MONEY AND THE SPACE ECONOMY

MONEY AND THE SPACE ECONOMY

Edited by

Ron Martin

Department of Geography,
University of Cambridge

JOHN WILEY & SONS
Chichester · New York · Weinheim · Brisbane · Singapore · Toronto

Copyright © 1999 by John Wiley & Sons Ltd, Baffins Lane, Chichester, West Sussex
PO19 1UD, England

National 01243 779777. International (+44) 1243 779777

e-mail (for orders and customer service enquiries): cs-books@wiley.co.uk

Visit our Home Page on http://www.wiley.co.uk or http://www.wiley.com

R. Martin has asserted his right under the Copyright, Designs and Patents Act, 1988, to be identified as the author of this work.

All rights reserved. No part of this publication may be reproduced, stored in a retrieval system, or transmitted, in any form or by any means, electronic, mechanical, photocopying, recording, scanning or otherwise, except under the terms of the Copyright, Designs and Patents Act 1988 or under the terms of a licence issued by the Copyright Licensing Agency, 90 Tottenham Court Road, London, UK W1P 9HE, without the permission in writing of John Wiley & Sons Ltd, Baffins Lane, Chichester, West Sussex, UK PO19 1UD.

OTHER WILEY EDITORIAL OFFICES

John Wiley & Sons, Inc., 605 Third Avenue, New York, NY 10158-0012, USA

WILEY-VCH Verlag GmbH, Pappelallee 3, D-69469 Weinheim, Germany

Jacaranda Wiley Ltd, 33 Park Road, Milton, Queensland 4064, Australia

John Wiley & Sons (Asia) Pte Ltd, 2 Clementi Loop #02-01, Jin Xing Distripark, Singapore 129809

John Wiley & Sons (Canada) Ltd, 22 Worcester Road, Rexdale, Ontario M9W 1L1, Canada

LIBRARY OF CONGRESS CATALOGING-IN-PUBLICATION DATA

Money and the space economy / edited by Ron Martin.
   p. cm.
   Includes bibliographical references and index.
   ISBN 0-471-98346-2. — ISBN 0-471-98347-0 (pbk.)
   1. Money. 2. Money market—Location. 3. Financial institutions—Location 4. Banks and banking—Location. 5. Space in economics. 6. Regional economics. 7. International finance. I. Martin, R. L. (Ron L.)
HG220.A2M5833 1998
332.4—dc21   98-29172
                                                                  CIP

BRITISH LIBRARY CATALOGUING IN PUBLICATION DATA

A catalogue record for this book is available from the British Library

ISBN 0 471 98346 2
ISBN 0 471 98347 0

Typeset in 9/12pt Caslon 224 from author's disks by Mayhew Typesetting, Rhayader, Powys
Printed and bound in Great Britain by Bookcraft (Bath) Ltd, Midsomer Norton
This book is printed on acid-free paper responsibly manufactured from sustainable forestry, in which at least two trees are planted for each one used for paper production.

Acknowledgements

I am grateful to all those who have contributed to this volume. Without their enthusiasm for the project, and their patience in responding to my various comments on their chapters, this book would not have been possible. I also wish to thank Tristan Palmer, formerly at Wiley, for his encouragement in helping to germinate a seed of an idea and his forbearance and assistance in seeing it through to fruition.

LIST OF ABBREVIATIONS

APCIMS	Association of Private Client Investment Managers and Stockbrokers
APT	Automated Pit Trading
CDS	Caribbean Data Service
DTB	Deutsche Temin Börse
EASDAQ	European Securities Dealers Automated Quotation
EBITDA	excess of current earnings before interest, taxes, depreciation and amortisation
EMU	Economic and Monetary Union
ETI	economically targeted industry
EU	European Union
FDIC	Federal Deposit Insurance Corporation
FTC	Federal Trade Commission (US)
GMA	grocery marketing area (US)
IBFs	international banking facilities
IET	Interest Equalisation Tax (US)
IFCs	international financial centres
IMF	International Monetary Fund
IPO	initial public offering (US)
IRR	internal rate of return
LETS	local exchange and trading systems
LIFFE	London International Financial Futures and Options Exchange
LBO	leveraged buy-out
MATIF	Marché à Terme International de France
MBIs	management buy-ins
MBOs	management buy-outs
NASDAQ	North American Securities Dealers Automated Quotation
OFC	off-shore financial centre
OPEC	Organisation of Petroleum Exporting Countries
OTC	over-the-counter
PQ	Parti Quebecois
REITS	Real Estate Investment Trusts (US)
S&L	savings and loans
SBIC	small business investment companies (US)
SEC	Securities and Exchange Commission

Contents

I Introduction 1

Chapter 1 The New Economic Geography of Money 3
Ron Martin

II The Changing Geographies of Banking 29

Chapter 2 The Stages of Banking Development and the Spatial Evolution of Financial Systems 31
Sheila C. Dow

Chapter 3 Globalisation, Regulation and the Changing Organisation of Retail Banking in the United States and Britain 49
Jane Pollard

Chapter 4 A 'Possibilist' Approach to Local Financial Systems and Regional Development: The Italian Experience 71
Pietro Alessandrini and Alberto Zazzaro

III Financial Centres 93

Chapter 5 The Development of Financial Centres: Location, Information, Externalities and Path Dependence 95
David Porteous

Chapter 6 Globalisation and the Crisis of Territorial Embeddedness of International Financial Markets 115
Leslie Budd

Chapter 7 Off-shores On-shore: New Regulatory Spaces and Real Historical Places in the Landscape of Global Money 139
Alan C. Hudson

IV Money and the Local Economy 155

Chapter 8 Financing Entrepreneurship: Venture Capital and Regional Development 157
Colin M. Mason and Richard T. Harrison

Chapter 9 Corporate Finance, Leveraged Restructuring and the Economic Landscape: The LBO Wave in US Food Retailing 185
Neil Wrigley

Chapter 10 Local Money: Geographies of Autonomy and Resistance? 207
Roger Lee

V Money and the Retreat of the State 225

Chapter 11 The Hypermobility of Capital and the Collapse of the Keynesian State 227
Barney Warf

Chapter 12 The Retreat of the State and the Rise of Pension Fund Capitalism 241
Gordon L. Clark

Chapter 13 Selling off the State: Privatisation, the Equity Market and
the Geographies of Shareholder Capitalism 261
Ron Martin

Endnotes 285

Bibliography 301

List of Contributors 330

Index 331

Part I

Introduction

CHAPTER 1

THE NEW ECONOMIC GEOGRAPHY OF MONEY

Ron Martin

INTRODUCTION: THE EMERGENCE OF A NEW SUBDISCIPLINE

This book is concerned with the economic geography of money. Since its inception, economic geography has been pre-occupied with the industrial landscape. To a large degree this reflects the modern origins of the subject, and in particular the emphasis of the early classic works on regional development and the space economy. However, interestingly, August Lösch, author of one of those early classic works, *The Economics of Location* (1939 and 1954), did draw attention to certain financial aspects of the economic landscape, in particular to geographical variations in interest rates, credit and consumer prices across the United States.[1] It was also was his intention to write a sequel dealing explicitly with money and location. Alas, because of his premature death, this was never completed (although a lengthy article 'Theorie der Wahrung' was published posthumously in *Weltwirtschaftliches Archive* in 1949, but has never been translated[2]). Had that book been written, the post-war development of economic geography and regional economics might not have neglected the study of money. As it was, his commentaries on interest rates and price inflation waves across space failed to ignite the passions of economic geographers and regional economists, who focused instead on industrial development and agglomeration.

Writing in the early 1970s, Richardson (1972, 1973) attributed the neglect of money in regional economics to the fact that regional economists borrowed too readily from neoclassical growth theory. With its assumptions of free and costless movement of capital and labour and perfect and ubiquitous information flows between regions, this theory essentially assumes away any regional role for money. This is not to suggest that regional finance was entirely ignored, however. For example, Gunnar Myrdal (1957) directed attention to the question of regional financial flows in his theory of cumulative, uneven regional development. He included the drainage of regional funds through a national banking system as one of the 'backwash effects' which have a negative impact on peripheral areas in an economy. However, unfortunately, although Myrdal's cumulative causation model of uneven regional development was subsequently to have a significant influence on both regional economics and economic geography, the financial aspect of his argument was not taken up. In fact, it was not until the mid to late 1970s that studies of regional and urban finance began to appear. A flurry of interest from US economic historians in the early development of the US national financial system between the 1870s and the 1920s, highlighted the distinctive regionalised banking structure that grew out of that process (see Carosso, 1976; James, 1976, 1978; Rockoff, 1977; Smiley, 1975), a regionalisation that persisted throughout the twentieth century. Regional economists and regional scientists likewise began to direct attention to certain aspects of regional finance in the late 1970s, focusing mainly on the regional impact of national monetary policy (Beare, 1976; Cohen and Maeshiro, 1977; Miller, 1978; Roberts and Fishkind, 1979). Although these various studies of the spatial dimensions of the financial system represented a welcome addition to the regional economics literature, they provided only limited insight into regional financial markets and the workings of financial systems across space.

For their part, geographers appeared even less interested in this field. Throughout the 1950s and 1960s there was little, if any, consideration of monetary issues by economic geographers. Kerr's (1965) empirical paper on the geography of finance and the rise and decline of financial centres in Canada was a lone, and largely unknown, exception. As in the case of regional economics and regional economic history, the 1970s did see a small surge of research by geographers into the circulation and impact of money in the space economy. Part of this focused on the flows of finance capital through the urban hierarchy (such as Conzen, 1977), whereas another stream of work sought to reveal the spatial biases (especially the discriminatory exclusion or 'redlining' of selected residential areas) in the allocation of mortgage finance within North American and British cities (Boddy, 1976 and 1980; Harvey, 1974 and 1977; Harvey and Chatterjee, 1974). However, innovative though these analyses were, they formed a partial and rather inchoate literature, and did little to stimulate other geographical work on finance.

Interestingly, during the mid 1970s a major work on finance appeared that combined the perspectives of regional economics and economic geography, namely that by the French banker-cum-academic, Labasse (*L'Espace Financier*, 1974).[3] This covered a wide sweep of topics on the geography of money, including banking networks, urban financial morphologies, national and international hierarchies of financial centres, regional payments systems and balances, the financial spheres of major cities, Eurodollar flows within the global financial system and offshore centres. In many ways, his *tour de force* anticipated the burst of interest in money by geographers that was to occur in the late 1980s, but because his study was never translated it was by-passed by Anglo-American geography and regional economics.

Thus, even as recently as the beginning of the 1980s, the 'geography of money' remained an underdeveloped subject. However, over the past decade and a half, and particularly since the late 1980s, both geographers and regional economists have begun to remedy that neglect, and at last there has been a growing wave of work on the spatial dimensions of money. In regional economics, research has moved ahead on a number of topics, including regional interest rate differentials (Faini, Galli and Giannini, 1993; Hutchinson and Mckillop, 1990; Mckillop and Hutchinson, 1990), the inter-regional flow of funds (Carlino and Lang 1989; Moore and Hill 1982; Short and Nicholas 1981) and regional credit availability (Alessandrini, 1992, 1994 and 1996; Amos, 1992; Amos and Wingender, 1993; Bias, 1992; Chick, 1993a;

Greenwald, Levinson and Stiglitz, 1993; Zazzaro, 1997). A substantial corpus of literature now exists on these and related themes (see the very useful survey by Dow and Rodriguez-Fuentes, 1997). Some contributions to the rapidly expanding literature dealing with international finance have also begun to consider geographical issues (Budd and Whimster, 1992; Cohen, 1998; Eichengreen, 1990; Eichengreen and Flandreau, 1996; Laulajainen, 1998; O'Brien, 1992). Equally important, economists have begun to make significant advances in theorising the geographical structure and spatial evolution of financial systems. One strand of this has focused on the geography of banking and credit allocation (Dow, 1987, 1988, 1990; Moore, Karaska and Hill, 1985), whereas another has been concerned with the growth of, and competition between, financial centres (Porteous, 1995).

Within geography, too, recent years have seen increasing recognition of the theoretical and empirical importance of finance and money for understanding the forces shaping the economic landscape. Thus, David Harvey has consistently assigned finance a key role in his evolving Marxist theorisation of the uneven development and crisis-prone tendencies of the capitalist space economy (Harvey, 1982, 1985, 1989b). At the same time, a growing number of more specific studies have appeared which deal with the spatial organisation and operation of particular financial institutions, services and markets, such as banking, venture capitalism, stock markets, and pension funds (Clark, 1993b; Gentle, 1993; Green, 1993; Lee and Schmidt-Marwede, 1993; Leyshon and Thrift, 1989 and 1997; Lord, 1987; Martin, 1989; Martin and Minns, 1995; Mason and Harrison, 1991 and 1995; O'hUallacháin, 1994; Tickell, 1994; Thompson, 1989). A third line of geographical enquiry has been concerned with explicating the economic, political and social dynamics of the world's major international financial centres (especially London, New York and Tokyo, but also off-shore centres), their role in articulating and moulding the global geographies of money, and the institutional cultures and practices on which their financial communities are based (Clark, 1997c; Corbridge, Martin and Thrift, 1994; Hudson, 1996; Leyshon and Thrift, 1997; McDowell, 1997; Roberts, 1994 and 1995; Sassen, 1991; Thrift, 1994). A fourth strand, still very much in its infancy, seeks to link regional financial flows and regional industrial development (Clark and Wrigley, 1997b; Courlet and Soulage, 1995; Schamp, Linge and Rogerson, 1993).

Thus, it seems valid to argue that the 'geography of money' is now firmly established as a new subdiscipline. Of course, we are still far from having a comprehensive synthesis encompassing the full range of financial geographies, from local money, regional finance and the national financial space to the global monetary system (see Corbridge, Martin and Thrift, 1994, for one attempt to encompass all these different geographical scales of monetary circulation). As yet, there has been relatively little integration of the work by economists and by geographers. As Leyshon (1995b) suggests, we have reached merely the 'end of the beginning' of the development of the subdiscipline. The task is now to take that development forward. The aim of this edited volume is to contribute to this endeavour, by bringing together a collection of original analyses which span the different interlocking economic geographies of money, written by leading researchers (both geographers and economists) in the field. The collection does not claim to be exhaustive in its coverage, nor does it seek to develop or impose any single conceptual or methodological framework. However, the different studies and topics covered here do demonstrate persuasively why economic geographers should take the study of money more seriously in their work, and likewise why economists who study money should direct more attention to the role of geography. My objective in the remainder of this introductory chapter is to sketch out some of the principal issues and themes which form the background to the rest of the book.

THE LANDSCAPES OF MONEY

Financial systems are inherently spatial. Initially, this statement might not be thought valid or important, since the very fungibility and convertibility of money enable it to transcend space more readily than any commodity. However, financial systems have complex institutional and organisational geographies that both reflect and influence their functioning. The various institutions and markets that make up the financial system, although they obviously vary in their specific activities and operation – from banks and insurance houses to stockbroking firms – continuously collect, receive and earn monies from all the localities and regions of a nation, and in many cases from other nations. These monies are then recycled, through various forms of investment, lending, trading and speculation, back across the regional system, and across the international system. Through this process, the various circuits of finance capital, such as banking, mortgage finance, insurance, pension funds, and stocks and shares, are constantly moving immense sums of money, credit and debt between the different regions and localities of a country, as well as between countries. These flows not only have their own intricate and ever-shifting economic, social, cultural and even moral geographies, but also shape the wider systems of production, employment, income and welfare of localities, regions and nations. The geographical circuits of money and finance are the 'wiring' of the socio-economy, as it were, along which the 'currents' of wealth creation, consumption and economic power are transmitted.

There are several different aspects to this geographical 'circuitry' of the financial system. One is the *locational stucture* of the institutions of which the financial system is composed. Financial institutions and markets are themselves distributed across space. Although in any advanced economy certain basic financial functions and institutional forms are widely distributed geographically (such as retail banking and mortgage-lending facilities), corresponding to the spatial distribution of the population,[4] other more specialised functions and institutions (such as stock markets, pension fund houses, merchant banks, bank headquarters, venture capital companies and foreign exchange markets) are typically much more spatially concentrated. Like other industries, financial activities are also characterised by economies of agglomeration, path-dependence and locational lock-in, and historically have tended to cluster geographically in particular urban centres and regions. Indeed, the urban hierarchy is, to a large degree, also a financial hierarchy. Thus, most countries have a number of major regional (or provincial) financial centres and, of course, typically a single national centre. Some of the latter, especially London, New York and Tokyo, are also leading world or global financial centres, serving and controlling not only their own domestic economies, but also wider continental regional economic blocs (Europe, the Americas and South-East Asia) and the global financial system as a whole.[5] The organisational and locational concentration of money markets in these and other global metropolitan financial centres has been associated historically with the development of distinctive urban built forms, architectural landscapes dominated by the key symbolic forms of 'high finance' (the City of London and Wall Street being the classic examples).

These interlocking locational structures, from the local, regional, national and international to the global, are at the same time *institutional geographies*. Although, like capitalism itself, the development of basic monetary forms and relationships has been similar across countries, just as there are different national (socio-institutional) variants of capitalism (Berger and Dore, 1996; Hollingsworth and Boyer, 1997), so historically different countries have evolved different institutional models of the financial system. Consider, for example, national banking systems. In the UK, as industrialisation proceeded in the eighteenth and nineteenth centuries, so a complex system of local and county banking developed alongside the growth of commercial banking in London. At their peak in about 1810, more than 900 of these small banks flourished. During the second half of the nineteenth century, however, with the onset of joint stock banking and the increasing dominance of London as the national centre of finance capital linked to the growth of the Empire, a wave of bank mergers and acquisitions led to the concentration and centralisation of banking activity into a much smaller number of national banks, many head-quartered in London. By 1914, the number of independent banks had fallen to 66. Of these, 20 had more than 100 branches, while the Midland, Lloyds and Barclays each had over 500 branches. Lloyds, one of the two major Birmingham banks, had taken over two London banks in 1884 and transferred its head office there in 1910. The Midland moved from Birmingham to London in 1891. Further concentration and centralisation followed, giving the final shape to the national branch banking system that came to characterise twentieth-century Britain.[6]

This is in contrast to US banking, which traditionally has been a local and state-based system. In part this can be ascribed to the way in which during the nineteenth century the westward spread of US economic development and the opening up of new regions encouraged the growth of local banks and related institutions. That these local banks were able to emerge and survive reflected the large distances between the different cities and their relative isolation from the older, established commercial and financial centre of New York. According to Odell (1992), the unfamiliarity with the new regions, and the primitive nature of communication between the older and newer regions, discouraged the early formation of a centralised national financial market. Also of critical importance was the development of a regulatory environment, including the McFadden Act of 1927, the Glass-Steagall Act of 1933, the Douglas Amendment to the Federal Bank Holding Company Act of 1956, and individual state banking laws across the United States, which controlled intrastate bank branching and prohibited out-of-state banking institutions from operating in-state branching subsidiaries or from owning banking or other financial subsidiaries through bank-holding companies. Only since the mid 1970s has this institutional geography begun to change, as individual states have progressively relaxed their own restrictions and as national legislation, such as the Riegle–Neal Act of 1994 (which effectively repealed the Douglas Amendment), has created a more national banking space. Since 1997, the Riegle Act has allowed bank-holding companies to consolidate their interstate banks into an office network and independent banks to branch interstate by merging with another bank across state lines. However, the establishment of *de novo* offices within a state by an out-of-state bank is allowed only where specifically authorised by state law, and most states do not permit it. Similar decentralised banking systems have been important in certain European countries, such as Germany, France and Italy (Carnevali, 1996), but in these, too, changes to the regulatory landscape are also underway.

The institutional geography of the financial system is important because it can influence how money moves between locations and communities. Thus, local and regional banking systems tend to be more rooted in and committed to the local economy and community than do local branches of centralised national banks. Other things being equal, in a local and regional banking system a higher proportion of locally originating monies are more likely to be retained and recycled locally; although this may mean that local banks are also vulnerable to localised economic decline, resulting in local financial crises. In a national branch banking system, local branches in less prosperous areas may adopt more cautious and restrictive lending policies than their counterparts in wealthy localities. In any case, they are likely to be constrained by head office in the degree of freedom they have to vary their lending policies at the local level. A national branch banking system may mean that branches in localities that suffer economic decline can be protected by profits made by the national network of branches as a whole. On the other hand, if a national bank is seeking to rationalise its operations, it is likely to be the branches in peripheral and economically depressed areas that are the first to be closed down. Thus, potentially at least, different institutional geographies can generate different geographical biases in the circulation of savings, credit and loans. As access to finance capital is fundamental to the process of socio-economic development and prosperity, such spatial biases in the circuits of finance may play a significant role in this process, within and between nations (as was demonstrated only too clearly by the Developing World debt crisis of the early 1980s).

Financial systems are also *regulatory spaces*. The history of money is also a history of regulation. Traditionally, nation-states have not trusted the financial marketplace to produce outcomes that are always in their interests; consequently, they have regulated their financial systems and financial institutions in various ways. At one level, governments set the key parameters of the domestic monetary system, for example interest rates, exchange rates, credit controls, money supply and restrictions on capital movements. At another level, they may also regulate the activities and functions of, and competition between, different sorts of financial institutions (e.g. by limiting the range of financial products and markets in which different types of institutions may trade). The spaces of regulation range from the local to the international. As an example of the former, as we have noted, the spatial-institutional geography of the US domestic banking system was largely a product of the state-based regulatory framework governing branching and entry in that system. At the other end of the geographical scale, regulation of the international monetary system is necessary to lend order and stability to foreign exchange markets, to encourage the elimination of balance-of-payments problems and to provide access to international credits in the event of disruptive shocks (Eichengreen, 1996; Leyshon and Tickell, 1994). The evolution of the international financial system over the last 150 years has involved a succession of regulatory forms, for example the International Gold Standard between the 1870s and the 1920s, the US dollar-based Bretton Woods System between 1945 and 1973, the post-war phenomenon of currency boards, the Basle Banking Accord of 1987, and the contemporary movement towards monetary integration, a single currency and a common monetary space among the member states of the European Union (EU). All of these international regulatory structures in fact have been characterised by distinctive geographies.[7]

Money has had a habit of seeking out geographical discontinuities and gaps in these regulatory spaces, escaping to places where the movement of financial assets is less constrained, where official scrutiny into financial dealing and affairs is minimal, where taxes are lower and potential profits higher. The growth of the Euro dollar market from the 1960s onwards was to a significant extent fuelled by US and other banks seeking to hold and invest their dollars outside the highly restrictive financial environment of the United States. For a while, London, with its more relaxed regulatory system, formed the centre of the Euro dollar market. In the 1970s, as the British government increased its regulation of foreign banks, so banking activity sought other less-regulated havens, in the form of 'off-shore centres' such as Hong Kong, Singapore, the Bahamas and Cayman Islands (Hudson, 1996; Roberts, 1994 and 1995). The growth of new information technologies, new financial products and global banks and related institutions since the mid 1970s has enabled money constantly to take advantage of the uneven and shifting geographies of financial regulation. At the same time, banks and related institutions have also been adept at overcoming the regulatory spaces governing their own domestic financial systems. For example, in the United States, by resorting to creative corporation rearrangements, such as holding companies and mergers, several banking, brokerage and insurance firms managed to slip out of the legislative restraints intended to limit their geographical reach and their permissible activities long before the US Congress acted officially to loosen them. Thus, through its parent corporation, Citicorp, which is not a bank under the law, Citibank could operate as a credit-card banker in all 50 US states, circumventing and rendering irrelevant the long-standing New Deal legislation that was supposed to keep banks serving their own local communities. To get around the legal requirement that banks lend only a certain proportion of their cash reserves, Citibank sold its loans to Citicorp, which is not subject to such requirements.

No less important than these three forms of financial landscape is a fourth associated with the *public financial space of the state*. Modern states not only set the regulatory structures that govern their domestic financial systems, and cooperate (and compete) with other states in regulating the international monetary system, they also redistribute vast sums of money across society and across regions and localities. Over the course of the twentieth century, public spending by capitalist states increased from about 10–15% of GDP in the 1920s to 30–50% by the end of the 1970s.[8] Much of this spending – funded by various forms of taxation and borrowing – has gone on the production of goods and services (nationalised industries and public infrastructure, such as utilities, roads and hospitals, and health and education), and the rest on various redistributive transfers associated with social welfare programmes. The public sector itself became a major circuit of money and finance. However, during the 1970s, states found themselves caught between rising social expenditure and growing social and business resistance to ever-increasing taxes. Commentators predicted the 'fiscal crisis' of the state (the *locus classicus* of this view being O'Connor, 1973). With the widespread shift to a neo-liberal, pro-market political ideology in the late 1970s and 1980s, in which state spending was seen as 'parasitic' on the market economy and low taxes became prioritised over extensive welfare provision, states throughout the world have been trying to reign back the scale of public finances and restructure their composition. Although governments have found it more difficult to retrench public expenditure than they envisaged (see Pierson, 1994), public finance is being recast in fundamental ways and with substantial social, economic and spatial implications (see Block, 1996, for a cogent critique of the political economy of the 'downsized state'; and Martin and Sunley, 1997, for the spatial effects of this new, post-Keynesian state form).

These four interrelated geographies of the financial system – the locational, the institutional, the regulatory and the public – shape the flows of money across space. Under a neoclassical theory of efficient capital markets and a nationally integrated financial system, provided financial markets function freely and there are no regulatory barriers to the movement of money, capital flows should respond readily to, and thereby function to reduce or eliminate, inter-regional and international differences in the rate of return. If a particular region experiences relative economic decline, local asset prices should fall as a result of declining demand. Lower asset prices and associated factor costs imply higher potential returns, thus stimulating inflows of capital from elsewhere in search of higher yields. Therefore, in a neoclassical world, not only would financial institutions be efficiently and optimally located across space, capital movements through the system would remove or certainly minimise any tendencies towards uneven economic development between regions. Such a scenario seems to have underpinned recent political moves to liberalise financial markets and regional development in several advanced Western economies, including the United States, the UK and the EU. It is also the ideology of those who advocate the pursuit of 'global neoclassicism', that is the complete liberalisation of financial markets on a worldwide scale (for critiques of this credo, see Banuri and Schor, 1992; Eichengreen, 1990 and 1996; Kapstein, 1994).

A somewhat different and more realistic view is that the financial system reproduces or even reinforces, rather than reduces, uneven regional development. In this model, financial institutions themselves centralise as they accumulate capital. As they centralise organisationally and spatially, local savers and depositors lose faith in, and outlets for, local investment. In addition, external investors become more reluctant as the cumulative effect of the exodus of finance and institutions comes into play. Local savers and investors prefer to construct more liquid portfolio structures for their money, rather than tie it up locally. These more liquid options are available centrally. However, this in turn exacerbates the pessimistic outlook in relation to the local regional economy, and increases the transaction costs which result from the peripheral character of such regions. In this scenario, financial integration – itself a feature of economic development – and the free movement of money and capital that accompanies integration will lead to spatial centralisation of the financial system, to uneven development between the centre and the periphery of the space economy, and to regional dependency. There will be a net flow of funds from the periphery to the core, and this flow may fuel the process of uneven regional development. It is in this context that the role of public expenditure is important as a regional stabiliser: in the post-war Keynesian welfare state model of economic management, various social and spatial transfers (through the tax-benefit system, as well as through the regional allocation of state investment in public infrastructures and the provision of nationwide education, health and other services) helped to ameliorate and compensate for the effects of geographically uneven development (see Martin and Sunley, 1997).

However, money is not just an economic entity, a store of value, a means of exchange or even a 'commodity' traded and speculated in for its own sake; it is also a *social relation*. Financial markets are themselves structured networks of social relations, interactions and dependencies – they are communities of actors and agents with shared interests, values and rules of behaviour, trust, cooperation and competition. Face-to-face contact, personal recommendation and informal word-of-mouth have always been central to the conduct of financial business and transactions, and remain so even in an age of advanced telecomunications – geography matters. These social relations are an important part of the embedded micro-regulation (accepted mores, norms, customs and rules) of business practice and behaviour in financial institutions and markets. Those practices in turn incorporate and reproduce specific structures of power, exploitation and domination. The highly gendered nature of employment opportunities in financial institutions is one such instance; see McDowell, 1997. Another is that in an era of progressive state deregulation of financial systems, intensifying pressure on individual and corporate performance, and the potential for speculation, the opportunities for unscrupulous 'rogue' activities have increased (as in the celebrated Maxwell and Leeson scandals in the City of London: see Clark, 1997c).

More generally, money is basic to the time-space distanciation of social interaction (Giddens, 1985). It allows – indeed provides a mechanism for – the simultaneous 'stretching' and 'compression' of social interaction across time and space, facilitating the storage, coordination and communication of the information and social power used in such interaction. Thus, money allows for the deferment of payment over time-space that is the essence of credit. Equally, money allows propinquity without the need for proximity in conducting transactions over space. These complex time-space webs of monetary flows and obligations underpin our daily social existence. In a capitalist system, money is a prime determinant of social inclusion and social exclusion in that it shapes our access to economic and social resources, our ability to participate in the wider community, and, in an era of hyper-consumerism, even the construction of our social and cultural identities. Thus, the geographies of money are also social and cultural geographies.

THE RE-MAPPING OF MONEY: CONTEMPORARY TRANSFORMATIONS

Without question, the growth of interest in the 'geography of money' in recent years has been stimulated by the profound changes and upheavals that have remapped, and are continuing to remap, the financial landscape. Since the mid 1970s, capitalism has been in the throes of an historic climacteric, a transition from one phase of development to another. Neologisms and interpretational schemas have proliferated in an attempt to characterise and explain the nature and trajectory of this transition.[9] The most prominent account – certainly within economic geography – sees this transition as a movement from a post-war 'Fordist' regime of accumulation and regulation to a new regime of 'flexible accumulation'. According to David Harvey, a leading proponent of this thesis, 'the extraordinary effloresence and transformations in financial markets' have played a key role in promoting and facilitating this transition of the economy:

> In the present phase, . . . it is not so much the concentration of power in financial institutions that matters, as the explosion in new financial instruments and markets, coupled with the rise of highly sophisticated systems of financial coordination on a global scale. It is through this financial system that much of the geographical and temporal flexibility of capital accumulation has been achieved. . . . I am therefore tempted to see the flexibility achieved in production, labour markets and consumption more as an outcome of the search for financial solutions to the crisis-tendencies of capitalism, rather than the other way round. This would imply that the financial system has achieved a degree of autonomy from real production unprecedented in capitalism's history, carrying capitalism into an era of equally unprecedented dangers. (Harvey, 1989b: 194)

Most commentators on the contemporary transformation of the financial system highlight three intersecting and mutually reinforcing processes of change, namely deregulation, technological innovation and globalisation. In many ways, the process of *deregulation* was set in motion in 1973, by the collapse and abandonment of the post-war Bretton Woods System of pegged exchange rates, dollar convertibility and capital controls. Nation-states were suddenly cut loose from the moorings which Bretton Woods had provided. By the end of the 1970s, in the context of high inflation, economic stagnation and the ascendancy of economic liberalism in much of the capitalist West, economic and financial deregulation began to assume increasing importance both in domestic policy and in the activities of international financial agencies such as the International Monetary Fund (IMF) and the World Bank. During the course of the 1980s and 1990s, a tidal wave of deregulation (and re-regulation) swept across the globe, beginning in the United States and the UK but quickly spreading to other developed and developing countries. Financial markets were dramatically redrawn as nations became locked in a process of 'competitive deregulation', in a 'race to the bottom' to free money and finance from the regulatory structures built up during the post-war decades. Once the process of liberalisation was unleashed, no country could risk imposing onerous regulations or controls on its financial centre for fear of losing financial business to other less-regulated (more 'competitive') centres. Restrictions on capital movements were relaxed or abolished, stock markets were deregulated, financial market and product boundaries were dismantled, controls on the operation of banks and other financial institutions were removed, and large sections of state-owned activities were sold off (privatised).

The deregulation of financial and stock markets in the United States and UK was part of the overall economic policy re-orientation of Reagan/Thatcher conservative ideology, with its blind faith in the 'free market'. In both cases, it represented a reassertion by the state of an underlying disposition towards financial interests after several decades of national welfare and industrial concerns (Bonefeld and Holloway, 1996). If the post-war Keynesian period had been one of 'managerial welfare capitalism', the 1980s and 1990s saw the advent and dominance of 'money manager capitalism' (Minsky, 1989). The crux of this change has been that financial institutions have demanded and been granted less regulation and more protection at the same time that nation-states themselves have prioritised the financial sphere and substituted monetary goals and policies for the employment and welfare priorities of the past. Deregulation was promoted by the same nations that, already containing the leading world financial centres, stood to gain most from encouraging a free market in money and finance. Yet, ironically, this combination of the state's inclination towards financial interests and its deregulation of financial institutions and markets has also served to reduce the room within which policy-makers can manoeuvre in terms of domestic economic policy.

At the same time that the regulatory boundaries of financial systems have been redrawn and rolled back, an historic process of *technological innovation* has been transforming money and capital markets (Carrington, Langguth and Steiner, 1997). The revolution in information processing and communication, centred on advances in computing, micro-electronics, telematics and related technologies, has not only transformed the way that financial transactions are conducted, but also the forms that money and finance take, and the nature of financial institutions themselves. We live in an age of increasingly 'dematerialised', electronic or 'virtual' money (Solomon, 1997), of 'smart cards', telephone banking, direct crediting and debiting, electronic fund transfer, screen-based trading and computerised arbitrage – in short, an increasingly cashless society in which money is becoming simply pieces of information transmitted via sophisticated telecommunications networks (Barnet and Cavanagh, 1996; Carrington, Langguth and Steiner, 1997; Castells, 1989 and 1996; Hepworth, 1991; Warf, 1989 and 1995). An essential feature common to all of these new forms of monetary transmission is that of 'speed-up'. Whether it be personal banking or foreign exchange dealing, the new information and processing technologies are dramatically reducing transaction and transfer times; in some cases these are already instantaneous. Money has always been mobile, but now it is truly hypermobile, and in this world of ultra-rapid exchange and circulation times, even seconds or fractions of seconds become important for making (or losing) profits on transactions and deals. Allied to these developments, the number of monetary and financial products has proliferated (including securitised instruments and various derivatives, such as options, forwards, futures and swaps), encouraged in part by deregulation, in part by the advances in transmission and transaction technologies. These new financial instruments and products allow larger, riskier and more easily tradeable financial investments. Alongside this expanding array of products, the institutional base of the financial system has changed, involving, for example, the rise of large investors and actors such as pension funds, insurance funds and equity funds; the emergence of telephone 'call centres' (for banking and insurance); and the growth of the retail-store financial business, that is the provision of credit, banking and loan services by consumer retail establishments.[10]

Together, deregulation and technological and institutional innovation have played a key role in promoting the third process of change in the realm of finance, namely *globalisation*. Like many other contemporary socio-economic concepts, the word 'globalisation' has gatecrashed academic and popular discourse without paying the entrance fee of a clear and agreed definition. Different authors use the word in different ways, and considerable debate surrounds its definition, measurement and alleged implications (see Allen and Hamnett, 1995; Amin, 1997; Berger and Dore, 1996; Boyer and Drache, 1996; Cox, 1997; Hirst and Thompson, 1996; Mander and Goldsmith, 1996; Michie and Grieve-Smith, 1995; Ohmae, 1990, 1995a and 1995b; Palan, Abbott and Deans, 1996). Typically, it is taken to refer to the increasing integration, hybridization, convergence and stretching of economic relationships across space, regardless of national borders and institutions. Arguably, it is in the realm of finance that this complex and multi-faceted 'time-space compression or shrinkage' is most advanced, by virtue of the ease of transmission and convertibility of money (Bonefeld and Holloway, 1996; Eichengreen, 1996; Kapstein, 1994; Martin, 1994; O'Brien, 1992). Deregulation and new information technologies are fuelling the movement towards integrated 'seamless' financial markets that ignore national borders and economic territories; the growth of 'homeless' or 'stateless' monies that move electronically around the globe at phenomenal speeds; the rise of truly transnational banks and financial institutions that have offices in all of the world's major financial centres, and which view themselves as global players in a global marketplace; and the transmission and increasing synchronisation of market price movements across the world.

The implications of these developments are generally agreed to be profound. Thus, according to authors such as Ohmae (1990, 1995a and 1995b), O'Brien (1992) and Kobrin (1997), financial globalisation renders national economic spaces largely meaningless and undermines the economic sovereignty and autonomy of nation-states. The result, it is argued, is the end of geography as far as monetary relationships and transactions are concerned. As O'Brien puts it:

The end of geography, as a concept applied to international financial relationships, refers to a state of econonomic development where geographical location no longer matters, or matters less than hitherto. In this state, *financial market regulators* no longer hold sway over their regulatory territory; that is rules no longer apply to specific geographical frameworks, such as the nation-state or other typical regulatory/jurisdictional territories. For *financial firms*, this means that the choice of geographical location can be greatly widened . . . *Stock exchanges* can no longer expect to monopolize trading in the shares of companies in their country or region . . . *For the consumer of financial services*, the end of geography means a wider range of services will be offered, outside the traditional services offered by local banks. (O'Brien, 1992: 1, original emphasis)

A similar prognosis is made by Cairncross (1997) in her thesis of the 'death of distance':

Distance will no longer determine the cost of communicating electronically. Companies will organize certain types of work in three shifts according to the world's three main time zones: the Americas, East Asia/Australia, and Europe. . . . No longer will location be key to most business decisions. Companies will locate any screen-based activity anywhere on earth, wherever they can find the best bargain of skills and productivity. (Caincross, 1997: xi)

In other words, in a world of globalisation, electronic money has annihilated geography, and location is rendered irrelevant and unimportant. As Kobrin (1997: 65) expresses it, 'The emergence of electronic cash and a digitally networked global economy pose direct threats to the very basis of the territorial state.'

However, as I have argued elsewhere (Martin, 1994), this view of financial globalisation exaggerates its spatial, political and economic effects.[11] It is certainly a one-sided account. It seems to portray financial globalisation as some sort of exogenous, 'top-down' phenomenon or force, and one that is unidirectional, leading ineluctably to the spatial decentralisation and dispersal of financial activities and services. But, this is to ignore the essential dialectic nature of globalisation, that it involves a constant tension between two opposing forces – of decentralisation and dispersion on the one hand, and centralisation and agglomeration on the other. Indeed, rather than leading inexorably to a highly decentralised and dispersed pattern of financial firms and activities, deregulation, new technologies and globalisation can equally reinforce the concentration of expertise and business within existing centres, as firms located in the latter can now access customers and funds, wherever these are located, just as easily from those centres.

In theory, both of these divergent forces – of decentralisation and dispersal, and of centralisation and concentration – are consistent with financial globalisation, and it is the dynamic interplay between them that will shape the evolving geographies of domestic and global finance (Laulajainen, 1998). Far from being an autonomous 'top-down' process, financial globalisation is inherently geographically constituted, the product of organisational, technological, regulatory and corporate strategies by individual firms, institutions and authorities in *specific locations*. Globalisation may well have annihilated *space* (Castells, 1989 and 1996; Harvey,

1989b), but it has by no means undermined the significance of location, of *place*. O'Brien comes close to acknowledging this himself:

> There will be forces seeking to maintain geographical control: regulators and financial firms will continue to protect and define their geographical territory, their 'turf' and market share . . . geography will remain one of the most powerful, evocative and obvious reference points. Identities are rarely given up if there are no clear identities emerging to take their place. Location will continue to matter while physical barriers exist, while travel still takes time, and while cultural and other social differences persist. (O'Brien, 1992: 2)

Yet, he concludes:

> . . . as markets and rules become integrated, the relevance of geography and the need to base decisions on geography will alter and often diminish. Money, being fungible, will continue to try to avoid, and will largely succeed in escaping, the confines of the existing geography. (O'Brien, 1992: 2)

However, escaping the confines of existing geography is nothing new – it is what money has always done.

What is clear is that, combined, deregulation, new technologies and globalisation are not obliterating the landscapes of money, but reconfiguring them in significant ways (see Table 1). One aspect of this reconfiguration is the increasing frequency of financial crises, at all geographical scales, from the local to the global. Financial markets have become increasingly volatile and crisis-prone. Examples include the Developing World debt crisis of the early 1980s, the failure of the US local savings associations in the late 1980s and early 1990s, the bankruptcy of Orange County in California, the crash on the world stock markets in late 1987, the failure of BCCI and Barings, the stock market collapse in the countries of South-East Asia in late 1997 and early 1998, and the recent failure of major Japanese banks. In today's integrated markets, what begins as a localised financial crisis can be propagated rapidly across the whole domestic financial system and in certain instances across the world to disrupt money markets in other nations. By reverberating though domestic financial systems, for example by promoting the restructuring and rationalisation of domestic banks and other financial institutions, these shocks can impact differentially across localities within nations far removed from the initial shock.

In a deregulated world, increased financial instability translates into increased geographical instability. Between 1984 and 1994, nearly 1300 commercial banks and more than 1100 savings associations failed in the United States – levels of failure not seen since the Great Depression (see also Hook, 1994; Warf and Cox, 1995; Avery, Bostic, Calem and Canner, 1997).[12] In the case of the savings associations, the number of offices declined by more than 8500 between 1985 and 1995, with the rate of closure being noticeably greater in lower income localities (Table 2). On a rather different spatial scale, although originating in factors largely specific to the countries of East Asia (especially ill-prepared and premature deregulation of capital controls, leading to excess inflows of capital, channelled

Table 1 The Contemporary Remapping of the Financial Landscape

Key processes	Structural changes	Geographical changes
De(re)regulation Abolition of exchange, credit and capital controls Removal of historic product and market boundaries Opening up of securities markets Payment systems opened up Privatisation International accords on capital adequacy standards	**Markets** Increasingly integrated Increasingly global Increasingly competitive Increasingly volatile and crisis-prone Growing financial disintermediation Shift towards non-bank institutions (pension and insurance funds, etc.) Rise of non-bank financial services	**Domestic financial systems** Some spatial decentralisation Growth of 'super-regional' banks through merger and takeover Decline of local banks Localised crises Growth of national 'call centres' Spatial consolidation (closures and rationalisation of local offices) Emergence of 'spaces of financial exclusion' Growth of alternative 'local monies'
Technological innovation Office automation Automated payment systems Paperless settlement systems Electronic fund transfer Screen-based trading Automated stock dealing and quotation New products and instruments (including securitisation and derivatives)	**Firms** Mergers and acquisitions Diversification Specialisation Demutualisation Transnationalisation Growth of non-bank financial firms New risk management initiatives Financial Business rationalisation driven by emerging overcapacity Bank failures Large labour shedding by banking and insurance sectors	**National financial centres** Integration into global network Increasingly internationalised Some decentralisation of activities Problems of 'territorial embeddedness' (e.g. urban property costs and constraints, labour costs and remote access to markets) Deregulation-induced trading instabilities (including trading scandals and speculative shocks, etc.)
Globalisation Integration of markets and exchanges Reduction of trade barriers in financial services Speed up of international financial flows Growth of 'stateless' monies Growth of transnational (global) institutions Rapid international transmission of crises and shocks		**Global system** Continued role of 'off-shore centres' Highly synchronised cross-national ('borderless') markets Competitive deregulation between global centres Shifts in trade and products between global centres Regional cooperation and integration Emergence of 'placeless' markets (e.g. NASDAQ and EASDAQ)

Table 2 *The Decline in Savings Association Offices in the United States, 1985–1995, by Relative Income of Local (ZIP Code) Area*

Relative income of local area	Number of offices 1985	Number of offices 1995	Change 1985–1995 Number	Change 1985–1995 %
More than 120	5319	3671	−1648	−30.9
80–120	13 934	9195	−4739	−34.0
50–80	4329	2509	−1820	−42.0
Less than 50	643	315	−328	−51.0
Total	24 225	15 690	−8535	

Source: Avery, Bostic, Calem and Canner (1997)

Note: ZIP postal code areas classified by relative average household income (US average = 100)

primarily into short-term banking instruments; loss of monetary control; asset price bubbles; and over-investment in risky and unproductive ventures, particularly unsustainable real estate markets; combined with malpractice amongst financiers and governments), the collapse of stock markets and currencies in that region in late 1997 sent shock waves across the global economy as whole.[13] Few Western stock markets, banks, investors and firms with holdings and interests in East-Asian markets and businesses escaped the effects of this crisis. In many instances, the depressive financial and trade effects were transmitted down to individual localities in Western countries, as were the decisions of East-Asian transnationals to cut back or abandon their direct investment plans in the West.

In addition to financial markets becoming more unstable, they have also become intensely competitive. Deregulation and new technologies are allowing financial institutions to diversify their products and markets, and are permitting non-financial firms to participate in financial activities. Increasingly, financial institutions are seeking to compete and become significant players outside their traditional market spaces, in wider national and global markets. This competition, together with the large cash reserves held by many financial institutions and the emergence of over-capacity in some sectors of activity, has stimulated a wave of merger, acquistion and takeover activity, as banks, insurance companies, accountancy firms and securities houses jostle for market share and greater profits. This merger and takeover activity is taking place on all geographical levels: locally, inter-regionally and internationally.[14] For example, in Germany, a merger movement is growing amongst local and regional banks (as in Bavaria and Berlin) to form 'super-regional' banks capable of competing across the European market as a whole rather than just within their local areas (Covill, 1997).[15] Similarly, in the United States, a 'merger mania' has broken out as the removal of the long-standing legislative barriers to cross-state banking has encouraged a growing number of acquisitions and mergers across the US states to form 'super-regional' groupings (Lord, 1992; Blanden, 1997).[16] Virtually all of the top 20 regional banks have been involved with mergers and acquisitions, not just with other banks but also with securities houses and credit-card companies. If the mania continues, it will not be

Table 3 *The Number of Foreign Banks in Selected Banking Centres, Early to Mid 1990s*

On-shore	Number	Off-shore	Number
London	520	Cayman	>600
New York	340	Luxembourg	220
Paris	170	Singapore	220
Frankfurt	150	Hong Kong	130
Shanghai	120	Bahrain	40
Tokyo	90	Labuan	40

Source: Laulajainen (1998)

long before the United States finds itself with truly nationwide banks with branch networks stretching from coast to coast, in other words with the sort of banking system that the UK has had for the past century. Likewise, within the UK, deregulation, demutualisation (of building societies and insurance companies), and increasing competition have stimulated merger activity across the financial sector. The process is not simply confined within national borders – substantial international mergers and acquisitions have been made by US, European and South-East Asian institutions seeking to secure a competitive position within global financial markets. At the time of writing, a $166 billion (£100 billion) merger – the biggest in history – has just been announced between two US financial companies, Citicorp and Travellers Group (both already global companies), to form Citigroup. This will be the world's largest financial services entity which will have assets of US$700 billion, serve 100 million customers in 100 different countries, and provide the full range of financial activities, including banking, insurance, fund management, securities trading and investment banking. This new global enterprise, head-quartered in the United States, will be the undisputed global leader in financial services. It will redraw the map of global financial competition and force other major companies to rethink their global business strategies.

Indeed, this wave of acquisitions and mergers is only part of the internationalisation of the world's leading metropolitan financial centres that has occurred in recent years. Numerous banks and other related financial institutions have expanded beyond their home bases and have 'gone global'. This process has been especially marked in London, New York and Tokyo. At the beginning of the 1970s, London contained 159 foreign banks, New York had 75 and Tokyo was host to 60. By the mid 1990s, the number in London had expanded to 520, that in New York to 340 and that in Tokyo to 90 (see Table 3). The global wave of financial deregulation in the 1980s and 1990s made it possible for footloose financial firms to set up directly in the world's leading metropolitan centres rather than serving them from a distance. Ironically, this burst of internationalisation undermines the sort of 'end of geography' thesis posited by Ohmae, O'Brien, Cairncross and others. International firms want to locate in these different metropolitan financial centres because each has a different financial specialisation, and each is the hub of a different (although of course overlapping and inter-connected) continental global

Table 4 *Investment in Information Technology and the Contraction of Employment in UK and US Banking*

	UK banking		US banking	
	1989	1996	1986	1996
Employment	436 000	348 000	1 560 000	1 346 000
Investment in IT	£1.85 billion	£3.82 billion	$9 billion	$20 billion

Sources: Mitchell Maddison Group and Federal Deposit Insurance Corporation

Note: IT Investment for UK in 1989 is 1988 figure

region. Foreign banks and related institutions have moved into these centres precisely *because* of geography, that is to expand their presence in or gain access to specific markets, to capitalise on the economies of specialisation, agglomeration and localisation (skilled labour, expertise, contacts, business networks, etc.) available in these centres, or to specialise their own operations and activities geographically (Laulajainen, 1998). How far this internationalisation adds to the efficiency of the world's financial systems is a subject of some debate. It may enhance financial competition, efficiency and technological innovation in host centres, but it may simply shift business and jobs from one global metropolitan centre to another. It may also generate inflationary pressures on local labour markets (forcing up salaries) and property markets (forcing up rental prices), imparting problems to the very territorial-institutional embeddedness that attracted firms to these centres in the first place.

A further process that has followed in the wake of deregulation, technological change, increased competition and merger activity within the financial system, is that of 'geographical consolidation', that is the closure, downsizing or rationalisation of local offices. This consolidation is typically justified in terms of needing to reduce over-capacity and duplication of local services, or because of labour-saving efficiencies achieved though investment in new information technologies. Geographical consolidation has been especially marked in the UK. In 1986, there were some 14 008 bank branches in the UK; by the end of 1996, this number had fallen to 10 334, a reduction of more than one-quarter. Over this period, Midland bank closed over 500 branches, while its profits soared from £434 million to £1.2 billion; similarly, its rival Barclays bank closed over 800 branches in this period yet doubled its pre-tax profit to £2.3 billion. As investment in information technology has escalated, so employment in financial services has fallen (see Table 4). Between 1989 (when employment peaked) and 1996, some 88 000 full-time equivalent jobs have disappeared from UK banking, a decline of 20%. A similar contraction has occurred in the US banking sector, where between the employment peak year of 1986 and 1996, 214 000 people (almost 14%) lost their jobs. What is particularly disturbing about these closures, and the similar run of closures of branches of demutualised building societies in the UK, and savings associations in the United States, is that consolidation has been spatially selective, in that, a disproportionate number of the closures have been of branches in low-income inner-city areas and rural communities (see Avery, Bostic, Calem and

Canner, 1997; Leyshon and Thrift, 1995). The withdrawal of banks and other related institutions from poorer areas may make commercial and shareholder sense for the companies involved, but it effectively produces areas of financial exclusion, in which local residents are left with little option but to resort to other local, non-bank circuits of finance (including private money-lenders), or else develop their own 'local monies' in the form of community-based local exchange and trading systems (LETS) (Lee, 1996; Meeker-Lowry, 1996).[17] In this repect, technological change, competition and globalisation are producing a similar process to the de-industrialisation that occurred earlier within manufacturing. Just as the restructuring and rationalisation of manufacturing has been highly uneven geographically, so it is proving with banking and related financial activities.

OUTLINE OF THE BOOK

In these and in many other ways, the geographies of money are being remapped and rescaled. The chapters that follow address various dimensions of these changing geographies in detail. They are grouped into four sections, each concerned with a distinct theme. Part II contains three chapters that examine different aspects of the changing geographies of banking. In Chapter 2, Sheila Dow builds on her earlier work on this theme by developing a 'stages of banking development' theory of the spatial evolution of the financial system. She uses post-Keynesian monetary economics combined with Myrdalian ideas of centre-periphery uneven development to show how the spatial structure and organisation of financial systems are intimately bound up with their economic and institutional development. Thus, historically, as banking has evolved from its early function of intermediation through to its present-day system of securitisation, so the geographical structure of the banking system has undergone successive phases of development (first local and regional banking, then the spread of national branch banking, followed by the concentration of financial activity in selected regional and national centres, and more recently the growth of global centres). The idea is not to suggest that each country has followed precisely the same sequence of spatial forms, but rather to use the theory to offer some clues as to why and in what ways banking and financial systems differ over space. The spatial evolution of the financial systems is shown to have implications for the flows of monies across the space economy, and hence for the process and pattern of regional development more generally. Different stages in the evolution of the banking system produce different geographies of the supply and recycling of funds between regions. Hence, this chapter combines a locational theory of banking with a monetary view of regional development.

In Chapter 3, Jane Pollard takes the issue of the spatial evolution of banking further by examining how retail banking is changing in response to the effects of globalisation, deregulation and technological innovation. The Californian example is taken as the main expository case because it is from this area of the United States that many of the technological and organisational innovations in banking are emanating. Within the United States, California's banks are influential players – the largest Californian banks are renowned as retailing specialists and, historically, have played a leading role in defining the 'state-of-the-art' in the US banking industry. On the one hand, the chapter highlights the differences between the US and British retail systems to illustrate how national and local regulatory structures, the changing nature of global finance and the proclivities of key individuals have produced a spatial organisation of retail banking in California rather different to that in Britain and somewhat distinctive within the United States. On the other hand, in examining how California banks have been responding to a rapidly changing competitive environment, the author also points to the growing similarities between US and British retail banking systems. In particular, she looks at how globalisation is affecting the 'organisational conventions' concerning best practice within retail banking, and how these conventions appear to be integral to the re-organisation of a more regionalised system of banking in both countries.

Chapter 4, by Pietro Alessandrini and Alberto Zazarro, is concerned with the regional structure, operation and re-organisation of the Italian banking system, in the context of the marked uneven regional economic development that has long characterised that country. The Italian banking system has always been typified by the presence of many small local banks operating in restricted territorial areas, combined with a few national banks. The peculiarities of this institutional framework and the high degree of localisation of many of the country's most successful industries (the so-called 'industrial districts'), have made the local bank a primary actor in the development of many local economies. Since the mid 1980s, however, radical legislative, administrative and institutional changes, such as the privatisation of many banks, the introduction of the 'universal banking' model and the nearly full liberalisation of credit markets, have been modifying the competitive environment of both local and national banks. The authors argue that the traditional coupling of 'local bank–small firm' will have to be completely rethought in the light of these changes, and they explore some alternative possible approaches to inter-regional financial integration and ways of facilitating the diffusion of financial innovations into peripheral areas.

Part III comprises three chapters dealing with financial centres and their role in national and global financial systems. In Chapter 5, drawing on the arguments in his recent book *The Geography of Finance* (1995), David Porteous seeks to answer the question of why large financial centres emerge and persist, especially in the face of space-time transcending information technologies. He focuses on the role of 'information externalities' in financial centre development, and on the concept of an 'information hinterland' akin to the hinterland of secondary and tertiary activity in conventional economic geography. His analysis is based on the intersecting roles of spatial asymmetries in information and the cumulative effects of path dependence in influencing the location of financial activities. Although once a primary national financial centre is established it is unlikely to lose its leadership, he shows how under certain theoretical and real-world conditions rival national centres can develop and how 'spatial switching' of financial dominance can occur between them. The chapter uses Montreal and Toronto in Canada, Melbourne and Sydney in Australia, and modern off-shore centres (OFCs), as empirical case studies to illustrate the theoretical ideas and arguments employed.

In Chapter 6, Leslie Budd turns attention to the problems that globalisation and new information technologies pose for international financial centres. Under the post-war regime of 'embedded liberalism' (national interventionist state policies combined with liberal policies towards the world economy), 'territorial embeddedness', 'the place-specfic regulatory, trading and associated economies of concentration', was critical to the stability and sustainability of international financial centres such as London, New York and Tokyo. The breakdown of the post-war regime, the wave of deregulation of money markets, and the impact of new communications technologies, have created the possibility of remote trading away from large international centres. One consequence of this regulatory shift is a changed relationship between the organisation of financial flows and geographical space. Sustaining the territorial embeddedness of large financial centres, by maintaining the benefits of the urbanisation, localisation and agglomeration economies found in these centres, may then have to rely on re-regulating the spaces in which they operate. In an environment where there is little constraint on cross-border economic and financial transactions, more interventionist regulation at a national level is no longer a possibility; nor can technological innovation be reversed. The question then arises as to whether, and in what ways, deregulated financial spaces can be re-regulated in order to combat threats to their territorial embeddedness. Drawing on the case of London, Budd traces the genesis and nature of this problem, and suggests that one possible way in which the 'crisis of territory' (the threat posed by new technologies to financial flows being concentrated in and articulated from particular places) can only be overcome by territory itself, that is, the degree to which a particular financial place can be transformed into financial flows through the agency of rent, so as to sustain its utility as a territorially embedded financial exchange.

The rise and impact of OFCs is the underlying theme of Chapter 7 by Alan Hudson. His interest is in the re-shaping of the financial regulatory landscape, a process in which OFCs have played a key role. The growth and success of these OFCs – essentially 'gaps' in the regulatory map of international finance and important elements in the globalisation of money and money flows – has stimulated changes in the regulatory spaces of the 'on-shore' (that is, established, regulated and metropolitan) financial centres by way of competitive retaliation. The impact of the Carribean OFCs on New York and the US financial system is used to illustrate this process. In the 1980s, a coalition of US business and government interests tried unsuccessfully to lure the off-shore dollar 'back home' through the development of international banking facilities (IBFs) in the United States; in effect this was an attempt to create an off-shore regulatory space on-shore. Hudson uses this example to highlight the socio-legal production of financial spaces, and the importance of trust in the financial landscape; moreover, trust which is backed up by legal regulations. The attempt to construct IBFs failed because it is very difficult to create afresh a new regulatory space, especially in places that, historically, have acquired regulatory reputations which to potential depositors have not been particularly reliable or trustworthy.

Part IV shifts the focus once more, towards the intersections between money and the local economy. This is, thus far, the least explored aspect of the economic geography of money, and the three chapters here provide rather different 'cuts' through this topic. In their contribution (Chapter 8), Colin Mason and Richard Harrison provide a detailed account of the rise of 'venture capitalism' and its role in financing new and small businesses in regional and local economies. Venture capitalism is the term used to refer to the provision, by specialised financial companies or by individuals (so-called 'business angels'), of (usually) equity capital for new and existing enterprises unable or unwilling to access established capital markets. In theory, venture capital is intended to help finance riskier, new and small enterprises, around which there is a long-standing debate about the existence of an 'equity gap', that is a shortage of start-up and expansion capital. Mason and Harrison demonstrate that the availability of venture capital within developed countries (the United States and Europe) is geographically uneven, and tends to be organised from and concentrated in selected regions. These spatial variations in venture capital investment activity are shaped by explicitly geographical factors in the venture capital investment process, such as a venture capital firm's network of offices and informants, the limitation on the geographical focus of investment activity as a means of minimising risk, and the need for spatial proximity to investee businesses for ease of monitoring and hands-on involvement. In addition, various demand-side influences also shape the geography of venture capital investments, including the nature of entrepreneurial activity in different locations and the knowledge of venture capital possessed by professional intermediaries which varies from place to place. The existence of 'regional venture capital gaps' in turn has adverse consequences for the economic and entrepreneurial prospects of those regions. In view of the link between venture capital and economic development, governments have responded with various initiatives which attempt to stimulate the supply of venture capital in regions where it is underdeveloped or absent. These policies are shown to have had little effect, and the authors outline some possible ways of promoting greater venture capital activity across the space economy.

A different dimension of the geography of business financing is taken up by Neil Wrigley in Chapter 9, in his study of the spatial consequences of capital structure transformations of the firm in general, and of leveraged buy-outs (LBOs) in particular. These are neglected topics in economic geography. Although economic geographers have long been concerned with issues of corporate restructuring, they have traditionally underemphasised types of restructuring which involve transformations of the capital structure and ownership configuration of firms. Likewise, the available literature on the geographies of mergers and acquisitions has to a large extent ignored issues of capital structure reconfiguration. Using the US food retailing sector as a case study (a sector that underwent a massive wave of LBO activity in the 1980s), Wrigley argues that corporate finance is not only a key issue for economic geography generally, but also for the emerging agenda of the geography of money and finance more specifically. He shows how, using information on the top 20 US food retailers, capital structure transformation affects the spatial organisation of the high-leverage firm, giving rise to a 'geography of divestiture', to local market concentration and to the avoidance of intra-market competition. In addition, capital structure transformation is shown to create interactive effects between the high-leverage firm and the decisions of rival firms, with important consequences for the nature of competition (exit, entry, pricing, etc.) in specific markets and locations. In short, capital structure transformations of the firm (as evidenced here in the form of LBOs) have profound implications for the economic landscape.

This section of the book concludes in Chapter 10, with Roger Lee's more philosophical-discursive essay on 'local monies'. The history of money is one of progressive universalism, the replacement of local means of exchange by national currencies, and in the current period, even the replacement of national currencies by stateless, global monies. These universal monies certainly facilitate the integration of local economies and communities, but at the same time render them subordinate to systems of valuation determined by wider, non-local conditions, markets and forces. However, the past 15 years have seen the growth and spread of local monies or community currencies. Local money is defined as money which is 'marked' in some way, but especially by relative geographical location, such that it operates alongside but is distinct from conventional official currencies. One such form of local monies is the local economic exchange and trading system (LETS), examples of which have sprung up in numerous local communities in the United States, UK, Australia, Canada and elsewhere. These systems enable local communities – especially those bypassed or marginalised by the economic mainstream – to re-establish local informal economies based on mutually beneficial exchange, thereby utilising local skills and human resources without the need for, or the constraints of, conventional money. Lee argues that such local systems of socio-economic exchange provide a means of re-asserting local autonomy and even resistance to the increasingly global imperatives of normal money forms. Not only do they help to retain more of the wealth and resources in the communities where they are generated, they also afford a means of community-building, of social inclusion and de-monetised welfare provision. The danger, however, is that the growth of LETS and other community currency schemes is simply seen by nation-states as facilitating the unwinding of their own publicly funded welfare and social services, an unwinding being driven, in part at least, by the dictates of global markets and monetary relations.

The final group of chapters, in Part V, focus on money and the contemporary reconfiguration of the state. In Chapter 11, Barney Warf discusses how the rapid and worldwide transition into a post-Fordist economic order has been accompanied by the steady, if erratic, peeling back of Keynesian social and economic policy, including worldwide deregulation, reductions in welfare spending and the removal of trade barriers. This process has simultaneously initiated a reworking of the relations between capital and space. Telecommunications have played a central role in facilitating the emergence of a new 'spatial fix' typified by a new global infrastructure built around fibre lines and satellite services, giving rise to hypermobile financial capital. The enormous quantities of financial funds that readily cross international borders have seriously weakened nationally based monetary policy. Although it is an exaggeration to claim the death of the nation-state at the hands of global money, in a world of hypermobile finance the autonomy of the state in economic matters has certainly been dramatically changed, and in important respects fundamentally circumscribed. At the same time, the hypermobility of finance capital has accentuated its bargaining position *vis-à-vis* localities, accelerating their competition with one another in a downward 'spiral to the bottom' typified by endless concessions and subsidies.

The 'retreat of the state' involves not only a loss of power and authority to global finance, but also a financial contraction of the state itself. Two different aspects of this financial downsizing are taken up by Gordon Clark in Chapter 12 and Ron Martin in Chapter 13. In his chapter, Gordon Clark focuses on the rise of pension fund capitalism and the simultaneous withdrawal of the state as an economic agent. He traces how many responsibilities and functions collected together under the banners of the 'welfare state' and 'social infrastructure' are either being returned to the private sector or are being systematically discounted in terms of their real value as governments underinvest or fail to maintain, budget-to-budget, the original value of those functions and entitlements. Set against these developments, there is another, equally compelling trend, namely the rise of 'pension fund capitalism', eclipsing all other forms of private savings and transforming the nature and structure of global financial markets. Clark suggests that, in part, the state has displaced responsibility for the financing and provision of urban infrastructure to the financial sector in general and to the pension fund industry in particular. This is a real and profound change in what we have believed to be the proper status of the state. As a consequence, we must better understand how and why pension fund trustees make investment decisions as well as the incentives and impediments to financial innovation in the urban and non-urban realms. He concludes that the urban fabric of Anglo-American societies will be systematically discounted by underinvestment over the coming generation with selective private investment replacing comprehensive investment by the state. Furthermore, in his view, there need be no connection between the goals of pension funds' investment strategies and the economic and social coherence of urban society.

In an increasing number of countries, the financial retreat of the state has taken an even more explicit and politically driven form, namely the privatisation (selling off) of large sections of its activities. Although privatisation programmes have been pursued for various economic and political reasons, one aim has been to promote what has been called 'popular capitalism', or 'shareholder capitalism', the spread of individual share-ownership across the population. Individual share-ownership had been falling progressively in most industrialised countries thoughout the twentieth century, and privatisation was seen as one way of restoring this segment of the capital market. In the final chapter, Ron Martin examines the UK experience in some detail, and shows that even with its extensive privatisation programme and well-developed financial system, the growth of popular capitalism in the UK has been limited. Moreover, the spread of individual share-ownership has been very uneven geographically, in that it has been disproportionately concentrated in the richer southern part of the country where, significantly, the individualistic, wealth-creating culture promoted by the Thatcher governments during the 1980s took greatest hold, and where the nation's financial system and its capital markets are based. He argues that the London stock market's disposition towards large institutional shareholders and the lack of a geographically decentralised, easily accessible retail capital market have hindered the spread of 'popular capitalism' across the regions of the country. The British experience is argued to hold some important lessons for other countries (including the ex-socialist states) where privatisation programmes are underway.

Part II

The Changing Geographies of Banking

CHAPTER 2

THE STAGES OF BANKING DEVELOPMENT AND THE SPATIAL EVOLUTION OF FINANCIAL SYSTEMS

Sheila C. Dow

INTRODUCTION

Monetary theory has conventionally presumed that banking systems are homogeneous except insofar as the particular types of deposit making up monetary aggregates may differ depending on the nature of the national banking system, that is the specific classification of assets making up so-called 'narrow money' or 'broad money'. The presumption of homogeneity is reinforced by open-economy monetary theory which incorporates international monetary relations in terms of changes in foreign exchange reserves, which are presumed to generate equivalent changes in monetary aggregates in different countries. Further, the mechanism for determining net capital flows is a 'world' rate of interest which, depending on the degree of international capital mobility, drives domestic rates of interest. The source of any capital immobility is not generally explored, even though such immobility may be suggestive of national differences in the nature and operation of capital markets.

In the banking literature, there is much more of a focus on the different characteristics of national banking systems and on the interrelations between them. As deregulation proceeds and international competition is encouraged (particularly in the European Union, for example), a central issue is the competitive structure of the banking system, and how that structure will be altered by increased global competition. Such analysis is inevitably spatial in character. Second, there is the question of the location of bank head-offices – will merger and acquisition activity concentrate banking so that head offices are concentrated in a few large financial centres, or will spatial dispersion be preserved? Similar questions arise with respect to capital markets, as governments promote the emergence of robust national capital markets to counter the strength of London and New York. Second, there is the question of the consequence of any change in the spatial pattern of banking and capital markets. Does the spatial concentration of banking generate a different pattern of credit creation from a more geographically dispersed banking system? To the extent that capital markets are not perfect, does the location of the national trading centre impinge on the accessibility of financial markets for smaller enterprises in non-central regions? What consequences might we expect for the spatial pattern of economic activity?

To date, there has been little research into these spatial dimensions of financial systems, even in the context of the profound changes that global deregulation and integration are having on money markets and relationships (for a review of the regional finance literature, see Dow and Rodríguez-Fuentes, 1997). For example, a recent survey of research into the consequences of European monetary integration (Commission for the European Communities, 1990) exemplifies the dichotomy between the monetary theory literature and the banking literature, whilst having little to say about regional and local finance. The banking literature, which is given only limited attention in the survey, primarily considers national differences in banking systems in terms of efficiency differences and differences in scope for local monopoly, which will eventually be eroded by competition. Little attention is directed at the connection between the aggregate features of a European monetary system and the implications of integration for the spatial structure of banking systems and financial flows across the regions of the EU.

The purpose of this chapter is to analyse the spatial composition of financial development in such a way as to integrate it with monetary theory. In particular, it will consider the spatial structure of financial systems and the effect of that structure on the pattern of regional economic development, in terms of the organisational evolution of financial systems. The monetary-theoretic framework employed here draws on Chick's (1986) theory of the *stages of banking development*. This theory is based on the view that any banking system or, more generally, any financial system, is a product of its history.[1] History determines the conventions employed, the state of confidence in the banking system, and the particular behavioural and institutional framework within which the state regulates the financial system. Therefore, history also determines which monetary theory is most appropriate for the analysis of the monetary system at any stage of development, and the form of monetary policy appropriate to that stage. More significant for the purposes here, by considering the evolution of banking and financial systems over time, some clues are offered as to why and in what ways banking and financial systems differ over space. In particular, issues arise as to the consequences of integrating financial systems at different stages of development (as in the case of the European monetary union).

The chapter starts with a discussion of the pivotal role of the banking system in relation to the wider financial system, and in relation to patterns of economic development. This is followed, in the third section, by an account of the stages of banking development framework. Implications are then drawn for the spatial pattern of financial development, and, in the fifth section, for the spatial pattern of financial flows.

THE PLACE OF BANKING IN FINANCIAL SYSTEMS AND IN REGIONAL DEVELOPMENT

By employing a theory of banking development, it might be argued that attention will be drawn away from more general considerations of financial development. Some now argue that it is inappropriate to single banks out as distinctive financial institutions, other than perhaps as an artifice of government regulation. However, the monetary theory we will employ explains the necessity of a distinctive banking system as the provider of money. This section is devoted to developing that argument.

The function of the financial system as a whole is financial *intermediation*, the pooling of financial resources among those with surplus funds to be lent out to those who choose to be in deficit, that is to borrow. The development of financial intermediation has clear potential for promoting economic development. Investment is the key to economic development. With financial intermediation, investors in new productive activities do not themselves have to generate a surplus to finance their projects; instead the projects can be financed by surpluses generated elsewhere within the economy. Further, as financial intermediation develops, efficiency gains in the financial sector itself can enhance the rewards for placing surpluses with the financial sector, thus encouraging the generation of surpluses. This in turn encourages productivity gains by the surplus-generating activities and increases the supply of financial resources to be lent on to deficit activities. This is the rationale for promoting financial development in order to promote economic development (see Drake, 1980).

This process of financial intermediation, however, requires another key element of any financial system, namely money. It is conventional to consider money's importance primarily as a means of payment, that is as a technical input into exchange the presence of which enhances productivity. Certainly, it is a necessary function of any asset classified as money that it be generally acceptable in payment. However, this payment function is interdependent on money's other primary functions as a unit of account and a store of value.

Prices are posted and contracts are made in terms of some agreed unit of account, and it is convenient for this unit also to be the unit of payment. It has been argued (by Heinsohn and Steiger, 1987; see also Wray, 1990) that the 'unit of account function' is the most important function for the economy. The distinguishing feature of a modern capitalist economy is that it rests on contracts. For example, production is organised by means of labour contracts, with payment in arrears because production takes time. More important in relation to the financial system, investment and production are often financed by borrowed funds which require debt contracts. In this way, the whole system of financial intermediation rests on contracts denominated in money.

Further, in order for money to fulfil its important functions of means of payment and unit of account, there must be confidence that it will hold its value relative to other assets. There may be assets the value of which rises periodically relative to money, but whatever is used as money will have the most predictably stable value. As the value of most assets is subject to uncertainty, the assets used for money are those in which there is least uncertainty. This has two important consequences. First, the banks and banking systems which inspire the greatest confidence will attract the greatest proportion of deposits. Second, the propensity to hold assets liquid (as in bank deposits) has a spatial dimension. Keynes's theory of liquidity preference suggests that the preference for holding liquidity will be greatest where confidence in expected yields of alternative assets is lowest. Given that spatial differences in economic conditions are common, we can expect a spatial character to liquidity preference.

Any modern economy requires an asset, or a collection of assets, to perform these functions. If the asset historically used as money (e.g. bank deposits denominated in a national currency) ceases to perform these functions, then some other asset takes on the role of money. Thus, in cases of hyperinflation, a foreign currency may take over as money. In other words, we need to think of money first in terms of its functions, and only then identify those assets which are performing them in any particular financial system.

Of course, what inspires confidence in any system depends on the historical development of that system, its institutions, conventions and habitual behaviour. This is where the banking system comes in. In modern economies, the bulk of the money supply consists of the liabilities of the banking system. It is bank deposits which form the basis for payments and for settlement of contracts. Non-bank financial intermediaries are built up on the basis of the banking system. They hold accounts with the banks which perform the function of reserves in which they make final settlement. The growth of the financial system is primarily a matter of growth of non-bank financial intermediation (Gurley and Shaw, 1967). However, that growth depends on the firm foundation provided by bank deposits as money. Thus, the success of a financial system rests crucially on the capacity of the banks to supply money. This requires that bank deposits be good stores of value, which in turn requires that banks honour their commitments, that is maintain sound portfolios. Ultimately, banks should be able to make payments in cash (banknotes issued by the central bank) if required. However, in sophisticated modern financial systems, if there is confidence in the soundness of banks, payment in cash is not necessary.

Some years ago, Goldsmith (1969) devised the notion of the 'financial interrelations ratio' (FIR) to capture the long-run relationship between financial development and economic development. The FIR relates the sum of the value of all financial assets to the sum of the value of all real (productive) assets. He showed that the FIR increases with *per capita* income (Y/N), but at a decreasing rate, as shown in Figure 1. This is a result which holds in comparisons between economies with different *per capita* income levels as well as in any one economy over time (Gurley and Shaw, 1967). There is a discussion below on the unusual economic conditions of the 1970s which may well have caused the FIR to increase at a faster rate. It can still be argued however, that if we consider any one class of assets, they follow the same pattern as the overall FIR, that is the total value of one class of assets increases relative to real wealth (W) as *per capita* income increases, but at a decreasing rate (Dow and Earl, 1982: 18–19), as shown in Figure 2.

Thus, coin relative to wealth (C/W) eventually was overtaken by notes (N/W), which were then overtaken by bank deposits (BD/W), then by non-bank financial intermediary deposits (NBD/W), and so on. Similarly, if we consider economies at different *per capita* income levels, the composition of their financial assets will differ according to a similar pattern. As far as banks are concerned, what is being argued here is that bank liabilities assume an increasing proportion of total financial assets as economic and financial development proceed, but that they are

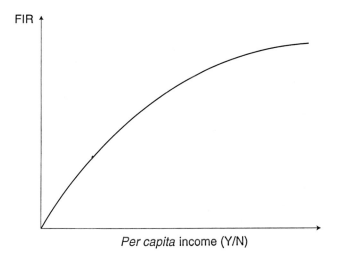

Figure 1 The Financial Interrelations Ratio (FIR)

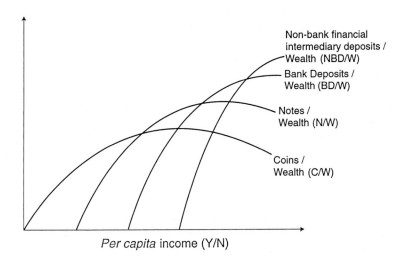

Figure 2 Ratios of Financial Assets to Wealth

later overtaken by other financial assets. Nevertheless, bank liabilities maintain their position as the core of the financial system; deposits falling as a proportion of a rising total wealth implies stabilisation relative to other assets, not extinction as some New Monetary Economists see it.

The next section considers how this role for the banking system evolves, and how the state develops the role of the central bank to promote confidence in the banking system. At the same time, it will become apparent that, as time goes on, the growth of financial intermediation, fuelled by the expansion of bank deposits, takes on a life of its own which is no longer clearly tied to economic development.

As financial development proceeds, therefore, questions arise as to its role in promoting economic development; beyond a certain stage of financial evolution, its contribution to economic production may become ambiguous. Financial development encourages saving by offering attractive returns, that is by appealing to the goal of *financial* accumulation. As a financial system develops, this goal may be better served by keeping financial resources ('churning') within the financial system rather than financing production and real investment. Not only may credit creation no longer be driven by requirements to finance productive investment, but financial conditions generated within the financial sector may actually have adverse effects on productive investment, and thus on real economic conditions. Through much of its evolution, the financial system may serve economic development needs well, but in the later stages of its evolution the financial system may become much less supportive of economic development.

As the financial system has real economic consequences, particularly in determining the volume and composition of credit creation, and as the financial system itself can be analysed in terms of its own spatial configuration, it follows that the real economic consequences may be different for different national and regional economies. The following section considers how banking systems at different stages of development influence the flow of financial resources among regions. It also considers the consequences of regions having differing degrees of financial development.

THE STAGES OF BANKING DEVELOPMENT

The stages of banking development framework is based on the view that there is an underlying logic to banking development. Different banking systems may go through the stages differently. Although the framework was developed with the English banking system as a model, it provides a common ground on which to discuss different banking systems where it is recognised that banking systems are not homogeneous. (This contrasts with the conventional view of homogeneity, which allowed Milton Friedman, for example, to draw conclusions for British monetary policy from the Chilean monetarist experiment.) The characteristics of the different stages are summarised in Table 1.

The first stage, that of *pure financial intermediation*, is where banks begin to emerge. Money consists of coinage issued by the state, or acquired in international trade from other countries. Financial intermediation develops, whereby a member of a community undertakes to pool savings from those with financial surpluses and to lend them out to those with financial deficits. Although paper IOUs may be issued as receipts for surplus funds deposited, final settlement is conducted in coinage. The financial system, such as it is, is constrained by the quantity of coin in circulation and the efficiency with which it can be used. Further, investment is constrained by the availability of borrowed funds, that is by the surpluses generated elsewhere and deposited with the financial intermediary.

The second stage is where banks as we know them emerge and *bank deposits come to be used as money*. To meet the needs of economic activity, where the

Table 1 *The Stages of Banking Development*

Stage 1: Pure financial intermediation
  Banks lend out savings
  Payment in commodity money
  No bank multiplier
  Saving precedes investment
Stage 2: Bank deposits used as money
  Convenient to use paper money as means of payment
  Reduced drain on bank reserves
  Multiplier process possible
  Bank credit creation with fractional reserves
  Investment can now precede saving
Stage 3: Inter-bank lending
  Credit creation still constrained by reserves
  Risk of reserves loss offset by development of inter-bank lending
  Multiplier process works more quickly
  Multiplier larger because banks can hold lower reserves
Stage 4: Lender-of-last-resort facility
  Central bank perceives need to promote confidence in banking system
  Lender-of-last-resort facility provided if inter-bank lending inadequate
  Reserves now respond to demand
  Credit creation freed from reserves constraint
Stage 5: Liability management
  Competition from non-bank financial intermediaries drives struggle over market share
  Banks actively supply credit and seek deposits
  Credit expansion diverges from real economic activity
Stage 6: Securitisation
  Capital adequacy ratios introduced to curtail credit
  Banks have an increasing proportion of bad loans because of over-lending in Stage 5
  Securitisation of bank assets
  Increase in off-balance sheet activity
  Drive to liquidity

---

coin-supply constraint is significant, the IOUs of the financial intermediary come to be used in payment rather than coinage. This occurs where there is sufficient confidence in the intermediary that the IOU would be honoured if necessary. Then title to deposits becomes a means of payment, a unit of account and a store of value circulating alongside coinage. The confidence in the intermediary will tend to rest on local knowledge, however, so that the IOUs will only perform money functions within a narrow local geographical area. Coinage is still required for settlement outside the local area.

This development is crucial to the development of banking as opposed to financial intermediation. Now that there is less call on the intermediary's reserves of coinage, funds may be lent out without a prior matching deposit. The intermediary can generate what we now know as the 'bank multiplier'. Deficit units can be financed without prior saving elsewhere in the system. The intermediary, now a bank, can itself create deposits as the counterpart to new lending which add to the local money supply and provide the basis for further financial intermediation. Before, financial intermediation was tied to the stock of coinage. Now, the financial system has the capacity to develop by means of increasing the stock of bank deposits.

This capacity is constrained by the extent of knowledge of the bank's soundness, and by the supply of coinage which is still required as the reserves of the banks. However, banks learn to evolve mechanisms for alleviating these constraints. In the third stage of banking development, banks learn that it is in their mutual interests to develop a *system of inter-bank lending*. When banks in one geographical area lose reserves to banks in another area, the former find their soundness threatened, while the latter may not have immediate alternative uses to which to put the newly acquired reserves. Further, they learn that confidence in the soundness of banks is subject to contagion. If a bank in one area fails and word spreads to neighbouring areas, confidence in other banks may suffer. Therefore, it is in the banks' mutual interests to limit competition to some degree in order to maintain the general confidence held in the banking system as a whole.

Maintaining confidence in the banking system is necessary for its growth and development. Banks can generate more deposits on any base of reserves of coinage the more confidence there is that those reserves will not need to be called upon. Confidence in the banking system is one of the concerns of the state. There may be a deliberate move to establish a central bank; but, if not, the banking system itself tends to generate a central bank to meet the needs of the banking system – Dow and Smithin (1992) express this argument in terms of the experience of the Scottish banking system. Thus, a central bank will normally exist by the third stage (as a creation of the state or as a result of market evolution). The liabilities of this central bank, in the form of its bank notes, or in the form of deposits, can add to the reserve of coinage, allowing further expansion of the monetary base on which the financial system builds.

The fourth stage is marked by a turning point in terms of the role of the central bank. Before this stage, the central bank may have lent reserves to banks in times of crisis. However, by the fourth stage it is recognised that a more general and institutionalised form of support for the banking system is required in order to promote additional confidence and, thus, further expansion. The central bank stands ready to act as *lender-of-last-resort* to any bank which finds itself in crisis. This development is crucial because it shifts to the banking system the capacity to alter the level of reserves of the banking system as a whole. If the banking system as a whole becomes overlent, there will be a general borrowing of new reserves from the central bank. Under these circumstances, the supply of reserves has become *endogenous*, subject to the influence of the banks themselves. In turn, banks can now respond to the demand for loans, rather than being constrained by an exogenous stock of reserves.

The expansion of the banking system, facilitated by enhanced confidence and the enlarged supply of reserves, in turn fuels a much greater expansion of non-bank financial intermediation with bank deposits as its base. At the fifth stage (which characterised the British banking system in the 1970s), banks find themselves facing fierce competition for financial services from new non-bank financial intermediaries. This competition has been the product of the banks' success. The banks are forced to become much more pro-active, seeking lending opportunities and the deposits to match them, that is to engage in *liability management*. This is where financial expansion starts to take on a life of its own, driven by the banks' concern over market share rather than the financing needs of borrowers in the productive sector. There will always be demand for credit to finance speculation. However, the banks' eagerness in the 1970s to extend credit and, thus, expand the level of deposits fuelled increased activity in speculative markets and increased interest rates, both of which left the financing of the productive sector behind (Chick, 1986).

The sixth stage represents a retrenchment from the consequences of the financial expansion of the 1970s, an expansion which had not been warranted by real economic activity. Monetary authorities, no longer able to control the supply of reserves, turned to capital adequacy requirements in an effort to constrain the volume of credit. As the banks found themselves saddled with bad debts, they found it more difficult to raise capital. Their response was to attempt to sell off assets in order to reduce their capital requirements and to increase the banks' appeal in equity markets. At the same time, the banks sought to avoid a recurrence of this situation of being caught with illiquid assets of dubious value. The outcome was the *development of securitisation* (Gardener, 1988). Banks turned existing loans into marketable securities and developed the provision of financial services in securities markets, facilitating borrowing by means of issuing securities, rather than lending directly themselves. At the same time, they encouraged the development of markets in *derivative* products which offered banks profit-making opportunities off the balance sheet, and thus not subject to capital requirements (although the requirements have since been changed to try to capture exposure to off-balance sheet risk).

Some argue that banks are now entering a seventh stage, or at least they would if they were not hampered by existing banking regulation. According to the New Monetary Economists (for example, Cowan and Kroszner, 1994) the distinctiveness of banks within the financial system is primarily the product of regulation which has singled banks out. Market forces, however, are eroding the basis for this distinctiveness. The securitisation of bank assets is reducing the peculiarities of banks on the asset side; historically, banks have specialized in making non-marketable loans. They were able to do so because of their distinctiveness on the liability side – banks could survive with an illiquid asset structure because bank deposits were used as money. However, it is argued that as transactions are increasingly being conducted with securities (such as bonds) rather than bank deposits, the uniqueness of the bank deposit is also being eroded. This trend is extrapolated to generate the conclusion that the financial system is developing by means of a *process of diffusion* such that banks will no longer exist as distinctive elements of the system. The financial system will then consist solely of financial intermediation; the function of providing money in the form of bank deposits will no longer be required.

The logic of the Chick framework suggests that this prognosis is wrong. Certainly, banking as we understand it (illiquid assets and liabilities-used-as-money) constitutes an ever-smaller proportion of financial activity. This is not surprising as wealth and economic activity increase, and it fits well with the

Goldsmith FIR, which suggests that any asset will eventually stabilise relative to expanding wealth. However, the logic of the Chick framework suggests that modern economies require an asset which is a good store of value to act as the denominator of contracts; the New Monetary Economics scenario implies that only assets of variable value will be available (with bank deposits varying in value with the value of the banks' assets, for example). The Chick framework, together with a study of banking history, suggests that, where a financial system itself cannot generate a money asset, an alternative will be sought from another financial system.

The conclusion, then, is that although traditional banking is declining in relative importance within the financial system in terms of volume of traditional banking business, it nevertheless retains its pivotal importance in providing the money base on which the rest of the financial system is built. This in turn continues to allow banks to play their distinctive role in providing direct loans to borrowers as an alternative to borrowing in securities markets. The importance of this source of finance for firms which do not have adequate access to securities markets is the premise on which the entire New Keynesian monetary economics is built (Greenwald, Stiglitz and Weiss, 1984). The different spatial characterisation of potential borrowers means that bank finance may be more important in some regions than in others. Similarly, portfolio behaviour on the asset side (particularly the desire for liquidity) may differ from one region to another. The following section focuses on the spatial constitution of financial system evolution – in particular on the banking system – in order to derive some conclusions about the implications of that constitution for regional economies.

THE SPATIAL EVOLUTION OF THE FINANCIAL SYSTEM

The stages of banking framework suggests that each banking system (whether regional or national) develops according to a common logic, whereby banks evolve to meet the needs of traders, borrowers and lenders, and devise similar strategies to promote confidence which allows greater growth. Increasing globalisation of finance may lead to the conclusion that international competition will promote homogeneity in banking development across economies.

However, a study of banking history reveals a diverse set of accounts of the development of different national banking systems. Further, it is clear that we currently experience spatial diversity, most notably in the form of distinctive financial centres in particular locations. This diversity can be seen to have arisen partly from differences in the origins of particular national or regional banking systems and in the role of the state. It was important, for example, whether the development of banking was initially motivated by private or state financial requirements (as was the case for Scotland and England, respectively) and whether segmentation was institutionalised to preserve regional balance (as in the case of the United States) or not (as in the case of Canada).

The second crucial feature which distinguishes the origins of particular banking systems (and thereby financial systems) is the presence of significant stocks of wealth. This factor may explain why the location of international financial centres has been noticeably persistent over time (Kindleberger, 1974). Stocks of wealth are of importance first in the provision of bank capital which generates confidence in bank liabilities, and second in promoting financial activity (borrowing and lending) which yields further surpluses. Therefore, there is a degree of inertia in the location of financial centres as those locations coincide with old wealth which originally inspired confidence in the banking system and which accordingly was able to generate further wealth and inspire more confidence, and so on in a cumulative, path-dependent process (see Chapter 5). Thus, for example, London has continued to act as an international financial centre almost independently of the performance of the rest of the UK economy. Within the UK, there are regional (provincial) financial centres, but they are clearly overshadowed by London (Gentle, 1993; Leyshon and Thrift, 1997). The inertia of financial centres might appear to have been challenged in the 1970s and 1980s with the emerging strength of oil-based wealth. New banks were formed in Western Canada, for example, encouraging predictions of a shift in the banking axis from East to West. However, these new banks could not survive the fall in confidence which followed subsequent reductions of the price of oil and went into decline (Dow, 1990). For the location of financial centres then, the overall picture is one of Myrdal's (1964) cumulative causation, based on the localised concentration of expertise, specialised institutions, networks of information and trust, and market confidence that these advantages help to promote. The central role of the banks as providers of money and the need for banks to inspire confidence mean that the financial centres which emerge early on in history can continually reinforce their advantage. More confidence means more business and the capacity to reap economies of scale, which inspires further confidence, and so on.[2]

The third crucial factor is the role of the state. For example, it was the need of the sovereign to have a bank to raise war finance which saw the beginning of the Bank of England, and allowed the Bank of England to have considerable market power. It was the absence of any significant role for the state in the Scottish banking system (from its inception in 1695) which meant that the private banks there evolved their own central banking system free from the controls of the English banking system until the state dictated otherwise in 1845. This illustrates that the stages of banking development may proceed without a state-sponsored central bank, but it still requires some institution(s) to perform the function of a central bank if the banking system is to develop. It was the political constraints on the US Federal Government which allowed the unit (state) banking system to persist, limiting the power of the New York banks; this was a system whereby banks were limited by state legislation to branching only within the state, or even prevented them from branching at all.

In modern times, the role of the state in the spatial distribution of financial centres is a matter of policy significance. Since the 1970s, there has been a widespread deregulation movement which has promoted increased competition in banking, which in turn has allowed increased concentration. Thus, deregulation in the United States is reducing the number of banks as they are allowed to increase their inter-state operations.[3] The European Union (EU) has been promoting the emergence of a single European market in finance with the European 'banking licence' at the heart of the policy. The emergence of this single market is encouraging merger and acquisition activity across the EU as financial institutions position themselves to compete in an expanded European market. While cross-border acquisition has been limited, the emergence of new large national universal banks, as the amalgamation of several national or regional institutions, is bound to have important spatial consequences. These new larger forces in European banking are located in the existing financial centres, which will have even more power to dominate the European market.

The relative capacities of different national banking systems in the EU to compete on a Europe-wide basis can be understood with reference to the stages of banking development framework. Some banking systems are at later stages of development than others. The British banking system is probably at the latest stage of development, with the new member states from Southern Europe probably the least developed. It should be emphasised that 'degree of development' refers to the capacity to create credit and make profits. As has been seen, banking systems at later stages of development are at the heart of a financial system which is increasingly distanced from productive economic activity. Thus, the London banks have been able to compete successfully in spite of bad debts from the 1970s and early 1980s because they evolved strategies for increasing the liquidity of their portfolios and profitability (securitisation and off-balance sheet activities). German banks have traditionally (although not universally) been regarded as being much better able than British banks to serve the needs of industry, because of their close relationships with local and regional industry (as well as because of the role of the state in creating a corporatist framework). However, the German banks are becoming increasingly aware of the illiquidity intrinsic to these relationships, and are attempting to employ similar strategies to those developed by the London and New York banks.[4]

Regardless of the relative contribution to productive activity, the banks in member states which are at earlier stages of banking development find it much harder to compete. They are adopting the new technology and products developed elsewhere. However, as long as there is not the foundation of a long and successful history of banking with a supportive state to ensure that confidence is maintained, these banks cannot create as much credit and attract the same rate of redeposit as the banks in Europe's financial centres. In other words, they are less competitive, and will face the threat of loss of market share and/or take-over in a single European market. Although this may be the case for banks in new member states such as Spain, Portugal and Greece, it is bound to happen to an even greater degree in aspiring member states in Central and Eastern Europe, such as Poland, Hungary, the Czech Republic, Slovakia and Slovenia. They not only have to contend with adopting new technology and the financial behaviour it accommodates, but also with a legacy of bad debts and a lack of experience in credit risk assessment.

There is apparently a recognition among member governments of the possibility of increasing concentration in banking in Europe, leading to domination of the market by a few financial centres. Accordingly, there are strenuous efforts being made to build up other national financial centres in the hope that they will be able to compete successfully with London as the European market opens up. These efforts range from more or less opaque barriers to entry for banks from one member state into another member state to active promotion of the development of equity markets in various member states such as France and Germany (Hawawini and Jacquillat, 1990). Yet there is a limit to what the governments of the member states can achieve in terms of enhancing the competitiveness of their national banking systems given the timescale available compared to the time it would take to build up international confidence and expertise in particular banks.

It is impossible to summarise the spatial pattern of banking development purely in relation to the stages of a banking framework because regulation has often been a major formative or complicating factor, as for example in the United States where banking was highly dispersed as a result of regulation. Central and Eastern European banking systems, too, are accelerating through some features of the stages of development as a result of competition with more advanced systems and state encouragement of banking development. The following section, however, explores the implications of the spatial character of banking and the different stages of banking development for the spatial distribution of economic activity.

THE SPATIAL DISTRIBUTION OF FINANCIAL FLOWS

What has been suggested here is that there is an inherent tendency in financial systems to concentrate in particular locations. This tendency relates to the capacity of financial-centre banks to inspire confidence and, thus, to generate more growth and more profits. This has been a pattern at all stages of banking development, with new waves of competition as a result of institutional developments (e.g. the emergence of joint-stock banking) being followed by a period of concentration (Dow and Smithin, 1992).

To the extent that financial centres are important employers and generators of income, this locational argument is of considerable importance. In the case of the UK, for example, the London financial centre has long held political power such that policies have been adopted to enhance the city's international earning power at the expense of domestic industry (Aaronovitch, 1983; Coakley and Harris, 1983; Ingham, 1984). This argument could be applied to the current policy which has maintained sterling at historically high levels despite the protests of employers and unions. The economic significance of the financial sector in this sense also lies behind the policies of developing countries to set up 'off-shore' banking in order to attract financial earnings irrespective of the developmental needs of the local economy.

However, the economic significance of the financial system extends well beyond its capacity to generate income. The banks alone can create credit without prior saving. In the process, they create the money which is the ultimate means of settlement for most (real and financial) transactions. The success of the banks in inspiring confidence in their liabilities-as-money fuels the success of other financial intermediaries which use bank deposits as reserves. As financial systems

Table 2 *The Spatial Implications of the Stages of Banking Development*

	Banks and space	Credit and space
Stage 1:	Serving local communities Wealth-based, providing foundation for future financial centres	Intermediation only
Stage 2:	Market dependent on extent of confidence held in banker	Credit creation focused on local community because total credit constrained by redeposit ratio
Stage 3:	Banking system develops at national level	Redeposit constraint relaxed somewhat, so can lend wider afield
Stage 4:	Central bank oversees national system, but limited power to constrain credit	Banks freer to respond to credit demand as reserves constraint not binding and they can determine volume and distribution of credit within national economy
Stage 5:	Banks compete at national level with non-bank financial institutions	Credit creation determined by struggle over market share and opportunities in speculative markets. Total credit uncontrolled
Stage 6:	Deregulation opens up international competition, eventually causing concentration in financial centres	Shift to liquidity by emphasis being put on services rather than credit; credit decisions concentrated in financial centres; total credit determined by availability of capital, i.e. by central capital markets

develop, the non-bank financial intermediaries expand relative to the banks; but that core of bank money is still an essential part of the system. Together, the banks and the non-bank financial intermediaries provide financial services to non-financial business. The nature, terms and availability of these services are of fundamental economic significance. To the extent that these services and their availability differ from one locality to another, the economic development of these localities is bound to be affected. These differences can arise from the local character of financial institutions, and from the degree to which local business is dependent on such local financial institutions (Chick and Dow, 1988). In other words, the financial sector also has an important indirect economic significance for the economies in which they are located.

It is possible to use the stages of a banking development framework to suggest some spatial patterns in the nature, terms and availability of financial services, particularly with respect to credit. The discussion is summarised in relation to the stages of the framework in Table 2. The stages of banking development can be characterised as a process whereby the banks increase their capacity to create credit. The excessive use of this capacity in the 1970s brought about a retrenchment (both to protect profitability and as a result of pressure from the monetary authorities) such that credit creation is now a smaller proportion of bank activity than in the past. It is, nevertheless, a distinguishing feature of banks that they can create credit, whereas non-banks cannot. Although New Keynesian economists maintain that it is the asset side of the balance sheet which distinguishes banks from other financial institutions, it must be remembered that banks' distinctive asset structure is only possible because their liabilities are used in payment. It is only because bank deposits stay within the banking system as payments are made from one account to another that they can engage in illiquid lending. Unlike other institutions, banks can enter into loan contracts with borrowers for periods of years, where the bank bears the full risk of default (although this can be mitigated by collateral requirements).

Those who focus on the banks' asset structure see the banks' distinctiveness in terms of the expert knowledge they bring to the credit-allocation decision. Certainly, the banking sector's long history of entering into these illiquid contracts has required banks to build up expertise in assessing default risk. Indeed, this has been a feature of banks since their initial emergence (the first stage). The key to banks, then, is *knowledge*. It is knowledge which determines not only the allocation of credit among potential borrowers, but also the total of credit insofar as this can be determined by the banks themselves.

The allocation of credit among potential borrowers depends on the banks' knowledge of their credit risk. As default risk is ultimately indeterminable, estimation of default risk must rely on expectations based on available information, supplemented by conventional judgement. The availability of information and the nature of conventional judgement depend on the history and location of the bank concerned. This is clearly an important issue when considering the structure of the banking system and the allocation of credit to different local and regional economies. The allocation of credit will differ depending on whether the banks concerned have a long history in an economy, and thus good knowledge of potential borrowers, and whether they are head-quartered in the financial centre – so that their knowledge will be biased towards large companies also head-quartered in that financial centre – or whether they are small local banks.

The Commission for the European Communities (1990) has sought to evaluate the likely effects of monetary union, a major plank of which is the single market in financial services. In fact, little attention was given to the effects of financial integration on the structure of the banking system and its effects on different economies within Europe. However, it was concluded that, although small local banks might suffer a competitive disadvantage initially, eventually a two-tier banking system would emerge with one tier consisting of international banks and the second tier consisting of local banks (where 'local banks' includes national banks devoted to the domestic market). The suggestion was that bank customers would enjoy a wider range of financial services (including improved access to credit) at a lower cost as a result of this competition.

However, the growing literature on regional finance suggests that the credit-allocation activity of large banks differs from that of small local banks; there is a parallel difference between lending to large companies and to small and medium-sized companies (Fazzari *et al.*, 1988). These differences stem from the differences in knowledge base between large banks and small local banks, and with respect to large companies and small and medium-sized companies. The concentration of financial activity in large banks expected to occur as a result of increased competition within Europe is likely on balance to reduce credit availability for small companies and companies head-quartereed in regions remote from financial centres (Dow, 1996a).

If banks at later stages of development are relatively unconstrained in their capacity to create credit, it might be argued that there should be no credit availability constraint. The only difference might then be in the terms on which credit would be offered, which might be higher if risk were assessed as higher due to poor information on small borrowers in remote or peripheral regions, and/or because of market segmentation (McKillop and Hutchinson, 1990). There are two particular implications of the stages of banking development framework. One is that banks at later stages of development have a competitive edge over other banks. In particular, banks at the sixth stage have developed sophisticated tools for protecting their liquidity, whereas banks at earlier stages are still constrained by the illiquidity of their assets. The second implication is that banks at different stages of development have differing capacities to determine the volume of credit. Not only does this increase the competitive disadvantage of banks at earlier stages of development, but it also means that borrowers in regions dependent on banks at earlier stages of development are more likely to be credit-constrained than borrowers in other regions. Thus, while the supply of credit is endogenous, the degree to which it accommodates demand may differ significantly in different regional economies and for different classes of borrower (Dow, 1996b).

It is also important to note that classes of borrower do not necessarily correspond to physical areas (Porteous, 1995). The social distance between those districts 'redlined' as representing unacceptable credit risk, and financially 'approved' districts physically adjacent to them, may be just as great as the physical distance between small communities and financial districts which are geographically remote. To the extent that these poor districts are dependent on informal suppliers of credit which are far less sophisticated than the advanced banks (namely money-lenders), they will be more credit-constrained and be offered less advantageous terms because of the poor competitive position of their credit suppliers. (Formal credit suppliers to such communities, such as consumer finance companies, may be competitive in financial markets, but offer poor terms because of their local monopoly position.[5])

The spatial pattern of credit creation may be reinforced by the spatial pattern of liquidity preference. Lower-income peripheral regions which have more limited access to credit from national and international banks are those most vulnerable to unexpected falls in asset values (whether in the form of real, financial or human capital). Given the dependence of peripheral economies on outside financial institutions and forces, confidence in expectations of asset prices will be relatively low. The rational response is high liquidity preference (Dow, 1993: chap. 10; Chick and Dow, 1988). This can take the form of holding liquid assets (often issued in the financial centre, in which confidence is high) and of abstention from capital expenditure and debt. Not only does this behaviour directly hold back economic development in such low-income and peripheral areas, but it has indirect consequences for local financial institutions. A propensity for holding liquid assets and for capital outflow limits the scope for these institutions to create credit and limits the demand for it to finance productive projects (as opposed to working capital). Low redeposit ratios are more limiting the less developed the banking system, so that this factor further impedes the competitive potential of local banks.

Finally, credit availability needs to be considered in terms of the effects of monetary policy. Chick (1993b) shows that different forms of monetary policy are appropriate at different stages of banking development. In particular, monetary controls are most effective up to the third stage of banking development, that is before the introduction of the lender-of-last-resort facility. Thereafter, the power of the monetary authorities to control monetary aggregates declines. Nevertheless, the form of monetary policy introduced in Western nations to counteract the excesses of the 1970s was direct monetary controls. This continues to be the approach to underpin plans for European monetary union. The mechanism may be the control of interest rates rather than money supplies, but the aim is still to control monetary aggregates. The stages of a banking development framework suggests that this form of control will have greater effect the earlier the stage of banking development. Where there is a single market which encompasses banks at different stages of banking development, there will be a differential effect of monetary policy. Monetary controls will be more deflationary in economies whose banking systems are at earlier stages of development.

Central banks have more tools of monetary policy at their disposal than control of monetary aggregates or particular interest rates. A traditional function of central banks, and one necessary to the development of banking, is to promote confidence in the banking system itself. This allows bank liabilities to be acceptable as money with minimal provision for reserves and, thus, maximum capacity to create credit. Again, this function derives from knowledge. Central banks build up knowledge of the banks in their systems and their operations, while banks build up knowledge of the supervisory practices of the central banks. The bank's customers in turn build up knowledge of the banking system and the central bank's efforts to ensure its soundness. Together, these processes generate confidence in the banking system which becomes conventional and habitual.

The stages of a banking development framework suggest that it is this function of the central bank which reasserts itself at later stages of banking development. The introduction of 'capital adequacy ratios' in the 1980s is a case in point, as central banks realised the need to build up more detailed knowledge of banks' risk exposure. This function could usefully be extended, given the argument made here about the differential effects of bank credit-creating activity on different regions and different classes of borrower. Central banks would be in a position to monitor the allocation of credit as the structure of the banking system evolves, and could use their regulatory powers to ensure that banks' capacity and willingness to extend credit accords with the best avaliable knowledge of risk. In other words, attempts could be made to ensure that credit is not denied potential borrowers simply as a result of poor knowledge, or because of the competitive disadvantage of local banks struggling to develop or even to survive.

Conclusions

The purpose of this chapter has been to develop the argument that the spatial development of banking has real consequences for spatial economic development. The first point made was that banking is the core of the financial system, in that it provides the means of payment which fuels economic activity and the reserves of non-bank financial institutions. This is what justifies a focus on banking. It was argued that the development of banking rests on the accumulation of confidence in bank liabilities which can then be used as a means of payment. It is the resulting high redeposit ratios which allow banks to create credit. The necessary confidence grows with the development of the banking system and the central bank's efforts to supervise and support it. If, with the aid of the state, the banking system encourages confidence, the banks enjoy a virtuous circle whereby they continue to attract deposits without the necessity of holding large reserves. This in turn allows them to create more credit. However, there is a problem of moral hazard if the central bank provides a lender-of-last-resort facility to maintain confidence in the banks. This allows banks to evade monetary control under circumstances of active credit markets, and to pay inadequate attention to default risk. There has been a backlash among central banks in the form of capital adequacy requirements, and on the part of the banks in the form of seeking a more liquid asset structure and alternative sources of income.

Deregulation of banking both within and among countries has opened up competition between banking systems at different stages of development, and therefore different capacities to compete effectively. The resulting concentration of banking in established financial centres will influence the spatial pattern of credit creation, with the emphasis in lending to peripheral regions being on credit for large companies head-quartered near the financial centre. This process will be exacerbated in economies with less-developed banking systems if they are subject to efforts at direct monetary controls, as envisaged for Europe in the Maastricht Treaty of 1992. Banking systems at earlier stages of development can be controlled more easily, and so are likely to experience more marked constraints on credit creation than more developed banking systems experienced in methods of evading monetary control.

However, we should be careful not to use the notion of a less-developed banking system as implying a lesser capacity to promote economic development. In fact, insofar as banks at earlier stages of development extend much of their limited lending capacity to local industry, they may well be serving the interests of local economies better than financial-centre banks whose priorities relate more to financial markets;[6] in this respect, there may well be a case for protecting these local banks against competition.

CHAPTER 3

GLOBALISATION, REGULATION AND THE CHANGING ORGANISATION OF RETAIL BANKING IN THE UNITED STATES AND BRITAIN

Jane Pollard

INTRODUCTION

Much of the contemporary literature on the globalisation of financial services has examined broad structural themes – the growing power of financial capital, time space compression and the homogenisation of financial space (Harvey, 1989b) – or the workings of key agglomerations of banking and other financial services in New York (Mollenkopf and Castells, 1991; Noyelle, 1989), the City of London (Coakley and Harris, 1983; Plender and Wallace, 1985; Pryke, 1991; Thrift, 1994) and, more recently, Tokyo (Sassen, 1991). This chapter examines the changing spatial organisation of retail banking, a part of the international financial system with which most of us engage, in some way, on a daily basis, and which is being affected by a series of changes in international financial markets. Retail banking in Britain has traditionally been dominated by the 'big four' (Barclays, National Westminster, HSBC-owned Midland and Lloyds–TSB), now joined by the likes of Abbey National, the Halifax, the Woolwich and other former building societies which have shed their mutual status. By contrast, in the United States in 1996 there were 9510 insured commercial banks and the state of California alone was home to 360 (Federal Deposit Insurance Corporation, 1996).

In thinking about how retail banking is changing in Britain and the United States, 'the California experience' of banking reorganisation is instructive in at least three ways. First, US and British banks are operating in an increasingly volatile, competitive global financial system and face many similar pressures. In the United States, however, a combination of deregulation and reregulation, a crisis in the savings and loans (S&L) industry, an overhang of LDC debt and the collapse of commercial real estate prices in the recession of the early 1990s, have forced banks – probably five to 10 years ahead of their British counterparts – to rethink how they 'do' retail banking. Second, within the United States, California's banks are influential players; the largest Californian banks are renowned as retail specialists (McGahey, Malloy, Kazanas and Jacobs, 1990; Pollard, 1995a) and have, historically, played a leading role in defining the 'state-of-the-art' in the US industry. Third, for all the differences in industry structure and the different histories and geographies of regulation in the two countries, the reorganisation of retail banking in Britain resembles, in many ways, that of banks in California.

This chapter focuses primarily on the *differences* between the US and British retail systems to illustrate how national and local regulatory structures, the changing nature of global finance and the proclivities of key individuals have produced a spatial organisation of retail banking in California rather different to that in Britain and somewhat distinctive within the United States. Yet, in examining how California banks have been responding to a rapidly changing competitive environment, the chapter will also highlight some of the growing *similarities* between US and British retail banking systems. Since the early 1970s, the US banking system has been undergoing a process of restructuring which is gradually eroding its distinctiveness *vis-à-vis* that of Britain and other advanced capitalist countries.

The following section describes the genesis of the spatial organisation of retail banking in California. This is followed by an examination of how structural changes and the worst financial crisis since the 1930s have influenced a reconfiguration of this spatial structure. Against this background, the next section considers some of the growing similarities between the reorganisation of retail banking in California and that now occurring in Britain, and comments on how such shifts in banking practice are formed and circulated within the industry. The final section draws some conclusions.

THE SPATIAL ORGANISATION OF BANKING IN CALIFORNIA

THE NATIONAL REGULATORY CONTEXT

The US retail banking industry, when compared with that in Britain, is very fragmented and decentralised. This industry structure bears the imprint of two key aspects of US financial history. First, the industry structure is a product of a dual or two-tier system of regulation wherein both federal and state laws govern the activities of banks. Second, the evolution of retail banking, and other financial institutions, has been shaped by the financial crises of the 1920s and 1930s; key pieces of federal legislation that have structured the geography of finance in the United States were introduced in the depths of the crisis which swept away almost 40% of US banks (over 9500 banks), and the savings of their depositors, between 1929 and 1933 (see Figure 1). The crises of the 1920s and 1930s are still benchmarks against which financial crises in the United States are assessed, although the S&L crisis of the 1980s now ranks as the most expensive financial crisis in US history.

Inter-war legislation in the form of the Glass–Steagall Act of 1933 was designed to segment different financial specialisms – investment banking, retail banking, mortgage lending and so on – in order to raise barriers to entry and prevent destructive competition between different financial institutions. Retail banks, investment banks and S&Ls (the equivalent of British building societies) were given quasi-monopolies in their respective product markets. Retail banks were to operate in money markets (where short-term credit monies circulated), investment banks were left to underwrite and sell longer-term equity and debt in capital markets, and the S&Ls were to provide deposit accounts (on which cheques could not be written)

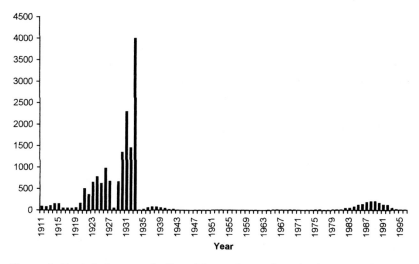

Figure 1 Bank Failures in the United States (vertical axis), 1911–1996. Sources: 1911–1933 US Bureau of the Census (1960), 1934–1996 Federal Deposit Insurance Corporation (1996)

and funnel those monies into mortgage lending. Further, federal deposit insurance was introduced to prevent bank runs, retail banks were prohibited from paying interest on cheque accounts and ceilings were placed on the interest rates banks could pay on savings accounts and term deposits (Regulation Q), thus limiting price competition between retail banks (Wolfson, 1993; Cerny, 1993).

These initiatives restricting the products retail banks could offer and the prices at which they could offer them were added to existing regulations constraining the geographic market areas in which retail banks could operate. The McFadden Act of 1927 allowed national banks to branch within their state limits if, and only if, state laws permitted branching. In the inter- and post-war period, many states operated unit banking regulations, wherein banks were not permitted to branch outside the state in which their head-office was situated. As recently as 1990, Colorado, North Dakota and Wyoming had legislation prohibiting in-state branching (Federal Deposit Insurance Corporation, 1996).

Such geographic restrictions affected the spatial structure of the banking industry in the United States from its beginning. Branching was regarded with suspicion because of the fear of bank failure and the difficulty of containing financial panics. Further, restrictions on branching were designed to prevent outflows of funds from small localities to larger urban centres and, of critical importance, to prevent the concentration of financial power in New York. Federal regulation, a strong pluralist tradition and a suspicion of the 'money trust' of Wall Street have produced a banking system in which power has been, and still is, much less centralised than in Britain (Fischer, 1968; Galbraith, 1975; Greider, 1987). So, inter-war regulation on banks' products and prices, together with geographic restrictions on banks' market areas has produced a comparatively fragmented retail banking industry in the United States (see Table 1) compared with that in Britain.

Table 1 Federally Insured Commercial Banks by Asset Size, United States and California, 1997

Asset size	United States	% of total	California	% of total
Total	9215	100	339	100
Less than US $100 million	5978	65	144	43
US$100 million to US$1 billion	2874	31	171	50
Over US$1 billion	363	4	24	7

Source: Federal Deposit Insurance Corporation 1997

THE REGIONAL REGULATORY CONTEXT

US banks are also affected by a second tier of state regulations. It is this second tier that has been especially influential in structuring retail banking in California. California has a history of bucking the national trend and operating a relatively permissive regulatory regime for its banks. Whereas east coast banks have traditionally been oriented to corporate lending, handling the accounts of Fortune 500 companies, California's banks are renowned for their retail focus, their extensive branch networks and their focus on small and medium-sized business lending (McGahey, Malloy, Kazanas and Jacobs, 1990).

> As a manager at one of the largest retail banks in California put it: You do not see the retail world as strongly in most States as you do in California. California has a bank on every corner . . . we consider ourselves *par excellence* in retail operations and the franchise we have out here in the Western US, that's something that is valued by everybody.[1]

Although it is true that the strong post-war economic performance of California has fuelled the development of retail financial markets and branch networks, the construction of dense state-wide branch networks was facilitated by early state banking regulations and, crucially, their interpretation by key bankers and regulators in the state.

The earliest banks in California were established in San Francisco following James Marshall's discovery of gold in 1848. Cross (1927) identified three phases of regulation for early banking in California. During the first phase of 1848–78, there simply was no effective regulation of banks. Hundreds of banking houses sprang up in and around San Francisco in the 1850s and 1860s, ignoring the state's Constitutional provision which outlawed their charter. These early banks were typically formed by private individuals and partnerships who held and traded gold dust and dealt in exchange drawn on other banks in California and, gradually, with the help of Express companies, that drawn on eastern banks. The largest of these early banks was the Bank of California, opened in San Francisco in 1864 – the first joint-stock corporation chartered as a commercial bank in California (Wilson, 1964). The first bank in Los Angeles was opened in 1868, closely followed by Isaias Hellman's Farmers and Merchants Bank (1871) which still exists today.

A second phase of bank regulation commenced in 1878 in the wake of a rash of bank failures and a month-long suspension of payments from the Bank of California in 1875. During this second phase of regulation, three state banking commissioners were appointed to undertake semi-annual inspections of all California banks. From 1878, banks had to publish information about their financial condition, and further amendments to the 1878 Act established reserve requirements and minimum capital requirements for state banks (Lister, 1993).

The third, and most important, phase of early bank regulation started in 1909 with the passage of the California Banking Act. This legislation – an early example of financial regulation trying to catch up with the workings of financial markets – was a response to financial panic. In 1907, the Knickerbocker Trust Company in New York collapsed, triggering a wave of bank runs starting on the east coast which gradually rippled westward to Oregon, Washington, Nevada and, eventually, California, wiping out 16 banks and more than 500 businesses state-wide (Cross, 1927; James and James, 1954). The 1909 Act appointed a state superintendent of banks, forced banks to separate the assets and accounts for their commercial, savings and trust banking functions, and imposed stricter supervision of banks' loans and investments. The most significant aspect of the law, however, was that it allowed bank branches to be opened, if, and only if, the superintendent of banks deemed a new branch in 'the public interest' (however defined). This clause of the legislation was crucial. State-wide branching had never been expressly authorised or forbidden in California; the 1878 legislation said nothing about the legality of branching. Other states, however, including New York, Washington, Nebraska, Utah and Colorado, prohibited branching, viewing it as a threat to the vitality and independence of small banks (James and James, 1954).

The bank that tested the limits of the 1909 legislation and, in so doing, came to symbolise the distinctiveness of California banks *vis-à-vis* their east coast counterparts, was A.P. Giannini's Bank of Italy. Giannini, a first generation immigrant who worked wholesaling fruit around the San Francisco Bay area, opened a bank in 1904 to serve the needs of, primarily, the Italian population in San Francisco. Following the 1909 legislation, Giannini successfully applied to the state superintendent to open a branch in San José and, in 1913, headed south to look over Los Angeles. The population had trebled in the city between 1900 and 1910, and, by 1913, water had arrived from the Owens River Valley and the harbour at San Pedro was under construction (James and James, 1954). That year, in spite of protests from local banks, the Bank of Italy was permitted to open its first branch in Los Angeles (Johnston, 1990).

Through the 1910s and 1920s, building on his early successes with branch applications, Giannini expanded the Bank of Italy into the Santa Clara and San Joaquin valleys in order to tap into the market for agricultural credit. Branching allowed the Bank of Italy to shunt funds around northern and southern California, following the seasonal cycles and credit demands of livestock, fruit and vegetable production. By 1927, the Bank of Italy was the largest bank in California and the third largest in the United States after National City and Chase National. The bank operated 276 branches in 199 localities and had over a million accounts; 20% of all Californians had an account at the bank (James and James, 1954). In 1928, with the blessing of the House of Morgan, Giannini bought the Bank of America on Wall Street in New York and formed the Transamerica Corporation. The formation of Transamerica allowed Giannini to control banks in different states while he waited for the advent of national banking legislation to enable him to merge his California and New York interests. In 1930, however, Giannini consolidated his banks in California; the Bank of America of California and the Bank of Italy were merged into the Bank of America National Trust and Savings Association (Johnston, 1990; James and James, 1954).

The Second World War transformed California's economy, especially that of southern California. The Bank of America grew rapidly after the war, providing credit not only for agriculture and construction, but also for the booming entertainment and military industrial complexes in Southern California. By 1953, it was the largest bank in the world with assets of US$8.2 billion and over five million accounts. Transamerica's close ties with the Bank of America, together with its ownership of other banks and insurance companies, made it the subject of anti-trust litigation.[2] By 1955, the Bank of America was a separate legal entity in California and, in 1958, following a long struggle, first with the Securities and Exchange Commission (SEC) and then with the Federal Reserve Board, Transamerica released all of its other banking interests – 27 banks with almost 200 branches in seven western states (Doti and Schweikart, 1990). These banks later became the First Interstate Bank, one of California's 'big four' retail banks, until it was taken over by Wells Fargo in 1995.

During the 1950s, 1960s and 1970s, the Bank of America and others established dense branching networks across California and into neighbouring states. Given the legislated restrictions on price and product competition affecting banks, geographic expansion through branching allowed them to tap fresh deposits and to compete for customers by advertising the convenience of a bank with many branches. Branching on this scale was peculiar to California. For example, at the end of 1966, there were 2797 bank branches in the United States that were located in counties non-contiguous with their head-office county; 47% of these were in California (Fischer, 1968). Branching was a critical part of retailers strategy in California, enabling the Bank of America and others to amass deposit bases larger than those available to east coast banks whose geographic expansion was limited by unit banking regulations (see Figure 2).

During the 1970s and 1980s, numerous foreign banks tried to enter the lucrative California market. Lloyds, Barclays, Standard Chartered and Midland all established footholds in California in the 1970s, only to retreat in the 1980s. During the 1980s, California outperformed the nation in terms of output and growth in personal income, employment and population (State of California, 1990). Six million new residents were drawn to California in the 1980s as defence-related manufacturing,[3] entertainment, construction and other industries grew rapidly. As British banks pulled out of California,[4] Japanese foreign direct investment[5] and the growth of trade with the Pacific Rim countries precipitated an influx of Asian capital. Between 1982 and 1988, assets of Japanese-owned banks in California grew from US$34.6 billion to US$93.4 billion, representing over one-quarter of all bank assets in California in 1988 (Zimmerman, 1989). Japanese

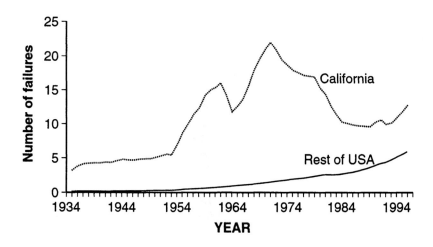

Figure 2 Branches per Bank, United States and California, 1934–1996. Source: FDIC (1996)

banks initially entered Los Angeles to follow their Japanese clients but have since expanded into retail and middle market business lending for a larger client base.[6] Reflecting their business orientation, Japanese banks held over 30% of all their assets in business loans in 1988, compared with 19.5% for domestic banks in California (Zimmerman, 1989). By 1989, four of California's largest 10 banks – Union Bank, Sanwa Bank of California, Sumitomo Bank of California and the Bank of California – were Japanese owned (see Table 2).

The booming California economy, the seventh largest in the world by 1989,[7] attracted not only Asian capital, but also a host of domestic financial institutions. In 1980, there were 281 commercial banks in California operating 4279 branches; by 1989, there were 479, mostly head-quartered in Southern California, operating 4532 branches (Federal Deposit Insurance Corporation, 1996). Competition was further intensified by the deregulation of the S&Ls. Sharply rising interest rates in the 1980s posed enormous problems for S&Ls that paid market interest rates to attract funds, but had a high proportion of their assets tied up in long-term mortgage loans repayable at low, fixed interest rates. In 1980, federal legislation removed the interest rate ceilings which restricted price competition between banks and S&Ls. By 1981, however, 85% of the nation's S&Ls, holding 92% of the industry's assets, reported losses (Litan, 1990). Further deregulation arrived in the form of the Garn St-Germain Act of 1982, which extended the asset powers and geographic reach of S&Ls, allowing them to invest up to 40% of their assets in non-residential real estate (Pizzo, Fricker and Muolo, 1989). California regulators went even further, passing the Nolan Act of 1983, which allowed state-chartered S&Ls to invest in real estate without limit.

Table 2 The Largest Banks in California by Asset Size, 1990 and 1997

	1990			1997	
Bank	Assets (US$ billion)	Employees	Bank	Assets (US$ billion)	Employees
Bank of America	96.2	47 000	Bank of America	224.3	64 916
Security Pacific	56.0	16 634	Wells Fargo	91.5	34 352
Wells Fargo	53.8	21 435	Union Bank	30.2	9381
First Interstate of California	19.1	13 590	Sanwa Bank of California	8.07	2726
Union Bank	16.3	7313	Bank of the West	5.4	1797
Bank of California	8.5	3809	Sumitomo Bank of California	5.2	1586
Sanwa Bank of California	7.1	3279	City National Bank	4.6	1599
City National Bank	4.9	2358	Imperial Bank	4.1	898
Sumitomo Bank of California	4.8	1686	Commercial Bank California	3.7	893
Imperial Bank	3.0	1009	WestAmerica Bank	3.6	709

Sources: Sheshunoff Information Services (1991) and FDIC databank web site (1997) (http://www.FDIC.gov/databank)

Table 3 *The Largest Savings Institutions in the United States, 1990 (by Assets)*

Rank	Institution	City	Assets (US$000)
1	Home Savings of America	Los Angeles	38 987 313
2	Great Western	Beverly Hills	34 949 576
3	Glendale Federal	Glendale	24 043 905
4	California Federal	Los Angeles	23 269 859
5	First Nationwide	San Francisco	22 088 222
6	World Savings and Loan	Oakland	19 337 776
7	HomeFed Bank	San Diego	17 752 680
8	American Savings Bank	Stockton	16 290 515
9	Great American Bank	San Diego	13 346 430
10	Crossland Savings	New York	11 609 177

Source: Thomson Financial Publishing (1991)

Behind this deregulation of interest rates and the expansion of asset powers was the notion that the S&L industry might grow or, more accurately, speculate its way out of crisis. If sufficient new deposits could be attracted, went the argument, they could be loaned out, with larger interest rate spreads, to offset the losses incurred on the long-term mortgage debt. To find those deposits and higher spreads, S&Ls had to be allowed to bid competitively for funds and then to invest those funds in assets with higher yields (read 'risk') than mortgages (Pizzo, Fricker and Muolo, 1989; Mayer, 1992). Southern California, already established as the home to the largest S&Ls in the country (Table 3), became the destination for new entrants seeking Californian state charters to invest in commercial real estate markets. In the sunbelt states of Texas, Florida, Arizona and California, S&Ls on the verge of insolvency could, and frequently did, pay above-market interest rates to attract deposits to fund their increasingly desperate forays into commercial real estate markets. The rewards of these endeavours, if successful, were privatised while the risk was socialised through federal deposit insurance.

During the construction boom of the 1980s, the average Californian house price increased from 60% above the national average in 1980 (US$99 550 compared with US$62 600) to more than double the national average by 1990 (US$196 120 compared with US$93 100) (California Association of Realtors, 1995). More significant for the health of Californian banks, however, was their participation in the commercial real estate boom of the 1980s. Like their counterparts in Florida, New York, Arizona and Texas, Californian banks, Japanese investors and the deregulated S&Ls participated in an office building boom which more than doubled office space in Los Angeles between 1979 and 1989, increased space in San Francisco by 75% and generated rapid growth in Santa Clara (Silicon Valley), San Diego and Sacramento (Centre for Real Estate and Urban Economics, 1991).

During the 1980s, competition in California's banking markets intensified dramatically with the influx of Japanese capital, the increase in the number of commercial banks and S&Ls operating in California, and the shift in the regulation and competitive strategy of the S&Ls. Despite the strength of the Californian economy in the 1980s, the banking industry in California produced lower returns on assets than the industry nationally (Federal Deposit Insurance Corporation, 1996).

A Changing Competitive Environment

California's economic performance during the 1980s fuelled the belief that the 'Golden State' was somehow different from the rest of the United States, somehow recession-proof (Cohen, Garcia and Loureiro, 1993). Yet by the late 1980s, California's military industrial complex and residential and commercial real estate markets were showing signs of weakening and, in 1990, California followed the nation into recession, experiencing its worst economic downturn since the 1930s. Between 1990 and 1994, California shed over 750 000 jobs, with the bulk of the losses occurring in Southern California (State of California, 1994). New office space was being completed in the early 1990s just as rounds of corporate restructuring reduced the demand for such space, while house prices fell more than 20% between 1990 and the end of 1994 (California Association of Realtors, 1995). The state's S&Ls were particularly hard hit by the collapse in commercial property markets.[8] Together with Texas, California was home to some of the worst excesses of commercial real estate speculation among S&Ls (Mayer, 1992), and was the recipient of considerable inter-regional transfers of federal funds to bail out its insolvent S&Ls.[9]

However, the problems facing Californian and other banks in the 1980s and 1990s were not just cyclical phenomena associated with real estate prices and the downturn of the business cycle. In the 1980s, banks across the country were starting to feel the effects of a series of broader structural changes in international and national financial markets, changes which left banks (and S&Ls) struggling to manage growing disparities between the requirements of their federal and state regulation, rooted in legislation designed in the 1930s, and the rapidly changing nature of the retail financial markets in which they were trying to compete. A host of changes in how financial services providers issued debt, how they dealt with risk and the systems they employed to identify and cherry pick profitable groups of customers were transforming the environment in which banks had to compete.

First, the inter-war legislation that protected banks from price, product and geographical competition had provided banks with a source of cheap funds (their depositors) and a source of demand for loans (large corporate borrowers). These two groups of customers effectively subsidised other bank activities. Yet in the 1970s, investment firms innovated by offering interest-bearing cheque accounts and depositors started switching their funds out of commercial banks into higher yielding mutual funds[10] and securities (Bryan, 1991). While banks' depositors started defecting to other institutions, banks were powerless to respond until their interest rate ceilings were deregulated in 1980 and 1982. Although the deregulation of interest rate ceilings helped to 'level the playing field' between banks and other financial institutions, banks now have to compete in global markets for their funds and pay market rates; gone are the days of cheap deposits.

Second, also on the lending side, US banks were gradually losing their position as key intermediaries in the provision of credit to corporate customers. In the wake of losses on LDC debt, and to appease regulators anxious about their balance sheets, banks increasingly turned to securitised forms of debt in the 1980s. By bundling up loans and selling them as securities in secondary markets, banks were able to move debt, and the risk associated with it, off their balance sheets and release the capital reserves held against such loans. Securitisation not only allowed banks to shed some of their LDC debt, albeit at a substantial discount,[11] it was also a means of circumventing changes in capital adequacy ratios introduced by US regulators who, in 1986, joined representatives from Canada, Japan, Britain and other European countries[12] to create a common set of capital adequacy standards in the wake of the LDC debt crisis. The securitisation of debt, however, was also a means by which some of banks' most profitable corporate customers could raise funds by issuing commercial paper, rather than borrowing from banks who had to charge interest rates that reflected the costs of their FDIC insurance premiums, reserve requirements, and, for many banks in the 1980s, deteriorating credit ratings. In addition to competition from newly chartered banks and S&Ls in the 1980s, investment banks, mutual fund companies and others were chipping away at banks' dominance of the corporate and industrial loan market. Once the core of banks' lending portfolios, commercial and industrial loans are now of declining significance in banks' lending portfolios (Boyd and Gertler, 1994).

Finally, there are two aspects of the spatial structure of the industry to consider. First, although the geographic restrictions on banks did constrain competition, such regulation also made it difficult for banks, and indeed S&Ls, to expand out-of-state to diversify their loan portfolios and supplement their deposit bases. Such geographic restrictions also denied banks a 'spatial fix' in the face of regional shocks and variations in real estate, energy and agricultural prices in the 1980s and early 1990s. Second, while banks' regulatory environment encouraged them to compete, especially in states such as California, on the basis of convenience and visibility, rather than by price or product, technological changes are reshaping the retailing of financial services. Throughout the post-war period, banks' branch networks constituted significant barriers to entry for firms wanting to enter retail financial services. This was because banks relied on the personnel and technological systems based in their branch network to generate the knowledge base – information concerning their customers, their products and their competitors – necessary to operate in retail financial markets. Physical proximity was important, allowing banks to obtain rich, interactive information about their customers and critically, their credit-worthiness.

However, during the 1980s, many non-banks made inroads into banks' markets by using telephone-based and geo-demographic information systems (and latterly the Internet), to gather customer information and to market and sell financial services. It is no accident that some of the major non-bank competitors moving into financial services are those already managing large relational databases. Thus, companies such as American Telephone and Telegraph (AT&T), using their extensive database of customer information and bill payment patterns, became one of the largest issuers of credit cards in the United States in the 1990s (Anonymous, 1993). AT&T sell their products over the phone without the need for a branch network for distribution. Similarly, innovations such as scanning and the universal product code, initially designed for logistical and inventory purposes, now enable grocery retailers and others to generate and maintain all manner of marketing information about consumers which is not available to banks.

The effect of these shifts in information technologies is that bank branches are no longer privileged channels for acquiring and distributing information necessary for the production, marketing and distribution of retail financial products. In addition to the regulatory differences that remain between banks and non-banks – in terms of FDIC insurance costs and reserve requirements – phone companies, retailers, insurers and others have illustrated their ability to separate the production of valuable information concerning customers' finances and spending patterns from its traditional association with the physical infrastructure of bank branches. As such, bank branches, throughout the 1980s, came to represent relatively expensive channels through which to generate marketing information and to deliver financial products.

Rethinking Retail Banking: From California to Britain

Any discussion of how Californian banks responded to their difficulties in the 1980s and 1990s has to focus on the four dominant retailers (Bank of America, Security Pacific, Wells Fargo and First Interstate) that accounted for over 70% of state banking assets in 1990. The erosion of their deposit bases and the depressed demand for domestic credit in the 1970s, led the Bank of America, Security Pacific and, to a lesser extent, Wells Fargo to move into overseas markets, lending to LDCs and expanding their operations in Europe. In 1986, reeling from losses on LDC debt, the Bank of America announced a second-quarter loss of US $640 billion (at that time the second largest loss in US banking history). To appease regulators and fend off a take-over bid from the Los Angeles based First Interstate, the Bank of America wrote off bad debt and raised money by selling some of its prized assets, including its discount brokerage, Charles Schwab. Nevertheless, the Bank of America's financial difficulties in the mid-1980s did prevent it from participating fully in California's commercial real estate boom and bust. Californian banks were some of the biggest investors, nationwide, in commercial real estate (Warf, 1994), tempted by the rapidly inflating asset prices of the mid 1980s and feeling the effects of disintermediation and securitisation which were squeezing the volume and profitability of their core businesses. Within the industry, the Bank of America is now regarded as something of a 'bellwether case' (Rogers, 1993: 260), in that it survived near failure and has, sooner than many of its competitors, reorganised its operations to focus on its core, domestic retailing business in California. These four retailers also had something of an advantage over their smaller competitors in that they all, to differing degrees, had a relatively diverse earnings base with some presence in Arizona, Nevada, Oregon and Washington which helped cushion them from the worst of the Californian recession.

The largest casualty of the recession in California was the Los Angeles based Security Pacific Bank which merged with Bank of America in 1991. Otherwise, and despite the severity of the 1990–4 recession, California's banks fared reasonably well compared with the banking industry nationally. California's banking woes wiped out less than 2% of the industry's assets in California, while Connecticut,

Illinois, New Hampshire and Texas suffered bank failures involving 20–44% of their banking assets (Federal Deposit Insurance Corporation, 1996). Between 1980 and 1994, over 1600 banks failed nationally; 87 of these were in California but were mainly small community banks or new entrants chartered during the 1980s.

THE REORGANISATION OF PRODUCTION IN BANKING

Throughout the 1980s, the four big retailers introduced a series of changes to reorganise the geographical anatomy of their production systems at an intra-regional and intra-urban scale (Pollard, 1996). One the driving forces in their reorganisation has been an attempt to rein in the costs associated with the operation of their branch networks, given the proliferation of competition from non-banks. To this end, the major retail banks all started stripping labour-intensive cheque and loan processing work out of branches in the 1980s and moving it into centralised facilities. For cheque-processing operations, this meant developing larger, regional data-processing centres and investing in larger, faster machines to reap economies of scale. In 1982, one of the largest retailers consolidated parts of their cheque-processing operations, shifting work out of branches and into 85 separate centres state-wide. These 85 centres have since been reduced to 11, with plans to move to just two centres, located in Los Angeles and San Francisco where their branch networks are at their most dense (Pollard, 1995a). In California, data-processing centres in Los Angeles and San Francisco are handling ever greater proportions of banks' processing work as banks close the smaller data centres in places such as Sacramento, Fremont and San José (see Figure 3).

Californian banks have also been at the forefront of banks' attempts to use technologies to reduce routine transaction handling in branches. In 1992, for example, the Bank of America, Wells Fargo, Security Pacific and First Interstate filled four of the top five slots in the *American Banker*'s national survey of Automated Teller Machine (ATM) investment. Although, since the early 1970s, banks had experimented with using telephone systems for simple bill payment services (Groh, 1986), full 'tele-servicing' – allowing customers to use telephones for a range of account enquiries – was an innovation of the early 1980s. In 1982, the Los Angeles based Union Bank piloted telephone banking for its premier customers and, in 1983, Security Pacific extended the concept, offering telephone banking to all of its customers prepared to pay a monthly charge. Telephone banking is now standard practice for the major retail banks and is being supplemented by various versions of on-line and Internet banking. Banks, then, like their non-bank competitors, have been trying to expand the number of different channels through which they deliver their products; routine account enquiries can now be dealt with via ATMs, telephones, PCs and, as a last resort, bank branches.[13]

On the loans side, centralisation has meant automating credit analysis by investing in credit-scoring systems for standard products such as car loans and home loans.[14] As transaction handling and credit analysis have been taken out of

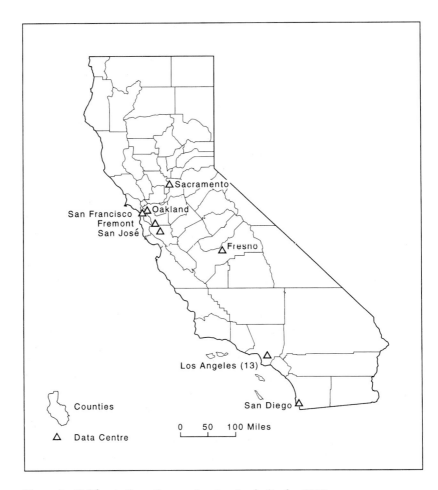

Figure 3 California Data Centres Serving Study Banks, 1970

branches, banks' data centres and loan centres have become increasingly important foci for information gathering, while branch-based local knowledge, with all the trappings of 'relationship banking', are of declining importance for many mass-market products. There is a good deal of variation, however, in how banks handle more complex kinds of loans. Arguably, the large retailers may have lost out on some business lending, particularly to small companies, because their scale interferes with their ability to be perceived as a 'local' bank and their impersonal, centralised lending authority encourages them to reject loan applications with any marginal elements (Pollard, 1995a). Three of the largest retailers are trying to maintain some specialised lending staff in branches, however, by operating a 'hub and spoke' system in which staff in designated branches, specialised in certain kinds of debt and investment products, take referrals from other branches.

Once designed as microcosms of an entire bank, branches have gradually relinquished their role in the transactions handling aspects of operations and lending and, since the early 1980s, have been adapted to function as sales outlets. Banks have tried to introduce more of a 'sales culture' into their branches, in part by offering a broader range of financial products such as unit trusts, that are not backed by FDIC insurance,[15] and by discouraging customers from using branches for routine transactions.

Another indicator of the increased emphasis on sales generation has been the extension of branch opening hours in high-volume locations. Banks' ability to decide where and when to extend their opening hours has been enhanced by their ability, developed in the late 1980s, to use branch-based technologies to measure their customer flows (Pollard, 1995b). Banks are also making greater use of information technologies to segment their customer base by age, income and postal code in order to customise financial products for, and target their marketing efforts towards, their most affluent customers. The logical extension of such a strategy is the provision of 'private banking facilities', such as those opened by the Bank of America in Beverly Hills, to service their high-net-worth customers. The cream of banks' clientèle who maintain high balances and/or routinely conduct business which generates fee income for the bank receive the time and attention of a designated banking officer in their home branch.

What of the branches that are not in Beverly Hills, the branches that do not attract the deposits of high-net-worth individuals? At the other end of the spectrum from the process which produces private banking services in Beverly Hills is the re-evaluation of the viability of branches in low-income neighbourhoods such as South Central Los Angeles (see Dymski, Veitch and White, 1991; Pollard, 1996). With sales measured per square foot of floor space and product managers now competing for space in branches at the large retailers, banks, like their less-regulated[16] non-bank competitors, want to concentrate their resources in high- and moderate-income areas. The result is a pattern of selective branch closures in low-income neighbourhoods throughout the 1980s in Los Angeles, Chicago, New York, Boston and other US cities (Towle, 1990; Campen, 1993). This withdrawal of bank infrastructure is not a development of the 1980s, as community groups and residents of South Central have watched formal financial institutions being replaced by cheque-cashing outlets and pawn brokers since the early 1970s. However, there is evidence that branch closures have increased in the 1980s as banks have faced increasing competition.

THE CHANGING DEMAND FOR LABOUR IN BANKING

Finally, these shifts in how banks organise their production systems have generated some significant changes for those working in the banking industry. Banking employment in California grew steadily in the 1980s, peaking in 1987 and declining throughout the early 1990s as the recession hit and banks failed (Pollard, 1995a). The merger of the Bank of America and Security Pacific in 1991 was especially significant, generating an estimated 15 000–20 000 job losses (Moffat, 1991). Job losses, however, are only the tip of the iceberg in terms of how employment relations are changing in the banking industry in California.

Table 4 Bank Employment by Occupational Group, Los Angeles, 1970–1990

Occupation	1970	%	1980	%	1990	%
Managerial	10 800	25.87	18 780	27.09	26 529	30.9
Professional	1050	2.51	1780	2.57	4889	5.69
Sales	300	0.72	920	1.33	2112	2.46
Clerical	27 900	66.83	45 940	66.21	49 945	58.16
Craft/Lab/Service	1700	4.07	1940	2.8	2395	2.79
Total	41 750	100.0	69 360	100.0	85 870	100.0

Source: US Bureau of the Census (1970b), (1970c), (1980) and (1990)

Throughout the 1980s and early 1990s, banks in California overhauled their employment relations, significantly changing their recruitment, training, promotion and remuneration practices. Banks became ever more automated and sold increasingly complex products and this was reflected in the changing occupational structure of the industry (see Table 4). Professional occupations – which include the computer programmers and systems analysts responsible for banks' automation – are now growing in importance numerically and symbolically. The largest banks in California have entire divisions of their administrative structures devoted to managing their technology, whereas in the 1970s technology was viewed largely as an adjunct to the workings of different departments.

At the other end of the occupational spectrum are bank tellers. During the late 1970s and early 1980s, banks trimmed the labour costs associated with their branch networks by centralising their data-processing operations. In the late 1980s and early 1990s, as the centralisation of data processing approached its limits, banks turned to consultants to recommend ways in which they could reduce the costs associated with those workers remaining in the branches. The solution adopted by the largest retailers involved the replacement of full-time staff with part-time and hourly workers, allowing banks to shed benefit costs which accounted for 30–35% of the wage bill. In 1989 only 33% of bank tellers in Los Angeles worked full-time all year, compared with 50% in 1969 (Pollard, 1995b).

As the skill sets required by banks are changing, so are their recruitment and training regimes. During the post-war period, banking offered a relatively high status, white-collar environment in which workers with (predominantly) high school education, could receive training and work their way through the ranks from bank teller up to branch manager and so on (Jackall, 1978; Noyelle, 1987). This is no longer the case. As banks have started to seek more specialised skill sets, they are recruiting college graduates and opening up their internal labour markets, at all levels of the occupational hierarchy, to the external labour market. There are now multiple points of entry into the industry and bank branches are no longer key sites for recruitment, training and then feeding workers into other parts of banks' occupational hierarchy as they were in the 1970s. In essence, banks are now recruiting specialists with proven track records in other industries, rather than training generalists. By way of example, the first generation of data-processing personnel in California banks in the 1960s and 1970s were drafted from the ranks of banks' operations officers; in the 1990s, such technical occupations

Table 5 Employment Shares by Median Hourly Income, Bank Workers in Los Angeles, 1979–1989 (benchmark = 1979)

Los Angeles Banking Industry	1979 Total	1979 %	1989 Total	1989 %	Change in share
All workers					
Low wage	3920	5.65	4919	5.73	+0.08
Middle	57 620	83.07	64 882	75.55	−7.52
High wage	7820	11.27	16 069	18.71	+7.44
Males					
Low wage	1060	5.37	1613	5.93	+0.56
Middle	13 220	66.27	17 037	62.59	−4.38
High wage	5460	27.66	8569	31.48	+3.82
Females					
Low wage	22 860	5.76	3306	5.64	−0.12
Middle	44 400	89.48	47 845	81.57	−7.91
High wage	2360	4.76	7500	12.79	+8.03

Source: Calculated from US Bureau of the Census (1980) and (1990)

Note: Median hourly wages for all bank workers in Los Angeles in 1979 were US$8.74 per hour (in $1989) (US Bureau of the Census, 1980)

are filled by computer science graduates and/or individuals head-hunted from other banks.

These shifts in the occupational composition, distribution of hours and patterns of recruitment are making their effects felt on the wages of bank workers. Three trends are discernible in census data for Los Angeles, the largest banking labour market in the state. First, although the banking industry is not a high-wage industry when compared with all industries in Los Angeles,[17] it is becoming a higher-wage industry in real terms. Throughout the 1980s, real hourly wages grew more rapidly in the banking industry (over 18% between 1979 and 1989), and in all its occupations, than those in the Los Angeles economy as a whole. Second, many of the largest banks in California are now turning to performance-related pay and report more room for individual wage bargaining, especially in senior and technical[18] job grades (Pollard, 1995a). Third, as real wages rise in the industry, the distribution of those earnings is becoming more polarised.

Table 5 divides up the earnings distribution in the banking industry into three (high-, middle- and low-wage) segments. The high-wage segment is defined as the percentage of bank workers earning more than 200% of the median hourly earnings for all bank workers in 1979, adjusted for inflation in 1989; and the low-wage segment is that portion of workers earning less than 50% of the median (Harrison and Bluestone, 1990). Table 5 shows the combined effects of changes in the earnings distribution *and* changes in the level of real hourly wages. Overall, as hourly earnings rose between 1979 and 1989, the low-wage share of employment for all bank workers increased by 0.8% whereas that of the high-wage workers rose almost 7.5%. The changing contours of labour demand in the industry have not only stretched the earnings distribution and opened up greater distances in terms of real hourly earnings between low- and high-wage workers, they have also started to close the gap in earnings between men and women.[19] Despite this convergence, however, there are greater differences opening up between particular groups of men and women. Hispanic men and women, for example, are disproportionately concentrated in the low-wage segment of the earnings distribution, while white men and women have been the chief beneficiaries of the growth of high-wage occupations (Pollard, 1995a).

Banking labour markets have changed significantly, then, in the 1980s and early 1990s. The banking industry is gradually shedding its labour-intensive and 'low-tech' image as its occupational structure shifts, albeit slowly, towards more managerial, professional and sales[20] occupations. The on-the-job training that moved white males and, in the late 1970s and 1980s, women and minorities, from the teller line to management has not disappeared entirely, but it is no longer expected that workers will enter banks in junior positions and spend their career moving into managerial positions. Changes in recruitment and training, increasing wage rate dispersion and differentiated access to work hours, implied by the increase in part-time and hourly work in some parts of the occupational hierarchy, are generating new cleavages and lines of horizontal and vertical division within banks.

From California to Britain

Many of these shifts described for banks in California resonate with changes being introduced in British retail banks in the 1990s (Leyshon and Pollard, 1998), although the effects of increasing competition are being worked through a different social, cultural and regulatory context in Britain. Like their US counterparts, British banks are operating in a changing regulatory environment which is eroding the oligopoly they enjoyed in the post-war period and increasing the number of players allowed to compete in different retail financial markets. In the 1980s, for example, building societies entered retail banking and insurance markets following the passage of the Building Societies Act of 1986. The flotation of the Abbey National and the Trustee Savings Bank (TSB), together with the growth of the Royal Bank of Scotland, also marked something of an escalation in competition for the big four.

Further, in addition to making their fair share of lending mistakes,[21] and facing a reduced demand for credit in the recession of the early 1990s, British banks are now competing with insurers, clothing and grocery retailers, utility companies and even professional football clubs who have decided to move into retail financial services. Although the cast of non-bank characters entering retail financial markets in the United States and Britain are slightly different, their effects are similar. In the British context, it is perhaps the large grocery retailers that wield considerable market power, operate successful loyalty-card schemes and exploit the growing synergy in how they and banks collect, manage and use customer information, which pose the most serious threat to banks. Relying on advanced marketing skills and banks' reputations for poor customer service (Riley, 1996), grocery retailers are using their strong brands to persuade some major banks –

Abbey National, Royal Bank of Scotland and Midland to name only three – to sign up to grocery retailer's own-label projects, forming expensive alliances which may yet generate some conflicts with banks' own business strategies (Alexander and Pollard, 1998).

Partnerships with grocery retailers are one option for banks in Britain, especially those with advanced telephone banking and data-processing systems. As in the United States, however, branch networks are key sites for British banks' strategies for cost containment, the generation of fee income and changes in employment relations. Since the early 1980s, banks have upgraded computing facilities in branches, introduced ATMs and centralised transaction handling in regional processing centres, seeking scale economies in processing and freeing up space in their branches for sales work (Leyshon and Thrift, 1993; Cressey and Scott, 1992). The National Westminster Bank, for example, is consolidating its regional processing centres from 150 to 50 (Donovan, 1996). This is centralisation on a rather different scale to that in California, given that banks there support branch networks across the western United States with just two data centres (one in Los Angeles and one in San Francisco), but the logic behind it is very familiar.

Similarly, banks in Britain are looking for cheaper ways to sell their products and telephone banking – a Californian innovation – has become increasingly popular for standardised products. In 1989, the Midland Bank launched First Direct as a branchless, telephone bank (although customers could have access to Midland branches), and other banks and building societies have followed suit in extending telephone access to business and personal customers. The Abbey National and others are now using charges to discourage customers from using branches if they operate a telephone-based account (Mackintosh, 1997). As bank branches are transformed into sales outlets, a new geography of bank infrastructure and financial exclusion is unfolding in Britain (Gentle and Marshall, 1993; Leyshon and Thrift, 1995). The Banking, Insurance and Finance Union (1995) estimate that 2800 bank and building society branches have closed since 1989. On both sides of the Atlantic, private banking facilities for the wealthy are now juxtaposed with branch closures in low-income neighbourhoods.

With regard to labour market changes, again the British context contrasts with the almost entirely non-union context for bank workers in the United States (Bertrand and Noyelle, 1988), but the changes underway are familiar, albeit less extreme in some respects. Between 1989 and 1995, there was a net decline of 90 000 bank employees in Britain (Mintel, 1995) as banks closed branches, centralised operations and extended their networks of ATMs. There is evidence to suggest that career trajectories are changing, with diminishing possibilities for vertical mobility from clerical job grades (Cressey and Scott, 1992) developing alongside fast-track schemes to promotions and high salaries for qualified professional workers. Banks are also introducing more part-time and 'prime time' labour into their branch networks (Leyshon and Thrift, 1993), although it is not clear that this practice is as widespread in Britain as it is in the Californian banks. In the Bank of America's branches in 1993, for example, just 19% of branch staff were full-time employees (Howe, 1993a).

The Globalisation of Retailing Innovations

Many of the recent innovations adopted in British retail banks (telephone banking, private banking facilities for high-net-worth individuals, the increasing centralisation of transaction handling and credit analysis, and the replacement of full-time workers in the branch system with part-time and hourly workers) have been imported from banks in the United States. This raises all manner of questions about how banks, and indeed other firms, identify and define 'problems' and their 'solutions'. For example, how much do British banks know about the sources of some of the practices they are now introducing? Do such innovations represent seemingly obvious routes to profitability? Are they perceived as practices tried and tested in one of the most competitive banking markets in the world? How important is the financial media in influencing banks' perceptions of 'best practice' and their definitions of what constitutes 'a problem'?

These issues can be raised not just in an international context in which retail banking innovations are, apparently, being transmitted across the Atlantic, but also in a regional context. For example, an important innovation for California's banks in 1987 was Wells Fargo's introduction of a staffing model which provided a template for replacing full-time workers in bank branches with hourly and part-time workers, thus reducing staffing and benefit costs in branches. This model was the product of an Atlanta-based consultancy firm, commissioned by one of Wells Fargo's Executive Vice-Presidents, to develop cost-cutting strategies for their branch network. This staffing model was influential in several respects. First, the individual who commissioned the research left Wells Fargo in 1987 to join Security Pacific and, with the 1991 merger with the Bank of America, moved into the Bank of America hierarchy. Both Security Pacific and the Bank of America introduced similar staffing models on his arrival (Howe, 1993b). Second, this model of cost cutting was widely discussed within the industry, and became something of a touchstone for those managers who felt that bank workers, accustomed to full-time employment, stable hours and job security, needed the equivalent of a short sharp shock to awaken them to the changing realities of banking in the 1980s. Third, the debate surrounding the model generated some pressure for human resource managers at other California retailers to 'conform' with the media-defined 'industry leader' and introduce similar changes within their organisations (Pollard, 1995b). Yet, some human resource managers, including those who introduced more part-time and hourly work to their banks, expressed ambivalence about such changes, fearful of their effects on turnover, customer service and morale; others derided the Wells Fargo model and the herd mentality that they felt was overtaking some of their peers (Pollard, 1995a and 1995b).

This is just one illustration of how the priorities of an executive vice-president at one bank, mediated through consultants, can become the basis for debate and for identifying and legitimating a series of changes that affect branch workers in California and elsewhere. Why did this particular reading of how banks might cut branch costs, which defined 'the problem' in terms of full-time and full-year employment, become the model that circulated throughout the industry? To understand this, it is important to explore the sources of managerial discourse in the banking industry – quite literally, who says what, who is listened to and how ideas are communicated, interpreted and legitimated. In the United States and Britain there are networks of key actors (regulatory authorities, consultants and managers), texts (financial media and management journals) and artefacts (computer hardware and software) that play significant roles in producing and encouraging the circulation of managerial discourses about best practice within the industry. From the perspective of this actor-network theory (Law, 1994), it can be argued that understanding the globalisation of retail banking innovations is synonymous with describing a network of social relations that govern how the ideas and theories which now underpin and legitimise the transformation of the retail banking industry in the United States and Britain have been produced and circulated throughout the industry (Leyshon and Pollard, 1998).

Conclusions

This chapter has illustrated some of the ways in which the nature and spatial structure of retail banking in California has evolved as a product combining aspects of a global financial system – changes in the organisation of credit provision, the intensification of competition and so on – with the idiosyncrasies of US financial history, federal and state regulation, and the post-war development of the Californian economy. This banking structure has been reconfigured since the late 1970s with a series of structural changes and the worst financial crisis in the United States since the 1930s. From this reconfiguration some links can be made between what has happened in California and the reorganisation of retail banking in Britain. For all the differences that remain between the two systems, the crumbling of inter-war regulation and the on-going reorganisation of the US banking system are eroding the distinctiveness of the US system.

In California, retail banks are heading up-market, dedicated to generating fee income by selling a widening range of insured and uninsured products in rapidly changing, intensely competitive, retail markets. Banks' information-gathering, marketing and product-delivery capacities are now multi-channel and becoming increasingly divorced from the physical infrastructure, technologies and personnel of their branch networks, networks that once constituted significant barriers to entry for non-bank competitors. Banks are working harder to identify, and do business with, groups of consumers that can generate the most fee income, customising their services to cater for changes in how different socio-economic groups consume retail financial services. The Californian vision of retail banking is one of an increasingly high-tech, exclusionary business, managed by a diverse group of highly qualified and well-remunerated professionals, and supported by a population of relatively low-paid part-time and hourly staff with little or no prospects of career development.

These themes resonate with much of the recent British experience of bank restructuring. California, it might be argued, has set an example for bank restructuring in Britain, and further down the road perhaps, for restructuring in other European Union member states. Many of the innovations adopted by British banks have been imported from the United States and US banks, in turn, have imitated staffing practices developed by retailers, and watched US commercial airline companies develop and test the telephony systems that are now used in their teleservice centres. This raises the issue of how banks in the United States, Britain and elsewhere are making choices about technologies, branch closures and employment changes. Within the industry in California, there are particular managers and consultants who have been very influential in defining and promoting a vision of how retail banking should be conducted in the 1990s. Their chosen route to profitability is not without its contradictions or detractors, nor is it the only route to profitability available for retail banks. In looking at the reorganisation of retail banking and the growing similarities in how banks on both sides of the Atlantic are restructuring, it is important to consider the sources of managerial discourse and to understand how, and for which constituencies, ideas about best practice are constructed.

Acknowledgements

Some of the material presented is based on research funded by the Geography and Regional Science Programme of the National Science Foundation, grant number SES9205339.

Chapter 4
A 'Possibilist' Approach to Local Financial Systems and Regional Development: The Italian Experience

Pietro Alessandrini and Alberto Zazzaro

Introduction

Over the last 10 years, great strides have been made in the analysis of the regional structure and location of banking and financial systems from both theoretical and empirical points of view.[1] This chapter deals with the subject by focusing on the experience of the Italian banking system. There are at least three good reasons why the Italian banking system and its geographical distribution offer an advantageous viewpoint from which to study the role of banks in regional development. The first concerns the economic regional disparities of the Italian economy. Italy is still clearly divided into two distinct areas: the Centre North, which is industrially very advanced and one of the most developed areas in Europe, and the South (the so-called 'Mezzogiorno'), which suffers from serious structural economic problems and poor growth prospects. Given this context, the question might be whether this spatial economic dualism is matched by an analogous financial dualism, and, more generally, how the geographical distribution of industrial activities and financial systems interact and evolve.

The second reason is that the Italian banking system has always been characterised by the presence of many small local banks operating in restricted territorial areas and a few (not particularly large) national banks. This banking structure is the result of the institutional framework of the Italian banking system, which has been marked for a long time by a very restrictive regime in terms of the geographical mobility of banks and their operative sphere; and the structure of Italian industry, which is largely based on localised networks of small and medium-sized firms or what have become known (in neo-Marshallian parlance) as 'industrial districts'. Although the production and labour systems of these celebrated industrial districts have been the focus of considerable study in recent years, by comparison the role of local financial institutions (such as banks) in the development and reproduction of these districts has received scant attention.[2] Both of these elements – the peculiarities of the institutional framework of the Italian banking sysytem and the high degree of localisation of many of the country's most successful industries – have made the local bank a primary actor in the development of many local economies.

The third reason is that, since the second half of the 1980s, the Italian banking system has been affected by radical legislative, administrative and institutional changes, such as the privatisation of many banks, the introduction of the 'universal banking' model, the nearly full liberalisation of credit markets and the consequent intensification of competition.[3] These changes, still in progress, are modifying the competitive environment of both local and national banks. The traditional coupling of 'local bank–small firm' will have to be completely rethought in the light of the new institutional framework, a subject certainly deserving of careful analysis.

The chapter is organised as follows. First, there is a brief description of some of the key structural features of the Italian economy. The following section then deals with the geographical organisation of the Italian banking system. This moves on to an analysis of the different approaches to interregional financial integration, and then a discussion of alternative models of the structure of regional banking systems and problems of competition between local and national banks. The conclusion explores some possible ways of facilitating the diffusion of financial innovations into peripheral areas.

SOME KEY FEATURES OF REGIONAL ECONOMIC DISPARITIES IN ITALY

As is well known, the Italian economy is characterised by marked disparities in regional development. Taking the 179 regions of the European Union (EU) as a reference (see Table 1),[4] at the beginning of the 1990s Italy had six regions in the group of the most advanced 20, and 11 regions in the group of the 60 with levels of GDP *per capita* above the EU average. All these regions are located in the Centre North, which includes Umbria, with an income level *per capita* near the EU average. The remaining eight of the 20 Italian administrative regions are less developed and all of them are located in the South. With the exception of Abruzzo, they are eligible for financial support from the EU regional development programme (Objective 1) reserved for the areas whose GDP *per capita* is less than 80% of the EU average. Among several indicators of economic backwardness that could be shown, one of the most significant is the unemployment rate, which in the Mezzogiorno reaches levels that are significantly above the European average and much higher than those recorded in the Centre North (see Table 1). This development gap between the South and the Centre North has remained basically the same since the beginning of the 1970s, despite the fact that in the two preceding decades there were encouraging signs of the gap closing.[5] The continuation of the North–South dualism is the result of two sets of circumstances.[6]

On the one hand, the southern regions have been lagging behind on account of contingent obstacles and failures in public intervention (that is, the poor performance of state-owned enterprises). Moreover, since the 1970s, there has been a dramatic transformation of public policy for the Mezzogiorno, from a policy supporting investments to a policy supporting incomes.[7] On the other hand, there

Table 1 Regional Economic Development in Italy: Selected Indicators

Regions	Average GDP per capita 1989–1991	Per capita GDP European Union regional ranking	Average unemployment rates 1991-92-93
North West			
Lombardia	134.7	8	41.8
Valle d'Aosta	129.6	9	82.8
Piemonte	119.6	20	76.4
Liguria	115.8	27	94.9
North East Centre			
Emilia Romagna	127.5	10	46.2
Trentino Alto Adige	122.0	18	32.7
Friuli Venezia Giulia	121.6	19	57.5
Lazio	116.8	22	108.8
Veneto	116.6	23	47.5
Toscana	109.4	39	80.7
Marche	104.7	49	66.7
Umbria	98.9	66	94.7
South			
Abruzzi	90.2	100	116.1
Molise	78.8	128	156.5
Sardegna	74.2	139	196.2
Puglia	74.1	140	157.5
Campania	70.2	143	224.4
Sicilia	67.5	145	230.4
Basilicata	64.5	147	223.5
Calabria	57.9	152	206.4
Total EU	100	179	100

Source: European Commission, *Report on Regions in Europe* (1994)

is the striking success of the North East Centre (NEC) regions, which in a few decades have reached levels of development close to those of the already industrialised North West, although their pattern of development has been very different. The notable performance of the NEC regional system, as it has been called by Fuà (1983), is due to a peculiar combination of factors,[8] among which a widespread entrepreneurial attitude (due to historical traditions of self-sufficiency), organisational flexibility (due to the small size of local firms) and social cohesion (due to the small size of the towns) stand out as the most important. The result is a model of development based on a large number of diversified local economic systems (or 'districts') scattered throughout the NEC area (and often grouped under the label of the 'Third Italy').

Italian economic geography, therefore, is very diversified and complex. For the purposes of this chapter it will be sufficient to look at the three main areas of the country: the North West (an area of long-standing industrialisation), the NEC (more recently industrialised) and the South (still in need of reducing its development gap with the rest of the country). Table 2 shows some distinctive features of

Table 2 The Regional Characteristics of Firms in Italy, 1991 (% in Each Area)

Regions	North West	North East	Centre	South	Italy
Number of employees		Size			
Less than 10	38.4	44.1		61.1	45.4
10–49	21.0	21.9		20.4	21.3
50–199	11.6	10.1		8.7	10.4
200–499	6.7	5.1		3.8	5.3
500–999	4.3	3.2		2.2	3.4
More than 1000	18.0	15.6		4.1	14.2
		Legal type			
Individual firms	67.3	68.7		81.4	71.7
Companies	32.7	30.3		18.6	28.3
of which: limited	1.5	0.9		0.4	1.0
		Localisation			
Mono-localised	94.2	94.0		94.5	94.2
Multi-localised	5.8	6.0		5.5	5.8
of which: nationals	0.8	0.6		0.2	0.6

Source: ISTAT

firms in each of the three areas. Generally speaking, the Italian productive structure is characterised by a large number of small firms, run at a family level and operating from one plant. There are very few medium-sized firms and limited companies. This peculiarity is most evident in the South, where four out of five workers are employed in firms of fewer than 50 employees, in artisan workshops and where the number of limited companies is well below the already low national level.

Taken as a whole, the territorial and productive features outlined briefly here are sufficient to highlight four main sets of problems that local credit systems have to face:

1. *problems of interaction* between real and financial sectors in regions with different levels of development – this concerns the question of which reciprocal influence tends to prevail at the local level between banks and firms;
2. *problems of integration* between the global and the local level, or between centre and periphery – this concerns the financial division of work between central areas and peripheral areas, which is particularly relevant within a unified monetary system;
3. *problems of dimension* connected not only with the size of firms, but also with the size of banks – it is necessary to understand what kind of bank (small or large, local or national, specialised or universal) is best suited to foster the development of peripheral systems of small firms;
4. *problems of innovation* concerning the introduction of those financial innovations which are most suited to regional needs – the main point here is how to balance innovative and traditional financial instruments for local savers and investors.

Problems of Interaction Between the Banking System and Regional Development

In principle, a region could do without a regional banking system, as so often happens in other industries. Local borrowers could raise the funds they need from banks located elsewhere. In an ideal world, with perfect information flows and no transaction costs, the multiplication of local banks would actually represent a waste of resources. However, in such an ideal world the sheer presence of banks would not be justifiable. Investors could directly make over their IOUs to savers, underwriting complete contracts (i.e. contracts which contain clauses for any contingencies) whose fulfilment would be enforceable by a third party such as, for instance, a court of law. In this hypothetical case, of course, there would be no need for financial intermediaries, as the efficient allocation of savings among productive activities would be ensured by the market mechanism.

Yet, the real world is quite different. Transactions are costly and information about agents is poor, so that it is never possible to design complete and fully enforceable contracts. To raise funds on the market, that is to make tradable one's own liabilities, requires the trust of a number of savers. This can be very costly. The role of banks is that of certifying the credit-worthiness of investors, monetising liabilities otherwise not tradable.[9] Banks, therefore, following Schumpeter's arguments, represent a fundamental device for the selection of successful entrepreneurs.

From this point of view, the existence of banks and their geographical diffusion are two 'parallel' phenomena: either one cannot justify the existence of banks or one has to recognise that banks will inevitably spread throughout the regions. The same market imperfections that explain why banks exist are what make credit markets (partly) spatially segmented. In this context, the emergence of powerful economies of agglomeration and the existence of transaction and information costs explain the formation of financial centres and the organisation of the financial system on a hierarchical order from centre to periphery.[10]

Regional banking systems represent the link between local economies and financial centres. On account of the spatial segmentation of credit markets, banks operating in a region (that is both local and external banks) are indispensable for overcoming the isolation of those local agents who are either so small or so 'new' that transaction and information costs are usually too high to permit them to access financial centres. Thus, banks operating locally are the main channel (often the only one) through which the financial needs of small and medium-sized firms are met. In turn, however, the geographical distribution of the banking system and its performance are themselves influenced by the level of economic development of the different regions. So, in a sense, the structure of the regional banking system mirrors the local economic system.

This interaction is clearly confirmed in the case of the Italian banking system. As shown by Figure 1,[11] the geography of the Italian banking system tends to reflect the geography of the productive system. In Italy there are two financial centres. The most important is Milan, which operates mainly with private financial flows. Second, Rome, as the capital, is concerned mainly with the centralisation of financial flows associated with the public system. Therefore, Lombardia (for Milan)

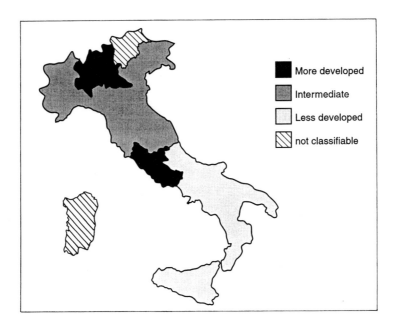

Figure 1 The Geography of the Italian Banking System, 1992. Source: Alessandrini, A. (1996). Note that the other areas are not classifiable by cluster analysis

and Lazio (for Rome) are the two regions with the most advanced forms of banking. The same indicators show an intermediate level of banking development for the other regions of Centre North and a much lower level for the Mezzogiorno, in line with the disparities in the productive system summarised above.

The provision of banking in the Mezzogiorno is significantly lower (Table 3). In that area, the population served by a single bank branch is 36% higher than in north-western regions and more than double that of the NEC regions. Only 60% of the local municipalities of the Mezzogiorno have an operative bank branch. This contrasts strongly with the NEC regional system, where diffused urbanisation and industrialisation result in a diffused presence of banks. Finally, in the Mezzogiorno deposits are smaller in relation to GDP than in the nation as a whole, and the difference is much wider when the percentage of loans to GDP is considered. The narrow gap from the deposit side is the result of the policy of income support adopted by the Italian Government over the last 20 years as one of the main kinds of intervention in the Mezzogiorno. On the other hand, the wider gap from the loan side is closely related to the slow industrial development of the area.

This geography is confirmed by other key behavioural and performance indicators, such as loan interest rates, the ratio of bad loans to total loans, the share of collateral on loans granted, and the gross operating income or the net earnings of the banks (Table 4). The low performance of banks in the South is attributable to several interacting causes. These are partly a result of the low profitability or wealth of southern borrowers, and partly of the internal inefficiencies of local banks. Lending to firms located in this area is more risky and this

Table 3 Indicators of Regional Banking Structure in Italy

	North West		North East Centre		South	
	1990	1996	1990	1996	1990	1996
Population per branch	2772	2680	2650	1659	5059	3751
Municipalities served by banks/Mun.	0.55	0.62	0.84	0.88	0.55	0.62
Deposits/GDP*	0.57	0.60	0.55	0.56	0.43	0.46
Loans/GDP*	0.48	0.54	0.42	0.48	0.27	0.31

Sources: Banca d'Italia; Svimez; and Banca d'Italia and ISTAT. * The data refer to 1990 and 1993

Table 4 Banking Performance Indicators (% Values)

	North West		North East Centre		South	
	1990	1996	1990	1996	1990	1996
Loan interest rates	14.1	12.4	14.7	12.8	16.2	14.8
Bad loans/total loans	2.9	4.7	5.1	8.8	8.0	26.6
Collateral/loans	33.9	29.8	45.0	2.3	74.8	55.7
Interest margin/total assets*	2.7	2.6	3.1	3.1	3.1	3.3
Intermediation/total assets*	4.1	3.7	5.8	4.3	4.1	4.2
Gross operating income/total assets*	1.7	1.4	3.1	1.6	1.5	1.2
Net income after taxes/total assets*	0.33	0.27	0.31	0.34	−0.24	−1.4

Sources: Banca d'Italia and Bilbank. * The data refer to 1993 and 1995

Notes: Interest margin = interest on loans, bonds + dividends − interest on deposits and other liabilities
Intermediation margin = interest margin + fees received + financial trading profits − fees paid
Gross operating income = intermediation margin − labour and other administrative costs
Net income after taxes = profits (losses) from ordinary activities + unexpected profits (losses) − taxes on income

could explain the higher interest rate and the higher share of bad loans as compared with other areas. However, empirical studies have clearly shown that southern banks perform a less efficient screening of loan demands.[12] Moreover, the labour and operative costs of southern banks are higher. The southern banks' ineffectiveness in screening investment projects and their higher operative costs are partly transferred to customers through higher interest rates[13] and partly contribute to lowering banks' income, which in recent years has been negative.[14] In turn, this behaviour adversely affects the quality of borrowing firms. All this creates a self-reinforcing vicious circle that can lock local credit markets into long-term situations of high interest rates and high default rates. Furthermore, issues of 'group reputation' can arise, creating discrimination in the credit market. Banks base their valuations of new firms, at least in part, on the *ex-post* average default rates for the same kind of firms operating in that area in the past. As a result, similar firms, simply because they operate in different areas, may have to borrow at different interest rates (Lang and Nakamura, 1990; Scalera and Zazzaro, 1997). Finally, southern banks tend to concentrate more on traditional banking activities (deposit collection and lending) than banks operating in other areas, as is shown by the wider gap between the interest margin and the intermediation margin.

PROBLEMS OF INTER-REGIONAL FINANCIAL INTEGRATION

THREE APPROACHES: PESSIMISM, OPTIMISM AND POSSIBILISM

Experience shows that the financial structure is not uniform throughout a country. It tends to be hierarchical with an advanced financial centre at the top and local credit systems becoming gradually less advanced depending on the local development of the peripheral regions in which they operate. It is essential, therefore, to consider the question of financial integration between the global and the local or, to be more exact, between the centre and the periphery.

There are three basic ways to approach this question.[15] The first two approaches present two well-known opposing views concerning the effects of integration: one is pessimistic and the other optimistic. A third point of view, which will be called 'possibilist', falls midway between the two extremes and offers a more realistic and accurate picture of financial integration problems. The first two theories, widely discussed in the literature, both emphasise a strong contrast between centre and periphery, but they differ substantially in their predictions of the effects that integration has on this division.

The pessimistic theory concentrates exclusively on 'localism'. It maintains that the centre would take advantage of asymmetries in information, behaviour and structure and would invest the savings drained out of the peripheral regions, slowing down their development. This could lead to some radical conclusions. First, local segmentation of capital markets is preferable to integration.[16] Second, the setting up of banks from outside the region – whether the opening of new branches, mergers or the purchase of a holding in local banks – is considered detrimental. However, this appears to be a misleading view. The idea of keeping the savings of a region within the confines of that region is not only illusory but could be also counter-productive. Inside a wider unified monetary area, savings must be free to move in search of the best portfolio opportunities and returns. Therefore, the main challenge for a region is to offer the best opportunities for investment, attracting funds from inside *and* outside its boundaries. Besides, statistical data show that often, but particularly in the less advanced regions, local banks have a greater ability to collect savings than to invest them in the same area. In contrast, branches of banks with headquarters in other regions tend to penetrate the loan market of a region more easily than its deposit market.[17] Thus, the platitude that local banks have a higher commitment to keeping savings in their region is not always supported by the evidence. Paradoxically, they tend to have a higher propensity to export capital out of the region than do local branches of external banks.

The optimistic theory, on the other hand, concentrates exclusively on 'globalisation' (or more exactly, integration). The idea is that the increased competition that follows integration will select the best enterprises (banks and firms) and standardise their performance at the levels of efficiency reached in the more advanced centre(s). This opinion leads to the argument that integration removes regional disparities in financial structures and capital availability. Clearly, this assumption of the spread of financial development and efficiency set in motion by the liberalised, integrated 'global' market is a simplistic generalisation typical of neoclassical theory. Experience shows that the advantages of integration and globalisation are not distributed uniformly among firms and regions. Instead, they tend to be more easily acquired by the strongest and best organised ones. So, in the absence of corrective policies, regional disparities could become wider rather than narrower.

In our opinion, neither of these two extreme scenarios is particularly accurate or helpful. A more realistic view is what is here called a 'possibilist' approach, which is based on the continual exploration of the possibility of co-existence, complementarity and interaction between different areas and between the centre and periphery.[18] The territorial integration of local financial systems is best carried out in a selective and gradual way, by trying to give preference to solutions that are best suited to the specific and varying characteristics and adjustment capabilities of local firms, institutions and agents. This preferred mode of reorganisation involves a combination of both passive and active inter-regional integration.

What is defined here as *passive integration* are those processes and developments that arise from outside the region. These are not simply flows of capital and monies into the region, but also the entry of non-resident banks by establishing new local branches, by incorporating (acquiring or merging with) local resident banks, or by buying controlling quotas of their capital holdings. If a region limits itself to this kind of externally imposed integration, it becomes a battleground between inward-looking local banks and agressive outside banks in an inevitably restricted market. Past experience shows that, under these conditions, the quality of credit does not necessarily improve, because the outside banks are most likely to find it profitable to adapt to the conditions prevailing in the local markets. *Active integration*, on the other hand, is when local banks are outward-looking in terms of financial investment and in their branching behaviour. It is important for a regional banking system to compete with other areas in order to gain the benefit of both regional and sectoral diversification. This process of inter-regional expansion increases the chances of benefiting from the gains of competitiveness. It also sets off that integration process called 'possibilist', which allows local banks, firms, savers and institutions to open up to the outside without abandoning their own territorial origins and roots.

Aspects of Inter-regional Integration in Italy

One way of analysing the integration process is in terms of the regional balance of payments.[19] Unfortunately, data limitations preclude detailed analysis of inter-regional flows. However, Table 5 gives an idea of the external constraint of each

Table 5 External Regional Disequilibria (Net Exports as a Percentage of GDP), 1980–1993

Regions	1980–1984	1985–1989	1990–1993
North West	7.5	9.4	9.1
Lombardia	10.0	11.4	11.4
Valle d'Aosta	−15.1	−19.3	−25.9
Piemonte	6.8	9.7	8.0
Liguria	−3.2	−1.1	1.5
North East Centre	0.6	1.9	2.7
Emilia Romagna	5.0	5.1	5.3
Trentino Alto Adige	−10.8	−10.7	−10.4
Friuli Venezia Giulia	−7.1	−4.7	−2.8
Lazio	2.4	4.1	4.8
Veneto	0.3	1.8	2.4
Toscana	1.2	2.9	3.4
Marche	−2.3	−1.4	−0.7
Umbria	−6.5	−11.2	−8.4
South	−20.3	−19.6	−18.8
Abruzzi	−17.1	−14.7	−13.3
Molise	−37.1	−31.7	−29.1
Sardegna	−22.6	−22.0	−21.3
Puglia	−11.7	−11.0	−10.2
Campania	−15.9	−15.7	−14.7
Sicilia	−23.6	−22.6	−22.0
Basilicata	−36.4	−36.1	−34.4
Calabria	−37.2	−38.0	−38.1
Italy	−2.4	−3.6	−4.9

Source: ISTAT

Italian region, showing the net exports of each region relative to its GDP. The existence of a wide and long-standing financial gap between savings and investment among the regions and also among broad geographical areas is clearly evident. This gives rise to the need for inter-regional financial compensatory flows by means of private and public capital movements and fiscal transfers. The Mezzogiorno is an area in substantial structural deficit and, thus, very dependent for financing on the rest of the country and, at the supra-regional level, on central government (and EU structural) funds. These transfers of financial resources allow the area in the long term to maintain a demand for real resources which is higher than the local internal supply. At the same time, the long-term financial inflow to the area has contributed to limiting outmigration (which in the past was very high). From this point of view, the integration of the south with the rest of the country is passive. In contrast, the other two areas of the country – the North West and the NEC – persistently export more goods and services than they import (see Table 5). Lombardia and Piemonte, which are the most industrialised regions of the North West, have the highest surpluses. In the case of the more recently industrialised NEC regions, either their surpluses have been increasing or their deficits decreasing over time.

Analysis of the inter-regional diffusion of the banking structure gives a clearer picture of the process of integration that is in progress. Since the beginning of the 1990s, the introduction of the new institutional model, which liberalised the options open to banks, has set in motion a series of important changes and inter-regional movements in the Italian banking system. These developments have taken various forms, including new branches, mergers, take-overs, acquisitions of holdings and agreements between local banks and those outside the area. Table 6 shows the distribution of bank branches in 1996 in the three areas and the changes from 1990 to 1996. During this period, there was clearly a marked increase in the opening up of local banking systems, which can be analysed in the light of the two aspects of active and passive integration mentioned above.

The process of active integration is set out in Table 6a. It becomes clear that the banks which have their headquarters in the Centre North and, above all, in the North West, are the most involved in this process of inter-regional projection by means of opening new branches. Lombardia, aided by the fact that Milan is the dominant financial centre there, is re-inforcing its position as region leader. One branch in four in Italy belongs to banks which have their headquarters in Lombardia. However, only just over half of their branches are in that region; the rest are to be found all over the country. Counter to this trend, only banks with their headquarters in the Mezzogiorno have been increasing the number of branches within their own area, but reducing them in other regions. This trend is indicative of a prevailing defensive strategy and is yet another sign of the weakness of southern banks; they are attempting to withstand the competition within their own area, but do not have the resources to extend their market horizons outside the region.

These tendencies are confirmed by analysing the second aspect of inter-regional opening which is concerned with passive integration. Table 6b shows market shares in terms of branches owned in each area by local banks (see data along the main diagonal) and by banks from other regions. Local banks everywhere have lost market shares to outside banks. This phenomenon is particularly marked in Lombardia, Lazio and the Mezzogiorno. This means that the strategy of territorial diversification in the setting up of new branches has taken two main routes. The first is directed towards the financial centres of Milan (Lombardia) and Rome (Lazio), where the most dynamic peripheral banks have established new branches in order to benefit, at least in part, from the advantages in information, competitive emulation and market relationships, which are typical of financial centres. The second route is directed towards the Mezzogiorno, where many banks of the Centre North have established new branches for two main reasons: first, because the Bank of Italy has urged them to be more active in the South, in order to strengthen the banking structure of the area; and, second, because northern banks have decided to acquire local positions in a less developed market considered (often erroneously) easy to capture. As a consequence, the southern banks, although they have expanded almost exclusively in the Mezzogiorno, have nevertheless lost market shares to branches of external banks located there, above all to banks head-quartered in Lombardia. In addition, as Table 6b shows, southern banks are also the only ones which have lost branch shares even in other areas of

Table 6 (a) *Active Integration in Italy (Distribution of Bank Branches by Location of Headquarters, % by Column)*

Headquarters Branches	Lombardia 1996	Lombardia 1990–1996	Other North West 1996	Other North West 1990–1996	Lazio 1996	Lazio 1990–1996	Other North East Centre 1996	Other North East Centre 1990–1996	South 1996	South 1990–1996	Italy 1996	Italy 1990–1996
Lombardia	55.7	−8.7	19.0	12.5	10.1	0.9	3.9	2	0.9	−0.2	19.2	0
Other North West	10.4	0.8	63.5	−16.3	7.4	−0.1	1.0	0.5	0.6	−0.6	12.1	−0.1
Lazio	5.7	0.9	3.2	1.5	42.0	−3.6	2.5	0.7	2.4	0	7.6	0.1
Other North East Central	15.5	2.3	7.5	1.2	19.4	0.1	87.5	−4.9	1.6	−0.7	38.1	0.4
South	12.7	4.8	6.8	1.2	21.1	2.6	5.0	1.5	94.6	1.5	23.0	−0.6
Total area	100.0		100.0		100.0		100.0		100.0		100.0	
Total branches	6167.0	1951.0	3159.0	1317.0	2594.0	777.0	8577.0	3099.0	3806.0	665.0	24 303.0	7809.0

(b) *Passive Integration in Italy (Distribution of Bank Branches by Region, % by Row)*

Headquarters Branches	Lombardia 1996	Lombardia 1990–1996	Other North West 1996	Other North West 1990–1996	Lazio 1996	Lazio 1990–1996	Other North East Centre 1996	Other North East Centre 1990–1996	South 1996	South 1990–1996	Total branches 1996	Total branches 1990–1996
Lombardia	73.6	−12.9	12.8	9	5.6	0.3	7.2	3.9	0.7	−0.4	4666	1529
Other North West	21.8	4.9	68.0	−5.1	6.5	−0.3	3.0	1.7	0.7	−1.2	2948	936
Lazio	18.9	3.4	5.5	2.9	59.0	−8.0	11.8	4	4.9	−1.2	1849	614
Other North East Central	10.3	−0.3	2.6	0.7	5.4	−0.2	81.0	−0.4	0.7	−0.4	9260	3042
South	14.1	5.5	3.9	1.2	9.8	1.2	7.8	2.8	64.5	−10.6	5580	1688
Italy	25.4	−0.2	13.0	1.8	10.7	−0.3	35.3	2.1	15.7	−3.3	24 303	7809

Source: Bank of Italy

Table 7 The Dimensional and Geographical Distribution of Mergers and Incorporations of Italian Banks (1990–1996)

Incorporating banks	Incorporated Banks												Total
	Large			Medium			Small			Other			
	NW	NEC	S	NW	NEC	S	NW	NEC	S	NW	NEC	S	
Large													
North West	–	–	–	3	1	–	3	1	9	–	2	–	19
North East Centre	–	5	–	–	1	–	4	4	4	–	12	–	30
South	–	–	–	–	–	–	–	–	–	–	–	–	–
Medium													
North West	–	–	–	1	–	–	8	2	–	2	–	–	13
North East Centre	–	–	–	–	4	–	1	8	5	–	–	–	18
South	–	–	–	–	–	–	–	–	–	–	–	4	4
Small													
North West	–	–	–	–	–	–	53	–	–	–	–	–	53
North East Centre	–	–	–	–	–	–	–	117	1	–	3	–	121
South	–	–	–	–	1	–	–	–	76	–	–	–	77
Other													
North West	–	–	–	–	–	–	–	–	–	7	–	–	7
North East Centre	–	–	–	–	–	–	–	1	–	–	18	–	19
South	–	–	–	–	–	–	–	–	–	–	–	3	3
Total	–	5	–	4	7	–	69	133	95	9	35	7	364

Source: Bank of Italy

the country. In contrast, the local banks in the NEC (with the exception of Lazio) are characterised by the fact that not only have they succeeded in containing the expansion of external banks in their area, they have also actively extended their presence into other regions.

The competitive capacity shown by NEC local banks can be partly explained by mergers and incorporation. Together with the opening of new branches, merger and incorporation are important aspects of the integration process which has characterised the 1990s. Table 7 shows all the mergers and incorporations that took place in 1990–6. First of all, the number of such reorganisations is striking: 364 in only six years. Second, the majority of these have been realised between banks of the same size and, above all, between small banks (68% of the total). Third, banks resident in the NEC played the most dynamic part in this integration process: in four transactions out of five they appear as incorporating banks, both of other banks within the area (60%) and of outside banks (40%). On the whole, the main trend has been towards mergers between similar banks (from the same area and the same size) rather than between different banks. This is understandable because mergers are undoubtedly a shortcut to achieving both operative and dimensional growth. At the same time, they present a difficult problem of internal integration. It is not straightforward to bring together personnel, procedures and organisational customs and cultures from different business backgrounds and direct them towards a common objective.

Problems of Dimension

Local Banks Versus National Banks

The fact that many banks have opted for opening new branches, or undergoing mergers or incorporation is indicative of the need to find the appropriate size and organisational form with which to confront the new institutional and competitive environment. In Italy two basic models of regional banking system have traditionally co-existed. The first is the model based on large national banks with branches all over the country (the national branch-banking system). The second is the model based on small, independent, local banks which operate in restricted areas (the local unit-banking system). In the next few years, however, the co-existence of banks which are structurally very different is likely to be challenged by increased competition following the liberalisation of the banking system and the beginning of the Euro system.[20] It is important, therefore, to analyse the advantages and weaknesses of different models of the regional banking system.

The main advantage of the first model is clearly connected to the banks' average size, that is to the presence of scope and scale economies in banking activity. Even though the empirical literature on scope and scale economies in the Italian banking system has not reached unequivocal and conclusive results,[21] there is no doubt that in rapidly evolving financial markets the minimum size of banks tends to increase. In more sophisticated financial markets, the ability of banks to attract business is dependent on the ability to supply sets of strongly innovative, often customised, financial services. The wider variety of financial products and services needed by their customers requires banks to rely on operative structures and professional abilities which are rarely features associated with small size. Furthermore, large banks, as they are able to diversify their portfolios, can afford to be less risk-averse than local banks.

On the other hand, the second model of geographical organisation, based on small local banks, ensures a degree of embedding and commitment of the banking system in the local economy which a system of large bank branches cannot easily emulate. The informative advantages of local banks stem both from an historically developed presence in the community and from personal elements, linked to the character of bankers and managers who feel part of, or who are familiar with, the local socio-cultural environment. A direct and in-depth knowledge of local entrepreneurs emerges, and it is this which is the decisive element of the 'local bank–local firm' relationship in economic systems of small and medium-sized firms. National banks can only slowly acquire this kind of local embeddedness. The distance between decisional and operative centres within a national branch bank structure reduces the availability of information about local firms and local growth prospects. In addition, the local-branch management of national banks is often in the hands of directors only temporarily committed to that branch. They often use their time in the branch as a stepping stone for their advancement in their own careers.[22] They tend to be very risk averse, opting for safe large investments, rather than riskier smaller investments, even to the detriment of important (innovative) projects for the growth of the local economy.

The close connections between local banks and firms produce both advantages and disadvantages. On the one hand, such relationships tend to promote technological and organisational innovation in the existing firms. The standing and the prospective development of local banks obviously hinges on the performance of the local economy. Therefore, local banks are interested in sustaining the market competitiveness of local firms. As the success of local firms depends on their innovative capacities, local banks have a direct interest in favouring the introduction of innovation. On the other hand, there are some negative factors. First, a deep knowledge of a single and often specialised economic environment can reduce the response of banks to developments emanating from other economic systems, especially from new industries not hitherto present in the region. To assess the credit-worthiness of a new firm operating in an industry which is 'new' in that region often requires a competence and a knowledge which local banks do not possess.[23] Second, continuous and exclusive customer relationships with local firms can induce banks to limit the entry of new and possibly strongly innovative firms. Indeed, financing these kinds of enterprises can create difficulties for existing local firms and make the old credits of local banks more subject to defaults, reducing the expected returns on them.[24] Third, this kind of attitude from local banks tends to reduce the innovative efforts of existing firms, which 'protected' by the behaviour of the banks are less stimulated to introduce innovations (Çapoğlu, 1991). Finally, the informational rents of local banks can make the local credit market less competitive, thus discouraging the entry of outside banks (Sharpe, 1990).

What Model for the Regional Banking System?

Thus, the main advantages of one ideal type of geographical organisation of a banking system are at the same time the main weaknesses of the other. If one accepts this conclusion, then in principle it leads naturally to the implication that the most suitable regional banking system corresponds to an 'intermediate' model (quite similar to that prevailing in Italy), in which local and outside banks co-exist and compete. To the extent that the advantages of both types of banks are present, this mixed system has the characteristics and a diversified dimensional structure which best respond to the needs of local development, and is well suited to links between the centre and periphery.

Given the increasing sophistication and efficiency of modern information technologies, national banks generally seem better equipped to operate even in the most peripheral local markets. However, not all local banks will necessarily disappear. Their strongest advantage, and one which they must defend and build on, is their competitive edge, which stems from their being firmly established in the area. Yet local banks cannot risk remaining either too small or too isolated. They must also seek to expand their flows of savings and investments through active links with more advanced operators and markets in the main financial centres. The regional banking system must ensure an adequate support to local smaller firms not able to access national financial markets. To provide this

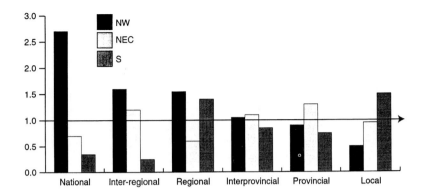

Figure 2 Banks Classified by Territorial Diffusion in Each Area, 1996 (Index numbers, Italy = 1.0). Source: Bank of Italy

support, it must obviously have a high degree of attachment to, and embeddedness in, the local community and economy in order to provide it with a deep knowledge of the needs and potentials of local entrepreneurs. Equally, the local banking system must be able to offer efficient and attractive financial services to local customers, and this requires close connections with the markets and instituitions in financial centres.

Despite the evident process of territorial integration taking place, the structure of the Italian banking system has not yet stabilised. More efficient forms of organisation are still at the selection stage. As a result, the intermediate model of coexistence between local and external banks has not yet reached its optimum level in all the different regions. In the Mezzogiorno, the banking system is still underdeveloped. Compared with the Italian average, the southern banks are mainly either small local banks or regional banks (see Figure 2).[25] There are hardly any national or inter-regional banks.[26] On the other hand, there are numerous branches of national banks with headquarters in other regions. But the most worrying aspect is that in the Mezzogiorno the local banks, unlike those prevailing in the Centre North, do not seem to be very dynamic and are geared towards the traditional collection and management of liquidity. This is confirmed by the ratio of loans to deposits, which in southern Italy assumes the lowest values for locally orientated banks (namely interprovincial, provincial and local banks; see Figure 4) whereas branches of national and inter-regional banks show values near to one. This suggests that in this area local banks play a marginal and secondary role in financing the economy. On the whole, these banks are small, isolated and inefficient. They operate in peripheral areas exploiting their isolation to carve out strong local monopoly positions *vis-à-vis* local firms and savers. Besides, given present trends, the southern banking system seems destined to be controlled by large banks from outside the area. Whether this will bring economic benefits to the South in the long term is difficult to say. In the short term, however, this appears to be the only way to stimulate an improvement in banking efficiency, financial resources and credit allocation in the area.

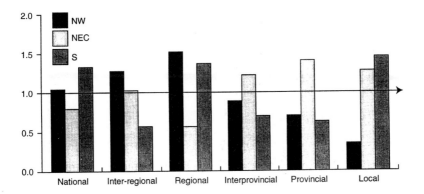

Figure 3 Branches of Banks Classified by Territorial Diffusion in Each Area, 1996 (Index numbers, Italy = 1.0). Source: Bank of Italy

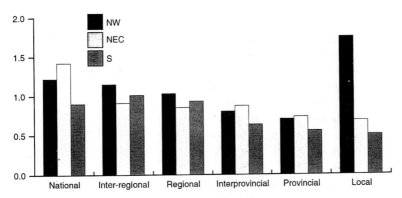

Figure 4 Loans/Deposits Ratio by Branch Localisation of Banks Classified by Territorial Diffusion, 1996. Source: Bank of Italy

In the Centre North, on the other hand, the banking system seems to be much more homogeneous both in terms of banks and branches. It is in this part of Italy that the intermediate model of regional banking is gradually developing. Of course, this evolution is occurring in different ways and at different speeds on account of the distinctive characteristics of local areas (see Figures 2 and 3). In the North West region there is a prevalence of banks which have a wide geographical coverage (namely national, inter-regional and regional banks). In contrast, in the NEC regions – characterised by widespread industrialisation and urbanisation – there is a slight prevalence of locally oriented banks (interprovincial, provincial and local banks). It is important, however, that in this area the local banks play a key role in the development of the local systems of small firms. Their intermediation capacity, measured by the loans/deposits ratio, is not much lower than that of larger national and inter-regional banks (see Figure 4). This reflects the fact that the local banks of the Centre North, and above all those of the NEC, have already proceeded down the path of 'active integration' through the opening of inter-regional branches, mergers and incorporation (as discussed above). This, obviously, does not mean that the banking system in these regions does not still have elements of economic backwardness to overcome and important modernisation steps to make. Especially if one looks at banking systems in the broader competitive context of the EU, it becomes clear that, on the whole, Italian banks are still too small and insufficiently innovative.

Geographical Diffusion of Financial Innovations: Some Concluding Remarks

The new institutional and competitive context of the Italian financial system, seen within the wider process of monetary unification in Europe, poses severe problems for the survival and adaptation of local credit systems. The most realistic and promising solutions are of an eclectic type and can be found, it is suggested here, by adopting a possibilist approach. This means favouring every possibility of co-existence between local and national-global systems, and between local and external banks, in order to make local markets arenas of interchange rather than dispute or isolation. The liberalisation of the operative choices open to banks introduced by the new Italian and European institutional framework admits the use of a plurality of financial innovations, thus making the achievement of this goal easier. These innovations concern not only banks, but also the introduction of new financial intermediaries and new types of financial market.

For one thing, the freedom to open new branches and to carry out acquisitions and mergers is an important organisational innovation which favours the growth in size and the inter-regional integration of banks. The effects of this process are not limited to changes in branches, the collection of savings and investments, but also involve changes in business methods and techniques. In the present context, no bank, whether large or small, can afford to get left behind using traditional forms of intermediation or loans which are too localised, not marketable, illiquid and, therefore, too risky. The survival of every bank, in general, and of small and peripheral banks in particular, is linked to their capacity to introduce innovations which will significantly improve the quality of financial services and market relations, as well as relations with financial centres and other intermediaries. The growth and success of local banks is strongly dependent on embracing such innovations.

The importance of being able to establish good business relations is something that has been highlighted by the experience of Italy's well-documented industrial districts. The same considerations can, in part, be applied to banks. They are able to remain small and peripheral, provided they are linked into a wider local network of collaboration and interchange. This explains why, as was the case with groups of firms in the industrial districts, there is an increasing number of groups of banks which pool their services, develop complementary specialisations and diffuse the improvements in quality and competitiveness necessary to withstand the impact of larger and more dynamic banks. In order to remain the chief intermediary between local firms and local savers, local banks will need to develop good relations not only

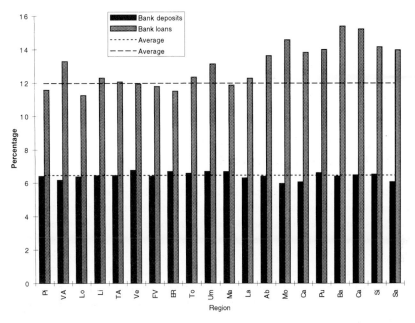

Figure 5 Bank Interest Rates by Region (September 1996). (Key given at the end of the chapter)

with other banks, but also with innovative financial markets, where marketable assets are exchanged.

An important example of the opportunity for the active integration of local banks in wider financial circuits concerns the two traditional instruments of bank intermediation: deposits and loans. Regional interest rates on deposits show very low differentials, much smaller than the regional differences in interest rates on loans (see Figure 5). This is not a consequence of greater competition between branch banks which operate only at local level. It is due, above all, to the growth of portfolios of marketable short-term bonds, such as treasury bills, which have forced banks everywhere to standardise the interest rate on deposits in order to keep them competitive with these alternative forms of investment of savings.[27] Bank loans, on the other hand, are more closely tied to the specific conditions of local debtors who, especially in local systems of small firms, typically have no other alternative than bank loans. This explains why regional variations in interest rates on bank loans continue to be substantial.

Even the typical constraints on bank loans, such as narrow localisation and strict bilateral links between banks and firms, can be overcome by using new forms of interrelation between marketable and non-marketable assets and, therefore, between local and national–global financial circuits. The first solution is securitisation, or issue of marketable bonds which can be quoted on the global standardised markets, backed by a certified set of the most reliable bank loans,[28] which maintain their characteristic of local non-marketable assets. A second solution is that of closed-end investment funds set up by banks in the form of specialised intermediaries, through which they can provide capital to local small firms for limited periods of time. These non-marketable capital assets are backed by the issue of quotas marketable on the secondary markets, being refundable only at the expiry date of the investment fund. In both cases (bank loans securitisation and closed-end investment funds), intermediaries act as the key link between marketable–non marketable assets and global–local financial markets. These solutions could also help to prepare the ground for the direct issue of marketable assets by local firms. This is a third solution which is obviously not suited to all the firms in the local economy. However, it does represent a goal towards which the leading medium-sized firms should strive. The issue of bonds and the stock exchange quotation of the leading firms would both reinforce their access to financial resources and, at the same time, widen their opportunities of financial diversification. Indirectly, the benefits would spread over the wider networks of small firms which operate as sub-contractors of the leaders and, therefore, would contribute to consolidating the development of their local economic–industrial systems.

Although it is quite easy to suggest possible solutions, implementing them is, of course, a rather different matter. There are lags, rigidities, inefficiencies and resistances to be overcome. These constraints are more rooted in the less advanced regions, and are both a cause and a consequence of the vicious circle of economic backwardness which characterises these areas. The most evident difficulties in the geographical diffusion of financial innovations stem from the unwillingness of firms (but also the incapacity of the smallest ones) to issue marketable assets, and from the parallel unwillingness of savers to invest in those assets as part of their portfolios. Italian households still opt for more liquid and, thus, less risky investments (see Figure 6). This tendency is most evident in the South, where there is a high incidence of households who hold bank deposits (which forms the basis of reference in Figure 6) and postal deposits. Even across the country as a whole, the proportion of savers who are brave enough to invest in more sophisticated and risky financial activities (such as shares and investment funds) is still very limited (in this respect, as in others, Italy's capital market is considerably less developed than that of, say, Britain or Germany).

All this reinforces a belief in the possibilist approach to inter-regional integration as the most suitable model for the future development of local financial systems in Italy. It offers a realistic perspective on what could be done to promote the gradual diffusion of financial innovations across areas in which strong elements of traditional, even backward, locally based banking structures and arrangements still persist. The modernisation of local banking systems in Italy is vital if they are to meet the challenges posed by the technological, institutional and regulatory changes that are transforming the world of finance (not least the opening up and unification of the European financial space), and if they are to serve the needs of local industry in an era in which the forces of global competition compel an almost constant process of restructuring and modernisation. In short, a new financial culture is required. Those banks that are at the forefront of developing that culture will also be the leaders in the difficult process of geographical integration and reorganisation facing the Italian banking system.

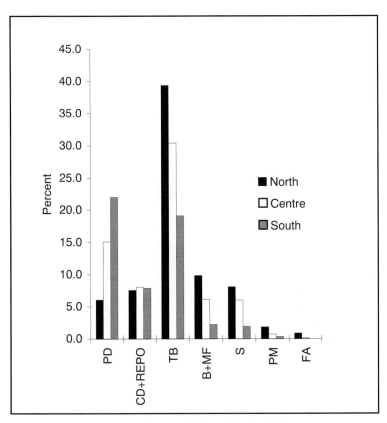

Source of Data: Bank of Italy

Notes
PD = Postal Deposits
CD = Certificate of Deposits
REPO = Repo Deposits
TB = Treasury Bills
B = Bonds
MF = Mutual Funds Shares
S = Shares
PM = Portfolio Management
FA = Foreign Assets

Figure 6 Percentage of Households Owning Financial Assets in Each Major Regional Area, September 1995 (Per 100 Households Holding Bank Deposits)

Acknowledgements

The authors wish to thank Ron Martin for his helpful comments, and the Consiglio Nazionale della Ricerche for financial support.

Appendix – Key to Figure 5

Pi	Piemonte	M	Marche
VA	Valle d'Aosta	La	Lazio
Lo	Lombardia	Ab	Abruzzi
Li	Liguria	Mo	Molise
TA	Trentino Alto Adige	Ca	Campania
Ve	Veneto	Pu	Puglia
FV	Friuli Venezia Giulia	Ba	Basilicata
ER	Emilia Romagna	Ca	Calabria
To	Toscana	Si	Sicilia
Um	Umbria	Sa	Sardegna

Part III

Financial Centres

Chapter 5

The Development of Financial Centres: Location, Information Externalities and Path Dependence

David Porteous

Introduction

Even a casual observer could point to two striking observations about the location of high-level financial services. The first is that specialised financial services are usually spatially concentrated in agglomerations called 'financial centres' which are typically large cities. Most people would identify London, New York and Tokyo as global financial centres, and they would probably also be able to name the leading national financial centre in their own country or even region. Second, there is apparently a remarkable stability in the ordering of financial centres over time. This may be called the 'Why London?' question, as London remains a leading global financial centre despite the relative decline in the global importance of the British economy. London was the first international financial centre of the modern age, and this 'first mover' advantage continues to underpin its leading position.

The geography of finance must provide theoretical explanations for these two observations. Recent developments in economics and geography, particularly in the theories of 'asymmetric information' and of 'path-dependent' processes, enable much better accounts to be given for financial centre development and change. The notion of 'path dependence' refers to the idea that chance events – 'historical accidents' – can have long-run cumulative consequences (David, 1988 and 1994). The localisation of economic activities provides a prime example of this process (Arthur, 1994). An initial locational pattern may simply be an accident of history, but, once established, that pattern can become path dependent and 'locked in' through mechanisms of cumulative causation. Recent developments in this theory allow for a more nuanced account, within which path-dependent localisation processes can be undermined or displaced by other forces. Without this possibility, it would be difficult to explain the rise *and* decline of financial centres at particular places over time, as the earliest to develop would stay dominant indefinitely. However, at the level of regional, national, and even global financial centres, processes of rise and decline do take place and provide useful insight into the forces behind financial centre formation.

One aim of this chapter is to develop a theoretical framework to account for the spatial agglomeration of financial activities over time and for the forces which cause financial centres to develop in particular locations. A second aim is to explore the processes by which 'spatial switching' of financial dominance may occur between centres within a given nation. Evidence relating to the rise and decline of the relative importance of national financial centres in two countries – Australia and Canada – will be used to illustrate the theory. In each case, the older financial centre (Melbourne and Montreal, respectively) has been displaced in the post-war era as the major national financial centre by a new city (Sydney and Toronto, respectively). There are further examples of this displacement process in other countries, such as Rio de Janiero and Sao Paulo in Brazil, and Cape Town and Johannesburg in South Africa. However, the parallels and comparisons are more direct in the Australian and Canadian cases, hence their usage here.

Definition of a Financial Centre

In order to understand the dynamics of financial centre development, a means of measuring and comparing financial centres over time must first be defined. At the outset, it is important to distinguish between a general spatial concentration of the financial sector and an agglomeration of the particular activities which characterise a financial centre. A financial centre is an area in which high-level financial functions and services are concentrated. This area is usually a city, but is often a more localised area within city boundaries, for example New York's Wall Street or the City of London. These services are not so much retail financial services, such as branch banking, serving residents of the immediate area, but specialised or high-level financial services which are in demand over a much larger area, covering the region, possibly the nation and even the global economy. A financial centre need not be the largest city in a region in terms of population or economic activity, although often this is the case, for reasons discussed below.

How does one measure the relative importance of financial centres? Clearly, indicators which measure high-level financial service activity are necessary. In his early (and somewhat neglected) paper on the geography of finance in Canada, Kerr (1965) suggested several such measures: employment in the financial sector relative to the total labour force; assets of financial institutions head-quartered in the city; the proportion of cheques cashed in a city; and the value of turnover on the stock exchange in the city (see also Laulajainen, 1998: chap. 7). Unfortunately, in practice, information on all of these aspects of financial activity may not be available at the individual city level.

Employment in Finance

Although a high proportion of financial sector employees in the labour force may be necessary for a city to be considered a financial centre, this is not a sufficient measure in itself. For example, in their study of New York City, Robbins and Terleckyj (1960) concluded that the 'money market core' of, in those days, 5000–10 000 specialists was at the heart of New York's financial sector. The rest of the

New York financial community (then some 312 000 workers) depended on the activities of these people. Similarly, Ryba (1974) in his study of Montreal (cited in Higgins, 1986) voiced concern over the relocation of bank trading desks and other 'core' financial activities to Toronto, as he argued that these functions were at the heart of a financial centre. Historical accounts of the development of centres such as London have also stressed the importance of an 'institutional core' or 'nexus'. In recent years, however, the introduction of new technologies has reduced the labour needs of many routine financial activities, so that employment is no longer necessarily a consistent indicator of the relative importance of different financial centres.

Bank Head-office Locations

The second measure suggested by Kerr, the proportion of national bank assets controlled by financial institutions head-quartered in a particular city, may offer a better indication of a financial centre's importance than employment. This is because bank head-office location to some extent proxies for high-level financial sector employment; after all, the most important financial functions in a bank are performed at head-office level. This is in accord with a long-standing tradition in economic geography of considering the control function of a city in terms of the number or assets of large corporations head-quartered there (Cohen, 1979; Stephens and Holly, 1980; Semple, 1985; Dicken and Lloyd, 1991).

Even this measure is not without its problems, however. For one thing, occasionally, the registered office of a financial institution and its main operating office are separated. This is a particular problem in Canada. The Bank of Nova Scotia, for example, maintains its registered office in Halifax for historic (and possibly political) reasons, although its executive office moved to Toronto in 1910 together with all of its most important financial functions (Naylor, 1975). A second problem lies in the declining share of banks in the total financial assets of many economies, while the influence of other (non-bank) financial institutions, such as insurance companies and fund managers, has increased substantially. Ideally, data would be needed on all financial institutions with control over significant asset shares.

Cheque Clearings

The proportion of total cheques cleared in a city reflects both the volume of activity in a region and the structure of the bank clearing system which determines where payments are processed. There are economies of scale in centralising the payments system. Kindleberger (1974 and 1985) suggests that these economies were one of the reasons for the early rise of London as a financial centre. Country banks rapidly sought affiliations with London-based clearing banks in the nineteenth century, ultimately giving rise, through processes of merger, take-over and concentration, to the large London-based national clearing banks. An additional advantage of this

Table 1 Cheque Clearings/Bank Debits in Two City-Pairs (% of National Total), 1950–1986

	Melbourne	Sydney	Montreal	Toronto
1950	37.8	40.6	26.0	28.8
1960	36.6	41.0	26.1	36.3
1970	32.9	43.5	23.1	41.8
1980	33.3	44.6	12.8	58.3
1986	25.7	57.0	9.0	67.2*

Note: * 1983 figure

Sources: Canada: Cheques Cashed in Clearing Centres: 1950–1971: Statistics Canada, 11.505 Table 4. Post-1971: Statistics Canada, 61.001.
Australia: Debits to customer accounts by all trading banks, average of weekly figures, by state, excluding debits to government accounts. Sources: 1945–1970: Butlin, Hall and White (1971): Table 51(i); includes non-metropolitan areas in states. 1971–1986: *Year Book Australia* various editions, Private Finance Section, under Trading Banks

measure for empirical studies is that, unlike most economic data, from early on information on cheque-clearing activity has often been collected and reported at a city level. Also, a small number of cheques typically comprises a high proportion of the total value cleared. These cheques are more likely to be for financial-sector-related payments, such as settlements for share purchases. Hence, the value of clearings figures may proxy to some extent for levels of financial activity more generally. However, in recent times, an increasing proportion of payments has taken the form of direct debits from computer files, as electronic and 'plastic' (credit-card) money forms are rapidly replacing the use of cheques. In part because of this, many countries no longer collect or publish regional clearing house data. Nevertheless, available cheque-clearing data clearly show the rise of Sydney and Toronto as financial centres over their respective national competitors (see Table 1).

Stock Exchange Volumes

A fourth measure of financial centre activity is the relative value of turnover on the stock exchange, or more generally, on all of the major financial markets in a city. Stock exchange data are useful for empirical work, in that they are generally available over a long period; and where cities have had their own distinct exchanges, it proxies for all the related specialised financial activities connected to it, such as stockbroking and fund management. Table 2 highlights the rise of Sydney and Toronto, although the comparative growth of the Toronto Stock Exchange is much more marked than that of Sydney.

In addition to stock markets, it would be desirable to have a measure of turnover in other financial markets too, especially in the key domestic money and foreign exchange markets which are closely associated with high-level financial functions. However, the data which exist are usually provided through surveys of foreign exchange dealers in particular locations; and these surveys are normally undertaken regularly only for first-order global financial centres such as London, New York and Tokyo. For example, the average shares of these three cities in the

Table 2 *Turnover on Stock Exchanges (% of Two-city Total Value of Annual Turnover), 1950–1986*

	Melbourne	Sydney	Montreal	Toronto
1950	NA	NA	40.7	59.3
1960*	49.3	50.7	28.1	71.9
1970	50.5	49.5	25.3	74.7
1980	47.9	52.1	11.8	88.2
1986	46.2	53.8	17.3	82.7

Note: The percentages are of a two-city total, hence add up to 100
* Australian figures in row are for 1966

Sources: Canada: Value of all shares traded (monthly average) 1945–1959: Historical Statistics of Canada (1983), H631-H650; 1959–1990; Canadian Economic Observer, SC11-210 90/91; Table 10.5 Australia: Melbourne Stock Exchange Annual Report 1973, 1981, 1986; annual value (year ending 30 September) of share securities traded; available from 1962–1993. Sydney Stock Exchange Annual Report 1965, 66, 67, 68, 71, 78, 82, 86

daily value of world foreign exchange trade in 1989 were 42%, 29% and 26% respectively (Gardener and Molyneux, 1990: 146). Information relating to trade in other securities and derivatives markets would clearly also be desirable (for example, any attempt to assess the significance of Chicago which excluded these activities would seriously understate its financial importance).

Foreign Bank Presence

A further useful institutional measure is the (relative) number of foreign bank offices in a city. This is often used to compile a rank ordering of international financial centres, but it may also be applied to national financial centres. Presumably, foreign banks are able to choose the optimal location at which to set up in a country at the time of their entry; they are not constrained, as domestic firms may be, by locational inertia from their start-up location. Their choice of location is likely to be based on a number of factors, including international and domestic accessibility, but perhaps most importantly on the volume of information flows in the city as this is an indicator of potential business. Therefore, their entry decision provides a useful pointer to the financial hierarchy at the time of entry; the setting up of a foreign bank office in a city indicates that city's relative importance at the time and, through information agglomeration processes, in turn promotes its importance as a financial centre. Table 3 shows the dramatic rise in the number of foreign banks in Australia since the late 1960s, and in Canada since the late 1970s. In each city pair, the centres with the earlier preponderance of foreign banks – Melbourne and Montreal respectively – very clearly lagged behind by the 1980s.

Other Characteristics

In addition to the five features considered above, other characteristics may help to measure the financial importance of a city. Several metropolitan features are

Table 3 The Number of Foreign Banks Present, 1950–1986

	Melbourne	Sydney	Montreal	Toronto
1950	4	3	1	0
1960	4	5	2	1
1970	7	13	4	1
1980	18	51	6	29
1986	34	90	11	78

Sources: Canada: 1950–1965: *Bankers' Almanac and Yearbook*, based on entries for city for each year. 1966–1986: *Rand McNally Bankers' Directory*. Note that this includes representative offices, branches and head-offices of (more than 50% owned) subsidiaries of foreign banks; if there is more than one office, each is counted unless one is designated main office or branch.
Australia: Includes all banks whose head office is outside the country, or which are wholly or mainly owned by foreigners. Hence, this includes affiliates/representative offices/wholly owned subsidiaries of foreign banks, as well as the ESA Bank and ANZ Bank, which had a London head-office until 1976. Source: 1950–1971: *Bankers' Almanac and Year Book*, 1972–1986: *Rand McNally Bankers' Directory*

closely associated with the financial importance of a city, even if they do not measure it directly. One of the most important is the role of *communications*. Castells (1989) and Warf (1989), among others, have emphasised the key role that communications are increasingly playing in financial markets. This has led large global banks to develop their own communications systems, such as Citicorp's Global Telecommunications Network. Measures which track high-level communication, such as the proportion of international phone calls, are usually not publicly available at a city level. However, Wheeler and Mitchelson (1989) have used volumes of Federal Express mail items in the United States to identify the informational and control functions of particular cities. Express mail provides a useful measure of high-value information flow which is more likely to be related to business purposes than the volume of ordinary mail. In addition to remote telecommunications, financial sector activities require frequent in-person contact. The transport system of a country, most importantly its airline routes, will shape the accessibility of a city and, hence, its appeal to the financial sector. However, data on air passenger traffic do not usually discriminate by type of passenger or reason for travelling (Irwin and Kasarda, 1991).

The head-office location of large *non-financial corporations* may also provide information about the importance of financial centres. Several early studies, such as Pred (1973 and 1977) and Cohen (1979) in the United States, Taylor and Thrift (1981a and 1981b) in Australia, and Higgins (1986) in Canada, have tracked the agglomeration of head-office functions over time. However, there is no necessary connection. For example, cities such as Detroit, Pittsburg and Cleveland continue to be home to major US industrial corporations, but have not emerged even as regional financial centres (Conzen, 1977), whereas cities with a smaller base of large clients than these three, for example Charlotte in Virginia, have stronger financial sectors (Harper, 1987). In general, there seems to be a crucial endogeneity here, in that banks and financial service firms may indeed relocate or start up so as to be close to major clients; but, equally, major clients may choose to locate closest to providers of funding and important services. Hence,

The Development of Financial Centres

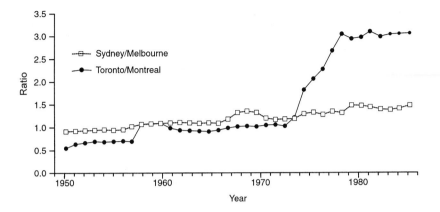

Figure 1 *Ratios of Financial Indices for Sydney/Melbourne and Toronto/Montreal.* Source: Porteous (1995: 111)

financial sector agglomeration both reflects and reinforces real sector agglomeration. This is one explanation for why financial centres are often in large cities.

These various indicators may be applied to any financial centre for which data are available, and may be assembled into a composite index which measures the relative importance of financial centres over time (see Porteous, 1995, for methodology involved). Figure 1 shows the ratio of such indices for the two case-study city-pairs considered here. The ratio is expressed as newer to older, hence a value greater than unity implies the dominance of the newer. The clear upward slope shows the rise of a newer financial centre in each case. Anecdotal evidence from city historians confirms these trends from quantitative indices, and the approximate timing of the change-overs or 'spatial switches' (for Australia, see Davidson, 1986; Daly, 1984; for Canada, see Higgins, 1986). Most importantly, these city-pair examples show that path dependence alone does not explain financial centre development; there must be an explanation to account for relative changes in position.

Forces Behind Financial Centre Formation

Economic geography has long recognised that the location of economic activities represents the outcome between agglomerating (centripetal) and dispersing (centrifugal) forces. These concepts have recently been revived as part of the so-called 'new economic geography' within economics, especially as exemplified by the work of Krugman (1991a, 1991b and 1995), and provide a useful framework for analysing the formation and evolution of financial centres.

There are three standard forces causing spatial *agglomeration* of any economic activity. These forces, originally recognised by Alfred Marshall, are generated by positive spatial externalities, or spillovers, at the level of the individual firm. These forces also apply to financial activities, but must be interpreted carefully to capture the particular features of this sector of the economy.

The first force can be called 'labour market externalities'. Financial firms are particularly dependent on highly skilled personnel. Therefore, they benefit from the presence of a large local pool of labour, as this may mean quicker filling of vacancies and better quality of staff because of a greater choice of applicants. As Krugman (1991a) notes, this externality promotes agglomeration not only from the side of firms as demanders of labour, but also from the side of workers as labour suppliers. Workers will benefit from a 'thick market' for their services, particularly as their skills become increasingly specialised. The general argument for labour market externalities rests on uncertainty in demand for goods (which generates uncertain labour demand) and increasing returns in production (which dictate that production be concentrated rather than evenly dispersed). In the financial sector, there is uncertainty in labour demand, as the cycles of financial boom and bust during the 1980s and the 1990s show. However, the evidence of economies of scale in production is more patchy in the financial services industry.

The second force is the demand for 'intermediate services'. Financial firms are users of specialised producer services, ranging from sophisticated communications hardware and software to complex legal advice. By locating near the source of such inputs, financial firms may enjoy better service and lower prices; all the more so as timely service is often vital for financial firms to be able to exploit profit opportunities.

The third force concerns 'technological spillovers'. In a sector which depends on rapid technological innovation, localised spillovers of knowledge among firms may also lead to agglomeration. However, Krugman cautions against giving this factor pride of place in explaining some of the most celebrated geographic concentrations of industry, such as computer hardware and software development in Silicon Valley. He suggests that labour market externalities, rather than technological externalities *per se* may be driving these spatial agglomerations. In the financial sector, localised innovation in financial products may clearly be a source of positive externalities as, for example, new derivative products or project financing options diffuse rapidly through spatially concentrated firms. However, technological spillovers are usually understood in the narrow sense of knowledge about 'how to do it'. In the financial sector, a broader type of information spillover – knowledge about 'what is out there/what others are doing' – may be more important than in the real economy. Hence, the concept of spatial externalities must be extended to address the information-intensive environment of financial services.

In addition to these standard types of agglomeration economies and externalities, there are three other types that can be distinguished, and which are especially important in the location of financial activities and services and the concentration of such activities in major financial centres.

First, there are *informational spillovers*. As financial firms benefit from their ability to process information more efficiently (meaning more quickly and more accurately) than other market participants, maximising access to information flow may increase firm profit. Of course, volume of information flow alone is not sufficient for ensuring its profitable use; instead it is the quality and timeliness of information which matters. This is where the inter-personal and other localised information networks in financial centres may be a vital means of rapid information diffusion. The Lund School of geographers has consistently stressed the importance of face-to-face contact for information exchange as a motive force in urbanisation (Tornqvist, 1979). Similarly, Pred (1973) has traced the pattern of information flows which shaped the early eastern US metropolitan system. He shows how the early dominance of New York City as an east coast commercial and financial centre developed in part because regular shipping packet lines brought news from Europe first to the city, and from there it diffused to other cities across the United States.

Second, there are *socio-institutional and cultural factors*. As Thrift (1994) has cogently argued, social and cultural factors play an important role in the formation and development of financial centres. Money is not only an economic commodity, a store of value and medium of exchange, it is also a socio-cultural relation. The rules and norms of behaviour of market participants, including issues of trust and reputation, are crucial to the conduct and operation of transactions between financial agents. These rules and customs both require and are reinforced by the spatial concentration of financial activities within a centre, and if successful (in the sense of giving the centre a competitive edge) become institutionalised, that is effectively embodied in and fundamental to the socio-institutional 'embeddedness' of the daily business practices of the markets in that centre. These social and cultural bases of financial activity also act as the conduits along which information flows occur.

Third, and particularly at the international level, the existence of permissive regulatory frameworks and structures (for example low taxes, minimal trading controls and freedom of entry) in a given location (centre) relative to other locations in other countries or regions, will reinforce the attraction of financial firms and activities to that location. This factor was important in the rise of off-shore financial centres (OFCs) in the 1960s and 1970s, and became an important element of the competition between major global financial centres in the 1980s and 1990s.

In opposition to these centralising or agglomeration forces, there are also *decentralising* or *dispersion forces* at work in the financial sector. First, there are the high costs of operations in financial centres. Office space is often at a premium; high costs of living feed back into high salaries. As a consequence, in recent years 'back-office' financial functions with relatively low value added have been increasingly decentralised to smaller provincial centres (for a British example, see Leyshon and Thrift, 1989). Second, congestion costs and the disamenities of life in large cities may affect the compensating wage differentials which must be paid over other more appealing places. Third, the same argument about information externalities generating concentration may be used to argue that firms located in one place lose the competitive edge on access to non-local information which firms in other locations may gather and act on first. If non-local information becomes increasingly valuable, then this may be a force for decentralisation. However, the rising value of non-local activity is less likely to cause a decentralisation of financial firms across a broad area than a re-attachment, or re-agglomeration, at another point, where the valuable information flows can be better accessed. Finally, for international financial centres, different time zones create a force for decentralisation. Different time zones give rise to a number of 'zonal' global financial centres, such as Tokyo, Sydney, Hong Kong, Frankfurt, London, New York, Chicago, Toronto and Los Angeles, among which trading activity (of foreign exchange, securities, futures and so on) moves in a 24-hours-a-day system of interlinked markets (see Harvey, 1989b: 162).

The balance of centralising versus decentralising forces at a particular time will determine the extent of agglomeration in the financial sector. The rapid development of communications technology appears to be strengthening the forces favouring decentralisation, as large volumes of information may increasingly be accessed – and transactions conducted – from remote locations at low and decreasing cost (*The Economist*, 1995). This is the so-called 'end of geography' argument made by O'Brien (1992). However, other than permitting back-office decentralisation, the latest technology does not yet seem to have promoted the dispersal of key financial functions to any large extent. One reason for this may be path dependence, in that financial firms locate where there have been good information flows in the past, and, by so doing, help ensure the continuation of good information flows. A further possible reason may be that the cost of relocation may be delaying the response to communications innovations. However, the financial sector is generally highly competitive and highly mobile, as firms have no bulky physical plant to move, so that opportunities to reduce costs without reducing profits by relocating are unlikely to be neglected for long.

A more plausible reason for the continuing spatial concentration of the financial sector is pointed out in the survey of financial centres by *The Economist* (1992). Although communications technology makes access to information more equal regardless of location, by creating more efficient markets it also has the opposite effect of pushing financial firms *closer* to informational sources in order to find unexploited information from which to profit. This explanation for continuing concentration may be further enhanced by considering the nature of information flows (Porteous, 1995). The quality of financial information is vital to its value. *Standardised* financial information (such as stock price quotations) may now be cheap and quick to transmit relative to its potential value. However, the quality of *unstandardised* information may decline sharply over distance between generator and user.

For example, a financial rumour about a particular firm may rapidly spread through global networks, but traders further from the source of the information may find it harder to confirm the information in order to act on it. Of course, they could contact local traders for verification, but this involves cost and delays their use of the information compared to the local firms. Local firms may have a better and quicker sense of the accuracy and value of new information as a result of having a variety of local feedback channels and enjoying localised information spillovers which enable them to receive information that cannot easily be standardised and transmitted over computer networks. This partly explains why most international banks establish local offices in countries in which they are active: to serve as 'listening posts' among other marketing and client service functions. Hence, although there are offsetting decentralising forces, even in the face of new communications technology, the forces of spatial agglomeration in the financial sector remain powerful. Instead of becoming less important, as O'Brien claims, the role of location continues to be critical in this sector (Martin, 1994).

WHY AGGLOMERATION IN A PARTICULAR LOCALITY

The above analysis has explained why the financial sector in general tends to give rise to spatial agglomerations known as 'financial centres'. The key question is why a financial centre develops in one particular location and not in another. Historically, the financial sector has developed in response to the need for financial services from the real sector – mining, agriculture and industry. At early stages of development, financial activity arises in a particular place because of information passing through that location due to trade flows or real sector activity, as Townsend (1990) has shown for Western Europe. Against this broad background, there are two levels of geographic analysis which can account for the growth of particular financial centres. The first is the immediate hinterland of the centre, meaning the area over which the centre is the dominant supplier or consumer of goods or services; and the second is the broader region comprised of various hinterlands across which there are processes of interaction which determine how particular city-hinterland systems fit into global or regional patterns.

LEVEL ONE: INFORMATION HINTERLAND

The concept of a hinterland is a familiar one in traditional economic geography. However, in addressing the special features of the financial sector, it is necessary to develop the concept of the 'information hinterland' of a financial centre, rather than hinterland in the general sense as previously described. The information hinterland may be defined as the space or region for which a particular financial centre provides the best access point for the profitable exploitation of valuable information flows. There are several parts to this definition which require explanation.

First, the criterion of *best access* implies a two-way flow of information – from the hinterland to the core city and from the core city to outside reference points. Such accessibility may be affected by natural (or 'first-nature', to use Cronon's, 1991 term) geographical advantage. However, accessibility will be shaped more decisively by man-made (or 'second nature') developments such as transport networks; after all, transport systems have managed to cope (more or less) with the huge natural disadvantage that several of the world's densest financial agglomerations (New York, Hong Kong and Singapore) are on small islands. Cronon's study of the rise of Chicago shows that although Chicago lacked first nature advantages, standing on a flat plain, it compensated for this by becoming a central market point in the Mid-west, a transportation hub with considerable second nature advantage.[1]

Second, the *value of information flows* is a function of the type of activity in the hinterland and its expected growth over time. This may be shaped initially by natural endowments (i.e. first nature), but the patterns of exploitation of natural resources and of secondary and tertiary sector development in the region are usually more important in the long run.

Third, the criterion of the *potential for profitable exploitation* is important. If this were excluded from the definition of information hinterland, there would be many small financial centres with small hinterlands. In order for information to be potentially profitable in a particular centre, there must be a financial infrastructure with sufficient trading opportunities. This requires financial markets which can support trading volumes high enough to overcome illiquidity risk. This is why there are typically no more than a few important stock exchanges in a country. Thus, there may be a trade-off between being in a location close to the source of new information and being in one where the information can be used profitably. The decision by foreign banks to establish a presence in a national financial centre, rather than service that national market from an existing global financial centre with its thick markets, may be the compromise outcome from such a trade-off.

The concept of an information hinterland does not rule out the crucial endogeneity between financial and real sector development alluded to earlier. Not only does the value of the information on the real sector in the hinterland shape the strength of the financial centre which is the 'gateway' to the region, but the strength of the financial gateway may in turn promote economic development in the hinterland by providing better financial services at more competitive rates (this is often cited, for example, as a factor behind the concentration of economic activity and prosperity in the South East of Britain, which benefits from having London as its centre).

LEVEL TWO: INTER-REGIONAL ATTACHMENTS

A financial centre is not only a gateway to a hinterland, but also the connection point or point of entry between different hinterlands. During a process of colonial development, a locality may become a 'point of attachment' through which foreign capital is channelled into the local economy. For example, Montreal's early dominance was attributable in part to it serving as the first point of attachment for British capital flowing into Canada (Code, 1971 and 1991). If it were not for the 'friction of distance' slowing the speed of information transmission or reducing its reliability, there would have been no need for a point of attachment as London banks could have serviced vast areas such as Canada from London without the need for local representation. However, because of the decay of unstandardised information over distance, there was a need to be closer to the source of generation. Precisely where such a point of attachment develops is a function of the information hinterlands of different centres within a country (the role of such information hinterlands in the rise of regional financial centres in the United

States is well illustrated by Odell, 1992). These concepts may be used to explain the development of our two case-study city-pairs.

The immediate natural information hinterlands of the respective cities are the provinces or states of which each is the capital: Quebec and Ontario for Montreal and Toronto, respectively; and Victoria and New South Wales for Melbourne and Sydney, respectively. However, their effective informational hinterlands extend over broader areas. For example, Toronto has also served as the financial centre for the western states of Canada; and Sydney was better positioned than Melbourne to service Queensland and other northern parts of Australia. These pairs of states differ considerably in their economic structure and their post-war growth patterns.

International links were particularly important in boosting the financial role of Sydney over Melbourne. Sydney's international orientation and accessibility at key times, such as during the mining booms, played an important part in enlarging its role as an international point of attachment. This in turn increased its economic weight and compensated for the relative weakness of the immediate hinterland in New South Wales (Cardew, Langdale and Rich, 1982). In Canada, the heavy volumes of post-war US investment in manufacturing in south-western Ontario also helped to pull the centre of Canadian economic gravity decisively southwards, with corresponding benefits for Toronto as the informational core city of the region. The general pattern of change in both city-pairs was a long, slow relative decline in the underlying economic hinterland of the older centres as the result of long-run changes in economic structure. However, particular technological shocks, such as the introduction of mass air travel, decisively pushed the balance of influence towards the newer centres. Both Sydney and Toronto were able to consolidate first nature advantages of a better physical location relative to new markets into second nature benefits by the construction of the first major international airports in each country.

This suggests a parallel in the financial sector to the forces which have determined the location and persistence of the US manufacturing belt in the Midwest (Meyer, 1983; Krugman, 1993). Mid-western manufacturers, with a base in strong regional markets, were increasingly able to exploit economies of scale in production. New technology, chiefly the railway, led to cheaper transport costs and opened up new markets in the South following the Civil War. Northern manufacturers could dominate these markets because of their established scale economies at a time when the South had little industry. Consequently, industrial development in the South was retarded and manufacturing in the South was slow to escape from the shadow of established Mid-western industry.

Code (1971) applies a similar argument to the development of Toronto, although with a different outcome. Toronto had the opportunity to emerge from the shadow of Montreal as a financial centre – in part because the critical innovations affecting rapid information flow, such as the computer, telex and fax, were post-war developments. Thus, Toronto was able to find its feet before technological advantage may have 'locked in' the established position of Montreal. In an industry heavily dependent on information transmission, technological changes affecting information flow can open space for new centres to emerge, or, equally, 'lock in' the advantage of those new centres, until the next significant wave of innovation changes the patterns of advantage.

Amid the similarities in the broad patterns shaping the city-pairs, one special difference deserves a mention here, namely the rise of the forces of Quebecois nationalism, in particular the Parti Quebecois (PQ), in the late 1960s. It is quite clear that, by then, the information hinterland of Quebec had already been in decline for two decades compared with that of Toronto. However, the nationalist policies of the PQ and others undoubtedly accelerated the decline of Montreal. Whereas prior to the PQ government of 1976, much of the growth of Toronto had come from being the start-up point for new business, after 1976 Toronto benefited from business relocations from Montreal as well. Such moves further bolstered the agglomerations in Toronto, and the advantages to subsequent firms of locating close to the growing financial community there. The nationalist climate of Quebec and the uncertainty regarding its future also meant that new foreign bank entrants were much harder to attract to Montreal. Hence, idiosyncratic political shocks also may play a role in shaping the map of informational advantage over time.

Modelling Path Dependence

It is clear from the examples of Australia and Canada that neither pure chance nor path dependence ('getting there first') can alone explain the processes of locational agglomeration in the financial sector. Brian Arthur has recently added important insights which provide a fuller conceptual framework for understanding apparently path-dependent processes. In doing so, he has provided some of the theoretical tools for modelling the financial centre development processes described here. This section aims to give an introduction to Arthur's insights applied to this topic. (For more detail, see Arthur's (1994) edited collection of his writings in this area applied to a wide range of observed path-dependent economic processes, such as technology development and industrial localisation.) Arthur (1988: 90) observes that firms which are not tied to raw material sources and which do not compete only for local customers are often attracted to a locality by the presence of other firms. Financial firms fit this description well. He proposes a model in which agglomeration economies from the presence of existing firms play an important part in the locational decisions of entering firms. Once there is sufficient agglomeration in a particular location, geographical 'lock-in' may occur in which one location clearly dominates the other.

The basic Arthur model is structured as follows. Firms receive a pay-off with net present value $\Pi_j + g(y_j)$ when they locate in locality j. Π_j, which Arthur calls the 'geographical benefits', are the pay-offs to firms from the inherent qualities of a place; $g(y_j)$ are the agglomeration benefits, which depend on the presence of y_j firms already located at j. Starting from no firms at any of a number of locations, firms are allowed one by one to choose where to locate so as to maximise profits. Arthur allows for different types of firms (i.e. with different $\Pi_j + g(y_j)$), where the probability of appearance of a particular firm type in the entry sequence is known. Arthur and others have shown that if the $g(y_j)$ agglomeration function increases

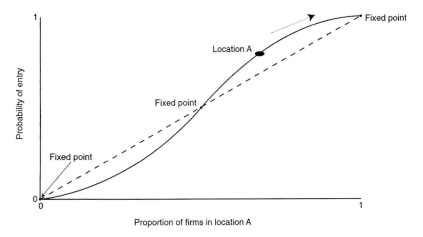

Figure 2 *Proportion-to-Probability Mapping*

without bound (i.e. the more firms at *j*, the greater the pay-off to each firm at *j*), then a certain region will eventually receive enough firms to become attractive enough to 'lock out' all other regions. This means that, after a point, all subsequent firms will enter in this region alone. Consequently, the share of this region in the total number of firms will converge to 100%, while that of other regions falls to 0%.

Therefore, the model allows for a mixture of chance and determinism. On the one hand, one cannot predict which region will become dominant, as this depends on the entry sequence of firm types. On the other hand, regions which are inherently more attractive have a higher probability *ex ante* of becoming dominant. Hence, the way in which the entry process evolves will determine the outcome. Path dependence is, therefore, no longer a 'black box' process, but one which can be analysed.

A 'proportion-to-probability' mapping, as shown in Figure 2, can be used to capture this general framework. This mapping shows the probability that a particular locality will be chosen by the next entering firm (on the y axis), given the current existing share of the locality in the total number of firms (on the x axis). Arthur has proved that the proportion of firms in a locality converges to a stable fixed point, at which the proportion of existing firms equals the probability of the next firm entering there. When the probability function is non-convex, there are multiple fixed points to which the outcome may converge, as shown in Figure 2.

APPLICATION TO THE CITY-PAIRS

The basic Arthur framework may usefully be extended to apply to the city-pairs described in this chapter. A financial centre is formed either by increasing numbers of financial firms locating there, or, in the case of foreign banks, by increasing numbers establishing representative offices there. The decline in one financial centre and the rise of the other may result from non-convexities in the underlying probability functions which would cause the proportional share of the declining city in total financial sector business to tend towards 0%, while the proportional share of the rival tends towards 100%. The shape of the probability functions is influenced by two factors in this model. The first factor is the nature of the informational externalities in each place. The second factor is that the 'inherent advantage' of a centre changes over time as well; that is, we need to consider $\Pi_{j,t}$, where t is the time subscript, rather than simply Π_j as before. What determines these changes? The above analysis has suggested that changes in the information hinterland of a city affect its attractiveness to financial firms. The value derived from the information hinterland depends on the value of information flows passing through it. This value changes over time, as the structure of the real sector of the hinterland changes. In addition, the growth of the real sector is not purely exogenous. A more dynamic financial sector may result in faster real sector growth, even as a larger real sector creates higher demand for financial services.

Financial firms located in one centre are not cut off from information generated in the hinterland of the other city. However, there is a friction connected to the use of non-local information. This friction is not so much from the cost of communication, which is now very small, but from the decay in the value of non-local information. As discussed above, this decay occurs because firms are not able to gather non-local information or to use it as quickly as the firms closer to the source, namely those in the closest centre. Hence, we may postulate an equivalent to the 'iceberg' transportation cost used in conventional real sector trade models: a certain proportion of the value of information 'evaporates' over the distance between source and use.

Consequently, we may redefine what Arthur calls the 'inherent attractiveness' of a locality as the expected discounted value of profits made on flows of local and non-local information. To this, we may add the profit made on information endogenously generated among participants in financial markets (the agglomeration effect). Hence, the expected discounted value of profit made by a firm entering centre A at time t may be expressed as:

$$\sum_t \rho_t \Pi_t^A = \sum_t \rho_t p_t \{\phi[I_t^A + (I_t^B/\varepsilon)] + \sigma_t(N_t^A)\}$$

where B denotes the other centre, ρ is the discount factor, p_t is the price received for the information, I^i is the volume of valuable information flows in location i, ε is the decay factor for non-local information, and N^i is the number of firms in location i. The subscript t in each case denotes time period t, while $\phi(.)$ and $\sigma(.)$ are functions which define the production of information outside the financial sector and within the financial sector, respectively. In the light of the discussion above, I^i is a function of both exogenous real sector forces and the endogenous financial factors, such as the size of the financial sector, in each location. Similarly, p will depend on the demand as well as supply of valuable information in each period, which will be a function of market size and the number of market participants. In each period, an entering firm would choose the location where expected profits are higher.

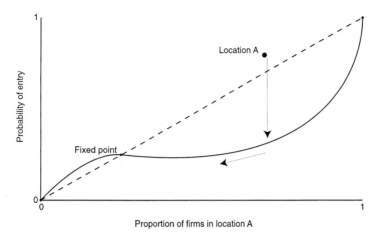

Figure 3 Probability Mapping After a Shock to the Function

As the relative volume and value of local and non-local information adjust with changes in the information hinterlands of the cities, so the proportion-to-probability mapping changes with time. It is the reshaping of this function which may explain the main phenomenon of interest in this chapter, namely the reversal of apparently path-dependent processes.

Let us assume that the probability function exists and has been defined for an economy with two entry localities. Financial centre A, which starts with an initially high proportion of the financial industry (for example, the position marked on Figure 2) may appear to be converging in the direction shown by the arrow to a position of total dominance, with all firms located there. However, a gradual reshaping of the information hinterland of the city, combined with external shocks, may change the position of the centre and the shape of the probability mapping so that the centre is now on a declining path, as shown in Figure 3. Correspondingly, the share of the rival centre in the financial sector increases in such a way that it may obtain dominance, at least until the probability mapping decisively changes again. The reshaping of the underlying probability mappings captures the effect described above of deeper changes in the attractiveness of each location to new entrants. In this way, the relative concentrations of financial activities in a location may change over time. The fact that there are reversals in the relative concentrations of firms such as those described in this chapter implies that there are considerable non-convexities in the probability mapping of the financial sector, and that these probability mappings change over time.

OFF-SHORE FINANCIAL CENTRES

This chapter has primarily analysed the growth of national financial centres and has not considered off-shore centres (OFCs), such as the Bahamas, the Cayman Islands, Gibraltar, Jersey and the British Virgin Islands (see Chapter 7).[2] These

OFCs usually have little or no informational hinterland, but exist because of specially created institutional advantages, such as flexible regulations or low taxes. Although financial sector employment may be created in such places, and indeed many OFCs have seen rapid growth in their workforces in recent years, the jobs are often not high level. For example, although fund administration is undertaken in centres such as Guernsey and the Isle of Man, many of the high-value-added functions of fund management remain located close to the major financial markets in London. Therefore, OFCs may suffer from the usual branch plant problems analysed by Dicken (1976 and 1986), namely little local control and high leakage as linkage into the local economy is typically very low. In his study of Panama, Johnson (1976) questioned the benefits of OFC financial sector policies in comparison with the costs in foregone tax revenues and in the neglect of indigenous entrepreneurship in other sectors. Nonetheless, countries as diverse as Botswana, Mauritius and Malaysia (Labuan) have been attracted by the experience of others and have embarked on setting up international financial service centres (IFSCs) in recent years.

Although the arguments of this chapter are not intended to account fully for the special features behind the growth of IFSCs, their rise does not undermine the explanations given here – in general, high-level financial functions seem to remain close to the information flows on which they depend, that is to an information hinterland. Furthermore, even though IFSCs are largely delinked or decoupled from local information flows, there is still evidence of a residual dependence on a geographic hinterland. For example, the OFCs around the British Isles receive most of their business from Britain, whereas US off-shore business tends to go to the Caribbean or Bermuda, and German off-shore money is concentrated in Luxembourg or Switzerland.

CONCLUSIONS

This chapter has applied the concept of spatial externalities to describe why high-level financial services tend to agglomerate spatially in 'financial centres'. Furthermore, the concept of the information hinterland of a centre has been developed to help explain why these agglomerations develop in particular locations, and why these locations may rise and decline in relative importance over time. The information hinterland of a national financial centre is a function of both intra-regional and international forces. The flows in this hinterland are shaped *inter alia* by patterns of intra- and extra-regional trade, by transport networks, by historical and institutional factors and by regulatory regimes.

The agglomeration effects may cause a locational 'stickiness' which slows the redistribution of relative financial sector power between two or more centres; but the examples used here suggest that the agglomeration effects in themselves are insufficient to prevent possible spatial switches in the long term. In the short- to medium-term, however, there may well be persistence on the basis of initial advantage. If an agglomeration is large enough to start with, then it may even outlast the decline in its immediate hinterland, as it may be able to shape and dominate a much larger hinterland. This may be the explanation for the continuing global financial role of London, as its wider information hinterland extends to most of Europe and beyond. In the long run, because the financial sector is increasingly mobile, the locational attraction of a financial centre can be protected or promoted only by securing and enlarging its information hinterland. The wave of competitive deregulation that swept across the world's major financial centres during the 1980s and 1990s represented, in effect, an attempt by the state authorities involved to do just that, as each sought to extend the information hinterland of its respective centre in an increasingly global market.

Technological, economic and institutional changes are constantly remoulding the information hinterlands of financial centres. First-nature locational advantage is soon undermined by second-nature advances in technology and institutional changes as the lessening of the friction of physical distance on factor and product mobility reduces the importance of first-nature factors in securing advantages for a particular location. Just as the arrival of air travel boosted Sydney over Melbourne, so new telecommunications technologies may reshape the comparative advantage of different cities (Hepworth, 1986). Changes in world patterns of economic growth may favour some cities over others, just as Toronto and Sydney have benefited from the reorientation of major national foreign trade and investment flows away from Britain towards the United States and the Pacific Rim, respectively. Institutional factors may also reshape advantage, particularly changes in political power which result in the favouring of particular regions in national policies and the choice of location of key institutions, such as central banks.

If local financial sector development brings local benefits, the remaining key question is whether financial centre development is amenable to locational policies of the type which are increasingly being implemented in places as diverse as Anchorage in Alaska and Dublin in Ireland. Other cities or regions have launched private sector marketing drives, such as Birmingham 2000 or Scottish Financial Enterprise, to develop the local financial sector (McKillop and Hutchinson, 1990: 163–5). To put the question another way, can Montreal and Melbourne do anything about their declining position; or can Toronto or Sydney do anything to bolster or assure their dominant national position?

In one sense, the analysis presented here is encouraging. Financial centres can and do emerge out of the shadow of other centres over time. However, the analysis also highlights the main force behind this emergence, namely the strong relative growth of the information hinterland of a centre and of its international attachments. This growth is not amenable to simple policy measures alone. Thus, even sustained efforts to promote the financial sector in Quebec have not shown marked success compared to Toronto, although they may have stemmed part of the decline of Montreal. Expensive relocation benefits or tax breaks for financial firms may be wasted, therefore, if development of the financial sector is seen in isolation. Indeed, some present-day incentive policies for the financial sector may be seen as the 1990s' equivalent of the 1960s-type growth pole strategies focused on the industrial sector, which have been widely criticised. The concept of an information hinterland suggests that complex channels of intra- and inter-regional information flows must be opened if financial centres are to grow. Tax incentives or de-regulation alone will not create a national financial centre, or even an OFC. This does not imply that nothing can be done to promote the financial sector in a new location, but that the process of cultivating a valuable information hinterland is a multi-dimensional and long-term project.

Chapter 6

Globalisation and the Crisis of Territorial Embeddedness of International Financial Markets

Leslie Budd

Introduction

The hallmark of the close of the twentieth century is one of a great transformation in the organisation of economic life. Since the mid 1970s, the Fordist 'golden era' (Lipietz, 1982), of rising standards of living, full employment and stable prices, that national economies enjoyed from the late 1940s has given way to an uncertain economic environment in which international market imperatives have dominated domestic priorities.[1] The 'crisis of embedded liberalism' (Ruggie, 1982; Keohane, 1984) that brought the Fordist golden era to an abrupt halt in the early 1970s mortally wounded international demand management. The term 'embedded liberalism' refers to the combination of interventionist domestic state policy and liberal policies towards the world economy. The two key components were the welfare state, in the widest sense, and a commitment to free trade whose benefits were to be distributed through the international division of labour, a combination sometimes referred to as the 'politics of productivity' (Maier, 1978). At the national level, organised labour accepted the logic of ever-increasing productivity in exchange for rising real wages and extensive welfare benefits. At the international level, the Bretton Woods system of international monetary regulation, underwritten by the world hegemon, the United States, provided a stable environment within which the politics of productivity could be pursued.

The reasons for the demise of the Bretton Woods system in the early 1970s have been well documented (Eichengreen, 1996) and need not be reiterated here. The pertinent point for the subsequent evolution of the monetary system and growth of financial markets was that the end of the Bretton Woods system brought with it a major change in the nature of regulating money and economic space. The environment of financial centres and the culture of their participants have altered radically in the last 20 years. The conduct of financial business by closed communities of conservative bankers and financiers according to arcane rules, has given way to a 'brave new world' of brash 'masters of the universe' drawn from all over the world, whose codes of conduct are often difficult to divine. These environmental and cultural changes have accompanied new concepts which account for the new business practices. They include digitised space, digitised money, 'globalisation' and the possibility of the 'end of geography' (O'Brien, 1992). These concepts essentially derive from the perception that new technologies have enabled flows of finance to become 'placeless', effectively existing in a 'virtual' world, taking off and landing at the major urban financial centres which comprise the hubs of the global network, in response to rapidly changing market signals about risk and return. However, the possibility of technology, deregulation and innovation generating placeless financial flows results in a paradox. The realisation of placeless flows has actually *reinforced* the locational embeddedness of the large international financial centres such as New York, Tokyo and London (Martin, 1994).

Financial innovation has created new financial products, whereby investor risk can be managed more efficiently. New communication technologies speed up market turnover and spread the net of market opportunities ever wider. These factors are underwritten by deregulation and the breaking of national boundaries in order to internationalise transactions. As financial markets derive advantages from economies of size, scope and agglomeration, the places where market activity occurs becomes an important consideration in the decision-making of large international financial firms and how they distribute their activities. So, despite the promise of a 'virtual' financial world, location remains an important consideration.

Drawing on the international relations literature, and using a critical-realist approach, this chapter argues that 'territorial embeddedness' has been a key factor in the stability and sustainability of international financial centres. However, a corollary of the crisis of post-war 'embedded liberalism' referred to above has been a crisis of territorial embeddedness. The combination of deregulation of financial markets (in order to create 'a level playing field') and new information and communication technologies, opens up the possibility of remote trading away from large international centres (essentially the argument put forward by O'Brien, 1992). One consequence of this regulatory shift is a changed relationship between the organisation of financial flows and geographical space. Sustaining the territorial embeddedness of large financial centres, by maintaining the benefits of urbanisation, localisation and agglomeration economies found in these centres,[2] will then rely on re-regulating the spaces in which they operate. In an environment in which there is little constraint on cross-border economic and financial transactions, more interventionist regulation at a national level is no longer a possibility. Neither can technological innovation be reversed. The question then arises as to whether deregulated financial spaces can be re-regulated in order to combat threats to their territorial embeddedness. The answer is a hesitant yes. Significantly, the 'crisis of territory' – the threat posed to financial flows being concentrated in and articulated from particular places by new technologies – can only be overcome by territory itself, that is the degree to which a particular financial place can be transformed into financial flows through the agency of rent, so as to sustain its utility as a territorially embedded financial exchange. In an era in which the friction of distance has been overcome and space has been re-shaped (Harvey, 1989b; Castells, 1996), it is the 'land question' which provides the basis for discussing the re-regulation of international financial centres.

The argument of this chapter proceeds as follows. The limits of embedded liberalism showed that the combination of a Fordist regime and the Bretton Woods international monetary system could not sustain the growth of economic welfare in a national economic setting. The subsequent growth and importance of financial markets gave rise to the notion that regulation of financial flows could not be contained within national spaces: market 'self-regulation' was the optimal allocative device. This growth and increased importance was stimulated by a combination of technology, innovation and deregulation, which is seen by many to be the basis of 'global finance' (Sassen, 1991). The advantage this combination brought was the possibility of virtual financial transactions, whereby financial flows would become placeless. Financial flows would no longer be contained within the network of large financial centres. International centres and their constituent markets, for example London, would then face challenges to their territorial embeddedness, leading to a possible crisis. If the financial territoriality of international centres such as London is to be sustained, their market participants cannot be sanguine about the threat of virtual financial trading technologies. The growth of such technologies threatens to change organised place-based exchanges into information systems which transcend place and annihilate space. Yet, before this imagined universe becomes the accepted wisdom, one significant element needs to be emphasised, namely territory – *place* – itself. The majority of net wealth of the advanced economies is accounted for by property holdings, both residential and commercial. Locational decisions of financial firms are heavily bound up with real estate issues, as rises in the value of these assets can be internalised to create greater liabilities.[3] Furthermore, the size and distribution of financial services in most countries is hierarchical, centred on a national capital because of the advantages of size and scope. The pools of specialised labour and expertise, and networks of contacts and association, found in such national centres remain vital to the daily business of the centres' financial markets and institutions. The spatial decentralisation of those markets and institutions would be unlikely to pull labour and expertise with them so long as highly renumerative employment opportunities were still available in the centre and the shared business and personal cultures that typify large, established financial centres were not easily replicated elsewhere. Territory, therefore, is likely to remain a crucial component in the sustainability of place-based financial centres even in a world of placeless financial flows. The shift in the nature of the regulatory space of financial transactions in the late twentieth century makes it critical to re-regulate the territoriality of large financial centres through the expedient of real estate. These are the themes of this chapter and, although the discussion focuses on London, the basic arguments also apply to other international financial centres.

THE LIMITS OF EMBEDDED LIBERALISM

The forces which undermined embedded liberalism were international. The assumption of Bretton Woods was that under its international monetary stabilisation and liberal trade regime, economic growth in one country would be transmitted to other countries, generating non-inflationary development. By the late 1960s and early 1970s, the transmission effects became less benign. The inflationary effects of the financing of the Vietnam War and the first oil shock of 1973–4, the slow-down in productivity growth, the growth of trade competition from the less-developed countries and rising national indebtedness in the West, all combined to produce widespread stagflation.[4] The increase in inflation and indebtedness may have slowed down economic growth, but it also opened up intoxicating possibilities for financial markets. In an unstable and uncertain economic environment, the nectar of financial innovation slaked the thirst for mitigating and managing risk. The demise of the Bretton Woods system and the huge increase in global liquidity in the 1970s damaged international demand management and the possibility of it being negotiated within national regulatory spaces. The excess liquidity and the impact of high worldwide inflation on assets and liabilities could only be regulated through the large international financial markets. By the mid 1970s, movements towards international monetarism and economic and financial deregulation were emerging. In 1975, the New York stock exchange was deregulated, and during the course of the 1980s other countries followed by deregulating their financial markets and relaxing their capital and currency controls. The hegemonic role of the dollar as the foundation of embedded liberalism gave way to the role of New York's financial markets as the leader of the new financial internationalism. The regulatory space of the world's economies and financial systems shifted from national boundaries to a few large (global) cities on which international financial flows were centred. These 'financial hot spots' experienced a two-way street of internationalism. They became the nodes in an evolving international network of financial flows, and, in accommodating these flows, became more international as places. Essentially, a number of developments that had been underway for some time grew both in size and importance.

During the 1960s and particularly in London, the Euro markets developed as global–local corollaries of the penetration of US production multi-nationals in Europe. The major players in the new Euro markets were initially branches of large US banks that re-cycled off-shore dollar deposits. The multi-nationalisation of banking corresponded to the shift in the regulatory space of the international economy. The end of the Bretton Woods system moved the regulation of the world's economies from a nationally embedded space to a less-embedded international one. The growth of off-shore financial centres (OFCs) and the attraction of financial innovation displaced production in favour of finance within the new regulatory spaces. The wave of deregulation and liberalisation policies introduced in the leading industrial nations from the mid 1970s onwards, stimulated the modernisation of the world's major stock exchanges in order to embed their markets into specific territorial and regulatory spaces (see Table 1 for a summary of the changes in the regulatory space of financial markets in the post-war period). Ironically, however, the very process of deregulation poses a threat to that territorial embeddedness. Likewise, current attempts to re-regulate the excesses of financial markets, particularly derivatives trading, by booking over-the-counter (OTC) to regulated exchanges, also threatens the territoriality of the financial marketplace. This is because regulated derivatives exchanges are becoming more and more like information networks whose computer-based services can be offered anywhere in the world (Laulajainen, 1998). In the case of the financial City of London, for example, the authorities face both increasing inter-place competition from other European centres (as the European Union deregulates the European

Table 1 The Changing Regimes of Financial Regulation

Period	Regime of regulation	Characteristics
Mid 1940s to early 1970s	Embedded liberalism	National regulatory system bound by fixed exchange rates, exchange and capital controls, underpinned by the Gold Exchange System, underwritten by the international convertibility of the dollar
Late 1950s to early 1960s	Regulatory challenge of Euro markets	Differences in national regulatory systems established concept of 'regulatory asymmetry' between financial centres; international demand for dollar deposits outside the control of Federal Reserve led to Euro dollar market being established in London, which had most liberal regulatory regime
Mid 1970s to mid 1980s	Successive international deregulation	Aftermath of the Bretton Woods system created the need for some international 'thermostatic device' to re-cycle excess, in particular dollar, liquidity; new technologies, financial innovation and the need for risk-managing financial instruments created the environment in which New York stock exchange was deregulated in 1975; the potential loss of London's international position led successive British governments to consider deregulation from 1978 onwards, culminating in 'Big Bang' in 1986 which increased London's international exposure; the other major centres in the European time zone quickly followed suit, albeit on a smaller scale; the locus of a few large financial centres embeds international flows in these few places
Late 1980s onwards	New re-regulation	Increased systemic risk from increased exposure to derivatives and fall-out from associated scandals, as well as the East Asian crisis, leading to demands for global re-regulation; in the European Union, the Single European Market programme of 1992 extends Europe-wide regulation to financial markets. Combination of electronic trading and moves towards regulatory 'level playing field' threatens locational embeddedness of large centres through increase in placeless financial flows; negotiation of real estate markets as strategy of re-regulating and re-embedding financial flows into specific financial locales

financial space) and intra-place competition from new financial spaces within London, for example Canary Wharf. The former is a product of the limits of embedded liberalism. The latter arises from opportunities for international rentiers to take advantage of the growth of international financial markets and the changed relationship between money and space. What connects them is the nature of property and land, and conceptions of territory which are strongly rooted in history.

THE GROWTH OF INTERNATIONAL FINANCIAL MARKETS

The importance and growth of financial markets has been treated as a relatively recent phenomenon. Indeed, at the start of the post-war period they were fragmented. They then began to grow through a combination of official policy and market responses to various opportunities arising from regulatory gaps between controls of essentially self-contained national financial systems (Harris, 1995). Where these national systems extended into currency blocs, profitable arbitrage opportunities were created.[5] Regulatory asymmetry between New York and London was a major cause of the growth of the Euro dollar markets in the 1960s.[6] In the 1970s, the crisis of embedded liberalism stimulated these kinds of 'off-shore markets' as the regulation of the world's economies moved from national Keynesianism towards 'global neoclassicism' (Schor, 1992). Global neoclassicism suggests that financial globalisation is the agency through which national economies become subservient to international markets. Globalisation of finance has been stimulated by innovations in financial intermediation, technology and financial instruments. Innovative instruments come under the rubric of derivatives whose conceptual and operational nature has caused much confusion. The growth in the type and use of derivatives has come from a simple logic, to manage risk in increasingly volatile asset markets. The increased globalisation of finance tends to equalise rates of return on financial assets.[7] Therefore, there is a constant search for above average returns through the agency of innovation. Moreover, in this search, risk management becomes a prime concern.

The conception of globalisation and its limits is beyond the scope of this chapter (for a more comprehensive coverage of the issues, see Featherstone, 1990; Dicken, 1992; Hirst and Thompson, 1996). However, if globalisation is a reality, then the law of one price should operate. That is, any differences between the prices of the same product in different locations are accounted for by differences in transactions costs (including transport costs). If the law of one price does operate, competitive markets function in a world of economic liberalism, and the tenets of global neoclassicism will hold. However, under such conditions, there will be little or no incentive for financial innovation, which itself thrives on volatility, because prices of financial assets will converge globally.

Globalisation appears to be the dominant parameter in the relationship of financial flows and space in the late twentieth century. The significant issue for the world's major international financial centres is whether their utility and role will be undermined by globalisation. The conundrum these centres face is that the key elements in their advancement also threaten to undermine their role. The advances in telecommunications technology, deregulation and innovation operating in a global environment suggest that place-centred financial flows will give way to placeless flows. According to O'Brien (a fervent proponent of this view), we are witnessing 'a state of economic development where geographical location no longer matters in finance, or matters much less than hitherto' (1992: 1). The argument is a simple one: the fungibility of money aided by the information revolution renders location increasingly irrelevant for financial markets. The argument is reinforced by two supposed tendencies. First, financial firms think of themselves as global long before markets are global. Second, the regulatory environment will shift from a national framework to a 'competition among rules' (O'Brien, 1992: 115) as the world becomes 'borderless' (Ohmae, 1990). To be fair to O'Brien, he does add caveats about the importance of geography remaining for some time. His account also points out that the territoriality of some international financial centres will be reinforced by deregulation and that some re-regulation will be necessary. Nevertheless, his major point is that a process is underway in which the global financial universe is becoming 'seamless'. The claim that the cost of capital is converging everywhere is advanced to support this contention. Yet, the fall-out from the East-Asian financial crisis of 1997 and 1998 shows that stochastic, region-specific shocks undermine the contention that risk-adjusted financial returns are equalising.

The contagious effects of this crisis may have taken on a global appearance, but its causes were a combination of old-fashioned current-account deficits associated with fixed exchange rates, overaccumulation of real capital, large speculative real estate ventures, and weak financial and regulatory systems. The 'tiger' economies were attempting to reposition themselves in the global order by changing the nature of money and space within them. Over 200 years ago, Marx noted that in attempting to reproduce itself, capital constantly seeks to re-structure the primal elements of space and time. The important lesson of the Asian episode is that geography still matters, as is confirmed by international policy makers seeking to re-embed confidence in the financial systems of the economies which have been most heavily affected. Although the internationalisation of financial markets may well have promoted the conditions for 'global neoclassicism', the change in the nature of the relationship between money and space is more complex than any simple 'end of geography' arguments allow for (Martin, 1994). It is this complexity which is explored and demonstrated in the remainder of the chapter.

Proponents of global neoclassicism argue that such a system allocates capital efficiently, so that any role for the state in the regulation of the domestic economy is residual. For regulators of financial markets, the focus has moved from the allocation of capital in the real economy to the systemic risks of the global financial system (Harris, 1995). It is becoming apparent that the global financial regime has brought in its wake a mode of regulation which has re-shaped the regulatory space of the real and financial economy. The mode of regulation of global neoclassicism links national and global conditions via a simple expedient – 'Interest rates, profit rates, wage rates, and commodity prices will equalise across borders. A country deviates from the logic of world markets at its peril' (Schor, 1992: 4).

Within economic theory, financial markets exist to manage time and uncertainty. In the first case, they intermediate between different rates of time preference between savers and investors.[8] In the second case, risk is a serious impediment to the optimal allocation of resources in economic life unless there is a set of contingent commodity markets. Where the number of available markets is smaller than the number of contingent commodities, an efficient allocation of resources is still possible if sufficient financial instruments exist. In other words, financial asset markets exist because there is an incomplete set of markets for commodities (Debreu, 1959; Arrow and Debreu, 1954).[9] Hence, there is a logic to the creation of new financial instruments which stems from the allocative functions within the real economy. However, the transmission mechanism between the financial and real economy appears to be out of kilter in the late twentieth century because the dominant mode of regulation generates financial instruments whose purpose goes well beyond the needs of allocative efficiency. Therefore, the claims for global neoclassicism are refuted by the behaviour of international financial firms, and globalisation is reduced to hegemonic ideology.

The utility of Keynesianism was that its transmission mechanisms were able to sustain prosperity and its distribution within mainly national boundaries. Disutility set in with the rise of stagflation, following the crisis of embedded liberalism. A similar logic can be divined in the development of global neoclassicism. Casual observation suggests that international stock markets have become more linked in the 1980s and 1990s than formerly, thereby appearing to support the notion of global neoclassicism. Cumulative and simultaneous crises of stock markets such as those of 1987 and 1997 have been explained by excess volatility:

> The Brady Commission's early report on the 1987 crash and the unsustainable boom which preceded it summarised a prevalent view of the causal role of international linkages between stock markets: *'investors made comparisons of valuations in different countries, often . . . using higher valuations in other countries as justification for investing in lower valued markets. Consequently a process of ratcheting up among worldwide stock markets began to develop'*.[10] (Harris, 1995: 203)

However, as Harris goes on to point out, there was a weak correlation between the large and rapid fall in the Japanese stock markets in 1990 and that of the New York and London markets. The explanation may be simple and concerns the specific operation and regulation of the Japanese financial system. A more comprehensive explanation would suggest that the convergence of international asset prices is not a direct consequence of the internationalisation of trading. Instead, it is via an indirect link of international responses to price changes in stock markets indices, which are taken to reflect changes in the fundamentals of domestic economies. If the allocative efficiency of global neoclassicism is to hold, one would expect a convergence of the costs of real investment. However, as stock prices are often driven by the market psychology of expectations, herd-like or casino behaviour follows, with its consequent volatility.

Technological innovation has meant that this volatility can be transmitted around the world almost instantaneously. As a result, counter-parties to financial asset transactions find themselves subject to a zero-sum game where one party's gains are immediately offset by another's losses. There are two possible routes to break out of this stasis. The first is to connect international stock prices to real economic fundamentals, so that international rates of return correspond to some form of basic datum, such as Tobin's q ratio (Tobin, 1969).[11] The second is to engage in product innovation that manages the risk environment engendered by volatility. The second route appears to be the most favoured option at present.

The opprobrium cast on derivatives because of the excesses of the German Metallgesellschaft, Orange County in the US, and the London Baring scandals has been misdirected because of a misunderstanding of the nature of these products.[12] In the same manner that innovations in information technology have allowed capital markets to manage one of their functions – time – more efficiently, so the creation of derivatives allows the other function – limiting uncertainty – to be managed more efficiently. However, a dual tautology arises from the twin innovations of financial information technology and derivatives. In the first case, the international convergence of returns on financial assets drives down transaction costs. Securities houses and banks then have to have commit more and more capital to take advantage of electronically transmitted price arbitrage opportunities. The instantaneous transmission of information across borders means any default or volatility becomes rapidly magnified and the consequences, as in the Baring collapse, amplified. Therefore, managing the time preference of savers and investors through time-saving technologies potentially leverages up inefficiencies. Using risk managing products to gain better returns potentially levers up the risk to the whole financial system.

In the second case, the change in the nature of financial intermediation allows investors to directly access capital markets.[13] In order to manage the risk environment, successive financial instruments are 'layered'. For example, a mortgage is secured against the value of a house. The mortgages held by a financial institution are then bundled together and the next layer of bonds is issued to investors against the value of the income streams from the mortgages. This is an example of the process of 'securitisation'. Another layer can be created, based on the different timing of the interest payments and changes in the nominal capital value of the bonds through options, futures and so on. At one level, technological and financial innovation is overcoming the constraints of time and space, but the layering of financial assets 'stretches' them away from underlying price changes. In order to break out of a potentially zero-sum game, banks and securities houses constantly search for new asset categories on which to 'layer' financial instruments. However, this search creates another tautology, a 'paradox of risk'. This is a variant of the classical 'paradox of thrift', which occurs where an increase in the desire to save causes a fall in actual savings because national income falls. By the same token, the paradox of risk occurs where the desire to hedge risk, through innovative financial instruments, actually causes an increase in systemic risk.

The rapid growth of financial derivatives has been accompanied by a bifurcation in their delivery. Derivative products are either traded through recognised exchanges, for example the London International Financial Futures and Options Exchange (LIFFE), or through bi-lateral contracts between counter-parties commonly known as the OTC markets. In 1986, the value of the world's outstanding derivatives contracts (technically known as open interest) was about US$2 trillion, split equally between exchange and OTC products. At the beginning of 1992, the total had risen to US$10 trillion, with the OTC share rising to 60%. At the end of 1995, the notional value of OTC derivatives market was about US$40 trillion out of a total of US$58 trillion (Bank for International Settlements, 1995). During the first half of 1996, the notional principal of OTC transactions rose by 20% as a result of market uncertainties and a growth in liquidity in the international financial system (International Monetary Fund, 1996). About one-third of this total was accounted for by six US banks (Kambula, Keane and Benadon, 1996). Most of the publicised debacles such as Proctor and Gamble, Orange County and Metallgesellschaft have involved OTC contracts. However, organised exchanges which deal in standardised products and have regulatory controls have not been immune to very public defaults, as the Baring case showed. OTC markets are more prone to credit risk than organised exchanges because the current position of contracts is opaque. The advantage is secrecy, but this makes unwinding an unprofitable position more difficult. Ostensibly the 'marking to markets' regulations generate greater transparency in organised exchanges. That is, the investor deposits a proportion of the total value of the contract and each day a 'margin call' is made to the investor to top up the deposited funds if the market moves against the investor, and *vice versa*.

The issue of the regulation of innovative financial instruments has vexed international authorities. Given the transitive nature of these instruments, it is not clear what should be the locus of regulation. The position was set out by the International Monetary Fund (1993):

> derivative instruments tend to strengthen linkages between market segments and between individual financial institutions in ways which are difficult to identify or quantify. Consequently, disruptions or increased uncertainty in one market may now be more likely to spill over into other derivatives markets and into cash markets.

The regulatory response to the perceived volatility of OTC markets has been driven by the fact that all the major US investment banks provide OTC products. In order to make these market transactions more transparent, the regulatory authorities for commodities and securities in the United States and Britain have agreed to a uniform set of disclosure rules for institutions under their supervision. In the United States, there have been changes in accounting rules and international proposals for the establishment of clearing houses for OTC transactions, similar to those in organised exchanges. Another proposal is that OTC trades be 'booked' (in an accounting sense) to a local exchange. The majority of OTC trades are cross-border transactions, accounting for 55% of interest rate and currency trades booked in the United States. There is also a dispersion of cross-border trades across the world. The three major trading areas of the United States, Japan and Britain account for only half of the global turnover volume. Furthermore, there is a decentralised structure to cross-border transactions, with traders entering into a contract in one location and then booking it to another. Therefore, the transmission effects of any disruption would be widespread across many national boundaries (Remolana, Bassett and Geoum, 1996). It is apparent that attempts to manage global OTC transactions through the localised regulatory spaces of national exchanges will not in itself combat the threat to the territoriality of international financial centres.

The dominance of a financial mode of regulation over a production one in the last two decades has, in fact, not substantiated the claims for global neoclassicism. The transmission mechanism of the financial sector to the real economy is only touched on here. However, in Britain the financial sector has become an important part of the regime of accumulation. In 1975, the financial sector contributed 7.2% of GDP. By 1996, this proportion had risen to about 24%. The market capitalisation of the London Stock Exchange represented 154% of GDP at the end of the 1996/7 financial year. Regulating the operation and performance of financial markets, therefore, has profound consequences for the regulatory space between the financial and real economy.

The globalisation of finance and the greater integration of stock markets have narrowed international rates of return on financial assets. In seeking more profitable arbitrage opportunities through technological and financial innovations, financial institutions have transmitted greater specific and market risk. This is somewhat ironic, because technological and financial innovation aims to make financial markets perform their twin basic functions more efficiently. Concerns over the threat of increased systemic risk has been engaging state authorities in reshaping the regulatory spaces in which these markets operate. It is apparent from regulatory concerns about OTC markets and accompanying re-regulation proposals, that attempts are being made to territorially embed these markets. However, paradoxically, trying to alter the nature of the regulatory space of an internationally deregulated centre through re-regulation might threaten that territorial embeddedness. It is the possibility of this crisis that challenges London's dominant position in the European financial time zone, as well as its status as a global monetary centre.

CHALLENGES TO THE TERRITORIAL EMBEDDEDNESS OF LONDON AS AN INTERNATIONAL FINANCIAL CENTRE

The position of London as a major international financial centre has been thought of as a recent phenomenon. However, there has been a spate of literature assessing London's historic international role (Porter, 1994; Kynaston, 1994). The power of the City of London to finance trade and wars goes back several centuries. In the nineteenth century, the City of London was a financial centre which promoted international trade and the development of the British Empire. In 1808, Thomas Baring wrote, 'London is the principal centre of the American Commerce, London houses acting almost solely as bankers for the American trade, receiving the proceeds of consignments' (quoted in Kynaston, 1994: 11). Later in the same century, foreign exiles such as Rothchilds became paymasters of British governments, arranging credits to pay for government expenditure and military adventures. However, the consequences of two world wars and Britain's relative and absolute economic decline meant that by the late 1970s London's international position was under threat.

The nature of London's regulatory space has allowed it to sustain and develop its international role.[14] In the 1950s and early 1960s, the Euro dollar market was established in London (Johnson, 1983). This local market played a global intermediary role in the creation of deposits in the currency of a third country. 'Euro' is a prefix which simply means 'off-shore', and markets in other currencies and accompanying bonds soon developed in the wake of the Euro dollar markets. A major reason for the establishment of the Euro dollar market was regulatory asymmetry between London and the United States. The demand for dollar deposits in Europe was constrained by monetary regulations which limited US interest rates and returns to international investors in the US markets. By setting up branches in London, US banks offered the prospect of significant arbitrage opportunities, conditioned by the interest parity theorem.[15]

Despite the liberal environment which had created the possibility of the Euro markets being set up in London, it was apparent that arcane practices were undermining London's international position. The major shift in London's regulatory space came with the so-called 'Big Bang' in 1986, which dismantled the old institutions and their market privileges. Dual capacity in the stock market gave way to single capacity, minimum commissions were abolished, and foreign banks bought up traditional stock brokers, jobbers and market-makers, creating financial supermarkets which served the range of financial markets.[16] It did not last long. The contagious effects from New York in the crash of 1987 exposed London's overcapacity of financial services and institutions. Perhaps what was more significant at the time was the way in which the built environment of the City of London and its environs was radically re-structured. Planning regulations were altered to accommodate a large increase in the stock of offices. A technological imperative determined a rapid growth of new offices with large dealing floors, cabling for computer-based trading and air-conditioning as well as the ubiquitous atrium. Real estate, as a factor of production and as an asset category, became an important component in the financial regime of accumulation and its mode of regulation. This issue of the role of real estate as part of the shift in the regulatory space of the City of London, and the basis of sustaining its territoriality is taken up below. The point to note here is that the project of deregulation and liberalisation, accompanied by the significant easing of planning controls was one of politically motivated modernisation. The government's aim was to embed London's position as the largest international financial centre in the European time zone and sustain and increase its global reach. By the same token, the smaller Big Bang in Paris and Germany's 'Finanzplatzdeutschland' were similar political projects of modernisation, with the same objective as London's. As exemplified by the link between the Paris and Frankfurt futures exchanges, supplanting London's dominant position has been a shared Franco-German goal. Furthermore, strategic alliances with other international markets allows smaller financial exchanges to gain access to greater shares of global transactions. It enables these smaller exchanges to become locales for international flows in large economic spaces, for example the European Union (Budd, 1995).

London's capacity as an international financial centre rests on three core activities: foreign exchange, international equities and exchange-traded derivatives (insurance is a fourth, but this is not discussed here). London has increased its share of international foreign exchange in the last five years. It remains the largest international centre with 21.5% of business conducted in US$/DM transactions, 17% in US$/Y and 11% in US$/£ (1995 figures). Its relative international position is given in Figure 1(a) and (b). Table 2 shows the performance of international stock markets in real and nominal terms. As a result of the difficulties of direct

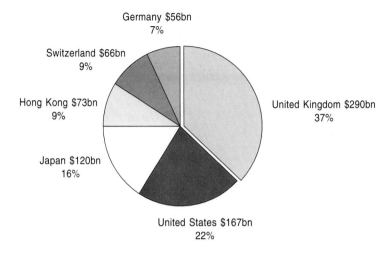

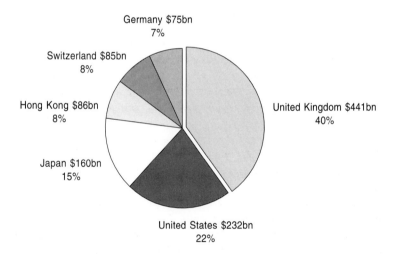

Figure 1 Foreign Exchange Turnover (a) 1992; (b) 1995 (Constant 1992 Prices)

comparison, Table 2 includes domestic business only. However, the significance of London in international equity business is amply demonstrated. In 1996, London accounted for 13.7% of the world's equity turnover on a sales basis, with Tokyo and New York accounting for 8.5% and 36.7% respectively. London's share of trade in foreign equities was 62% for the same year, with New York accounting for 23.5% and Tokyo 0.1%, compared to 66%, 16.6% and 3.3%, respectively, for 1992 (London Stock Exchange, 1996; O'Brien, 1992). Total business rose by 82.4% between 1992 and 1996. Table 3 shows the real value of foreign equity turnover in London between 1991 and 1996.

Table 2 International Stock Markets by Market Capitalisation and Turnover (£ billion), 1992–1996

	1992 Cap	1992 Turn	1993 Cap	1993 Turn	1994 Cap	1994 Turn	1995 Cap	1995 Turn	1996 Cap	1996 Turn
New York	3083	177	4628	1578	4162	1726	5251	1963	5860	2345
Tokyo	1943	314	3834	839	4650	878	5100	956	4801	824
London	1418	442	2844	1211	2831	1250	3096	1229	3217	1514
Germany	1109	869	1434	1412	1563	1580	1920	1844	1763	1842
Paris	662	644	783	997	757	982	916	906	905	954

Note: Calculated by taking the domestic deflator for each country and adjusted for different effective exchange rates

Source: Own calculations for London Stock Exchange (1994), (1995b) and (1996) and OECD (1997)

Table 3 The Annual Turnover of Foreign Equities Traded on the London Stock Exchange 1991–1996 (£ billion)

1991	1992	1993	1994	1995	1996
247.8	298.7	420.4	341.6	367.5	468.5

Source: London Stock Exchange (1993), (1994) and (1995b) and OECD (1996)

It seems that London has been increasing its international role in respect of foreign equities. However, this expansion carries with it an implicit threat. In 1990, the Dutch authorities discovered to their horror that over 50% of the shares of Dutch companies were being traded in London. They called in US consultants to advise on reversing this position. Whatever the reason, this total has since dropped to about 30%. The situation has been similar for the shares in French and German companies. The point about London is that the capital base of the British economy is not sufficient to sustain London's current equity market capacity. However, being an international off-shore entrepôt also makes London vulnerable to shifts in the regulatory spaces of the world's stock exchanges.

In 1996, the LIFFE was the fourth largest exchange in the world after the three Chicago exchanges, calculated on an open interest and half-yearly basis. The LIFFE also faces competition from the Marché à Terme International de France (MATIF) in Paris and the Deutsche Termin Börse (DTB) in Frankfurt. Their relative global performances are given in Table 4. Although until recently the contracts traded were not in direct competition with each other, differences in the ways of doing business, the effective franchising of electronic trading systems and prospective global alliances with other international exchanges have all intensified competitive strategies between the major exchanges. The LIFFE has franchised its Automated Pit Trading (APT) system to other exchanges around the globe, for example to the Sydney futures exchange. The German DTB agreed to an alliance with the Swiss SOFFEX exchange and the French MATIF at the end of 1997, to form the EUROEX. One factor behind this link was to anticipate the

Table 4 The World's Largest Futures and Options Exchanges by Outstanding Contracts ('Open Interest': Millions), Half-year, 1994–1996

	January–June 1994	January–June 1995	January–June 1996
CBOT	121.70	109.65	117.34
CME	117.21	115.15	103.06
CBOE	92.44	93.99	91.32
LIFFE	87.23	70.89	82.91
BM & F	42.84	72.08	70.42
DTB	33.81	28.31	37.63
MATIF	56.59	25.58	30.14

Notes: CBOT = Chicago Board of Trade; CME = Chicago Mercantile Exchange; CBOE = Chicago Board Options Exchange; LIFFE = London International Financial Futures Exchange; BM & F = Bolsa de Mercandorios & Futuros (Sao Paulo); DTB = Deutsche Termin Börse (Frankfurt); MATIF = Marché à Terme International

Source: *Futures and Options World* (1994), (1995) and (1996)

intense competition for pan-European markets following the introduction of the single currency in Europe. At the end of 1997 and beginning of 1998, the DTB had traded a greater number of futures contracts on German government bonds (known as 'bunds'), overtaking the LIFFE for the first time. The 10-year bund contract is the most heavily traded financial future in the world at present. At the time of writing, the LIFFE's response to this loss of market share had still not been formulated. What this process suggests is that the ownership, and thus the territory, of organised exchanges, through their member firms, may become more divorced from their role as information exchanges, providing price series around the globe. In this scenario, will the organised exchange become the residual location through which contracts are booked in order to conform to prudential regulation? It is a point which is returned to at intervals in the remainder of this chapter.

The growth of strategic alliances between futures and options exchanges ostensibly increases their global reach. There are more complex forces at work, however, with exchanges franchising their trading systems so that their products are transacted internationally. System franchising and strategic alliances are akin to the growth of multinational industrial companies in the 1950s, 1960s and 1970s. For example, the LIFFE is an exchange embedded in London, but it is becoming a trading network whose business is likely to become more rootless. The London embeddedness of the LIFFE may be reduced in the future as it becomes a regulatory site either for registration of OTC trades or acting as an OTC clearing house. This speculation appears to be a consequence of the international networking of other information systems, as noted elsewhere. It reinforces the notion that 'Telecommunications simultaneously reflect and transform the topologies of capitalism, creating and rapidly recreating nested hierarchies of spaces technically articulated in the architecture of computer networks' (Warf, 1995: 375).

Most of the alliances between exchanges have occurred in the last few years and are concerned with after-hours trading through remote electronic systems. The MATIF and the DTB agreed a link-up in 1994, whereby their products would be traded on the DTB's electronic system during the day and on the GLOBEX (an after-hours electronic exchange based in Chicago) during the night. This particular alliance is being wound down, with Reuters, one of the major partners, not renewing the GLOBEX agreement when it lapses in 1998, pointing to the instability of these arrangements. As an another example, the MATIF is switching to a French-based after-hours system which the Chicago Mercantile Exchange and the New York Mercantile Exchange will also join in exchange for trades cleared through the Chicago and New York Mercantile Exchanges' Clearing 21 system. Turnover on the after-hours trading system has been disappointing. The GLOBEX system trades only 1% of the Chicago Mercantile Exchange's and 9.5% of the MATIF's total volumes (1996 figures). The Chicago Board of Trade (CBOT) Project A after-hours system accounted for 1% of total volume. The LIFFE's APT system accounted for just 3%. Some of these alliances go back a decade or more. The logic driving them is internationalisation and economies of scale. International banks and securities houses have taken over smaller domestic specialists. In giving market access to their big customers, these international financial institutions are members of all the largest financial and commodities exchanges. Economies of scale cut dealing costs, and derivatives markets are no different from any other financial market in that external economies of scale and transactions costs are crucial to market players. Alliances also allow 'offset' agreements, in which liabilities are netted out against each other, thereby increasing the efficiency of the clearing and the settlement of contracts. Moreover, the greater growth of the OTC market and the regulatory demands for the pooling of clearing and settlement information add impetus to these set-ups. Furthermore, the derivatives market is becoming a mature one, and another branch of investment banking. According to the Bank of International Settlements, 602 million futures and options contracts were sold in the first half of 1994. In the first half of 1995, 581.5 million were sold. In Europe generally, the maturing of the market is occurring more rapidly, as shown by the 36% decline in the volume of the MATIF during 1995.

Despite these alliances offering the possibility of certain contracts being traded face-to-face on each other's exchanges, the prospect of 'remote membership' suggests difficulties in maintaining the places in which the exchanges are located in a 'nested hierarchy'. There is a trend for global alliances between derivatives exchanges to be concerned with efficient clearing rather than trading, reinforcing the trend towards remote trading with organised exchanges becoming information systems or nodes rather than places to trade. The other issue for the territoriality of certain financial centres in Europe is the prospect of economic and monetary union. This is encouraging the development of pan-European financial products and markets. A single currency (the Euro) will reduce the demand for instruments based on national government bond issues, so that a degree of consolidation will occur, with threats to smaller European exchanges. Furthermore, the OTC market may face problems with contracts which mature after the year 2002 (the supposed date for monetary unification), because the national currency benchmarks will disappear. As OTC contracts are not standard ones, as those traded on organised exchanges are, complex litigation may result. Ostensibly, the territoriality of organised exchanges should be reinforced. Yet, with the combination of remote trading and global strategic alliances more concerned with clearing than with trading, exchanges could be become rootless or placeless information systems. The instability of many global alliances also strengthens this argument.

Similar challenges face the London Stock Exchange as it is potentially squeezed between the modernisation of competitor European exchanges and European Commission deregulation. It is useful at this point to look at the combination of European-wide regulation and the prospect of electronic trading of stocks, which will allow financial agents to operate remotely from the physical territory of financial centres. It is these kind of challenges which the 'end of geography' literature has highlighted and which were reviewed above. The comparative advantage of the London Stock Exchange is its liquidity in regard to large volume trades, where it has the lowest transaction costs in the European time zone. A measure of liquidity is the 'spread' between bid and offer prices (buy and sell signals); the smaller the spread, the greater the liquidity. Given London's dominance in foreign equities, as detailed above, it is the place for large investors to transact their international (especially European) business. Moreover, the trading system in London was designed to maintain liquidity in the most straightened of circumstances. The 1987 crash, however, displayed the limits of London's trading system in respect of small investors who could not contact their brokers by telephone. As prices were quoted only on the screens of the privileged market-makers, small investors were denied access to the basic function of a mature stock market – transparency.

Trading on the London Stock Exchange occurs on a 'quote-driven' system. That is, market-makers quote particular bid and offer prices of stocks, then deals and orders are taken via the telephone. Market-makers are given certain privileges in the market by the London Stock Exchange for risking their own capital to maintain liquidity. 'Order-driven' systems, on the other hand, are electronically traded in that prices for particular blocks of shares are quoted, from which customers can order via electronic screens. The advantage of the latter is that it has lower dealing costs and the market conditions; that is bid and offer prices are more transparent.

In September 1995, a new order-driven system named Tradepoint was set up in London. There have also been proposals for trading via the Internet. The initial response was for the London Stock Exchange to post a warning to its members. It is now proposing to be the first international stock exchange with a hybrid system, combining both methods on its new electronic trading platform, Sequence VI. Part of the fall-out from proposals to change the trading system of the London Stock Exchange was the resignation of its chief executive. The establishment of alternative electronic order-driven trading systems, such as Tradepoint, pose potential competitive threats to the London Stock Exchange's privileged position and its territoriality. If trading can be done through electronic media and these media display prices transparently to all investors, what function is there for an organised large exchange in one place? By implication, is the stability and sustainability of the whole of London's financial territory threatened by these innovations, because trading and associated employment can occur anywhere?

These issues are tied up with competitive strategies to sustain London as a global market in the European time zone in the face of challenges from continental

European stock exchanges. However, meeting these potential competitive threats with new trading innovations carries the possibility of undermining London's financial pre-eminence in Europe and its position in the global triangle. Paradoxically, further modernisation of the stock market will stimulate placeless financial flows at the expense of London as a premier financial place. Technological information combined with a globally neoclassical world suggest that the concentration and centralisation of activities in large financial centres may be a thing of the past. However, that it has not happened to date and that territory remains a crucial ingredient in the dominance of these centres, also suggests that there may be some fundamental forces at work which tie large volumes of financial activities to certain locations. It is the process of *re-structuring time and space*, rather than the end of geography, which is at work here.

For a number of years, London dominated cross-border trading in shares through its Stock Exchange Automated Quotation International system (SEAQ-I). The extent of the domination was such that 90% of cross-border trading went to London in the late 1980s. Trading on SEAQ-I has started to decline as the other European exchanges of Paris, Frankfurt and Amsterdam have invested in modernisation. This has encouraged local members to trade locally with lower dealing costs and greater liquidity. Turnover on the Amsterdam exchange has doubled in the last few years, in Paris it has increased by 36% and in Frankfurt by 45%. The DTB, which runs the Frankfurt securities and derivatives markets, announced plans in 1995 to move towards full electronic trading. The aim is to establish the German exchanges as the largest and most attractive financial centre in Europe, through increased efficiency and transparency. Furthermore, European Commission regulations came into force in January 1996 which make it easier to trade on continental exchanges.

The European Union's Investment Services Directive allows 'remote membership' of any European exchange. Thus, an investment bank based in London can gain membership of the Finnish stock exchange without having a physical presence in Helsinki, merely by trading via an electronic network. This attempt to further regulate a 'level playing field' in Europe poses a threat to the locational embeddedness of London's international business. The proposed hybrid trading system is explicable in terms of this twin challenge. In this regulated environment, the French and German stock exchanges have proposed a joint trading system, mimicking the MATIF/DTB link. Banks and securities houses are unlikely to close foreign branches as they act as local agencies for a range of services. However, although there is likely to be an overall increase in capital and local market entry, local brokers are likely to see their market share reduced.

The proposed Franco-German merger suggests size and economies of scale are still important determinants of competitiveness, but also points to the prospect of more pan-European trading systems. For example, the European Security Dealers Automated Quotation (EASDAQ) system launched itself in 1996 as a European version of the North American Security Dealers Automated Quotation system, which itself holds a 5% stake in the EASDAQ system. There will be dual listing of the systems, reinforcing the attractiveness of the European trading system. At the end of 1996, two European 'footsie' indices, known as Eurotop 100 and Eurotop

300, were established between the London stock exchange and the Amsterdam futures and options exchange.

The core areas of London's financial dominance, reviewed above, are not so discrete. The largest category of futures and options are currency instruments, followed by interest rate instruments. The biggest contract by volume is the Euro dollar contract, traded on the CBOT. The fundamental point is that the re-regulation of London's financial spaces leads to a contradiction. The de-regulation and liberalisation of the City of London increased its territorial embeddedness. Re-regulation is leading to a crisis of embeddedness. Some of London's key markets such as futures and options are becoming multinational information networks with sites transplanted around the world. The fundamental shift in the regulatory space of the production economy is being followed by a shift in the financial economy. If London is to remain an important space in the world's nested hierarchies of financial centres, what agency is going to territorially embed it in a more internationally regulated environment? There are grounds for suggesting that real estate is that agency.

SPECULATIONS OF REAL ESTATE AS THE REGULATOR OF TERRITORIAL EMBEDDEDNESS

Real estate is both a factor of production and an asset category. This dual determination makes the social relations of real estate complex. Moreover, real estate or the 'built environment' has become an important determinant in the regulatory space of financial markets and their locales. The size of wealth held in the form of property is large in both the United States and Britain. In 1990, the value of residential real estate and land in the United States was about half of net wealth (US$8.35 trillion of a total domestic net worth of US$16.24 trillion) (Case, Schiller and Weiss, 1991). If commercial real estate is added, then most of domestic net worth is accounted for. A similar position holds for Britain in aggregate, but there are marked differences in portfolio preferences between the two countries. For example, in 1991 some 34% of US insurers assets were invested in commercial real estate compared with 6.7% for British insurers (Coakley, 1994; Warf, 1994). Both insurers and pension funds in Britain have reduced their exposure to property over the last 15 years. Between 1979 and 1996, investment in British property fell from 14.4% to 4.6% of pension funds' total assets. For the same period, insurance companies assets in property dropped from 24.1% to 9.2%. This decline in the asset base devoted to property correlates with the decline in returns. For the period 1980–92 annualised returns were 8.4% in property, compared to 18.5% for equities and 11.8% for government bonds. In 1995, these figures were 2.04%, 11.64% and 8.76%, respectively. Only in three years during the period 1980–92 did property outperform the other asset categories (Coakley, 1994). Between December 1986 and December 1996, monthly returns in Britain were 0.97% for equities, 0.73% for bonds and 0.17% for property. Given the poor performance of property as an asset, why do financial firms continue to invest in real estate? As property crises have become an international phenomenon, is there only the most tenuous link between real estate as an asset category and other financial assets in any one locale?[17]

The problems for investors in commercial property are large transactions costs, significant management fees and illiquidity. For many companies, the use of derivatives is to increase the liquidity of their transactions, but a fully functioning property-based derivatives market has not developed. Such a market would allow the benefits of internal income flows, arising from rises in capital values, to be released. Property returns are less volatile than those for equity and bonds; they are relatively uncorrelated with equities and are weakly inversely correlated with bonds (Robinson, 1996). This suggests that property is important in a well-balanced investment portfolio, but the lack of liquidity deters investors. A derivatives market based on broad property indices would provide the necessary liquidity. There is some evidence of this market developing. To date, most transactions have been OTC contracts. In Britain, the attempt to generate exchange trade futures lasted six months in 1991 on the London commodity market FOX. Dealing was suspended because of insider trading, as traders sold to each other to maintain liquidity. In Britain, the prospect for this type of market has grown as a group of large institutional investors, with Treasury approval, are offering OTC property futures contracts based on a Real Estate Index Market. In France, the first securitisation of a French real estate commercial lease was 50% oversubscribed in April 1997 (*Financial Times*, 30 April 1997). There are problems with OTC transactions, as examined in the previous sections. The lack of a formal intermediary to maintain liquidity, such as an organised exchange, may constrain the development of a fully functioning derivatives market. In the United States, this function is performed by the Real Estate Investment Trusts (REITS), whose investments have risen in popularity as the yield gap with US stocks has grown.[18] Although REITS accounted for only a very small proportion of the value of all US real estate (less than 1% in 1992), such instruments are an important hedge vehicle for real estate transactions, particularly for spread investing (whereby capital is raised at one rate and higher yielding real estate is bought). As the REITs provide hedging instruments, positions can be more easily unwound in the event of a downturn. Moreover, the tax treatment of REITS has increased their popularity with institutional investors. These possibilities suggest that property can become a financial asset like any other. A threat to the territorial embeddedness of a large financial centre can result where there has been significant investment in the built environment. Why then, should one be arguing that property becomes the regulator of territorial embeddedness?

It is beyond the remit of this chapter to examine the complex relationship between property and financial markets in detail. The financing of commercial real estate through bank loans has generated negative externalities in the form of higher cost of capital. This externality arises because of badly performing property loans in both Britain and the United States following the downturn in property markets. Weakened balance sheets of banks feed through into cost-cutting, redundancies and enhanced social polarisation. In Japan, the bursting of the 'bubble economy' and the scale of badly performing property loans has led to the collapse of certain banks and made the financial system vulnerable. As noted above, the recent East-Asian crisis appears to replicate these threats, but on a regionwide basis. There had been contagious effects in the rest of Asia, with banks in the 'tiger economies' falling over each other to provide loans for prestigious real estate developments prior to the crisis. These developments were financed by foreign currency loans which have to be covered by local currencies whose values have fallen dramatically. The increased internationalisation of the financing of real estate markets means that returns in major cities have become correlated. Therefore, downturns in one city can no longer be covered by upturns in another within a property portfolio. As a very illiquid asset category, property exposure cannot easily be hedged, despite recent developments. As a factor of production, its supply is bound up with a number of factors that extend beyond derived demand. It is apparent that the changed nature of the social relations of commercial real estate has become an important component in the regulatory space of international financial centres (Pryke, 1994; Warf, 1994).

The introduction of electronic trading and regulations which allow financial firms to be physically remote from their operation centres suggests that they should not buy large buildings in large, expensive cities. Why, then, do they still do it? Despite the changing relationship of money and space in the late twentieth century, old-fashioned agglomeration economies are still important. If the financial markets are information sets, then there are large thresholds between each set given the nature of portfolio investment (see Chapter 5). Externalities are generated which can only be captured if the information sets are in one place. Furthermore, given the fast pace of financial innovation and its accompanying technology, large financial centres create 'innovation environments' in which face-to-face communication or proximity is crucial. A more fundamental explanation of why large international banks and securities houses buy large expensive offices is the valorisation of space. That is, that the banks can internalise the rental flows of the real estate through the increase in the assets held on their balance sheets. The other explanation is cultural and historical. People have a strong sense of territority. The City of London has been a powerful territory for many centuries. Despite the image of 'masters and mistresses of the universe' populating financial markets, financial actors are not immune to culture or history (Thrift, 1994).

The present crisis of territorial embeddedness in the City of London has also been underlined by intra-place competition within London itself. The financial restructuring of Canary Wharf has led to the possibility of ownership returning to the original developers. As a result of a number of inducements and an improved transport infrastructure, the occupancy rate of the office developments there has increased quite significantly. Major international banks and securities houses have shifted their London headquarters to Canary Warf. Concern with the competitive threat to the territory of the City of London (the 'Square Mile') — not only from other European centres, but also from Canary Wharf — has led the Corporation of London to establish a local economic development agency. Its prime purpose is to persuade existing and new firms to locate within the City of London, rather than be tempted downstream to Canary Wharf. To this end, the agency is prepared to allow development in conservation areas. To date, the Corporation appears to have had some success in its intra-place competition: ABN-Amro (the Dutch financial conglomerate) is building its London headquarters in the City of London. Its direct national competitor ING (that took over Barings) is also looking at its property strategy in London. The LIFFE is to move to a new building being constructed on a City-fringe site at Spitalfields. The Corporation of London is attempting to negotiate a crisis of territorial embeddedness by re-regulating its territory. However, given the change in the nature of the social relations of commercial real estate, the outcome of this strategy remains uncertain.

Conclusions

The crisis of post-war 'embedded liberalism' led to a shift in the regulatory space of the world's economy. The move from production to finance which developed at the beginning of the post-war period culminated in the 'financial rentier' economy of the late 1980s and early 1990s. The globalisation of finance, stimulated by innovation and technology and by nation-states themselves, generated convergence of returns on financial assets, so that the function of global neoclassicism meant that the market could only be bucked at one's peril. Political projects of deregulation and liberalisation became projects of modernisation in order to embed major cities in the new 'global order'. In the case of London, de-regulation was the vehicle for sustaining London's premier international position in many markets and embedding them there.

The current challenges to London's position arise from the re-regulation of systemic risk at an international level. Cooperation between financial authorities on exposure of derivatives risk, capital adequacy rules by the Bank of International Settlements and the European Union, directives for investment services, and booking OTC contracts to organised exchanges or to new clearing houses, form part of the new re-regulation (see Table 1). This is aimed at better managing systemic risk at the same time as keeping intact the liberal sentiment of banks and securities houses managing their own unsystematic risk. The attempts at general re-regulation of derivatives exchanges, through strategic alliances and franchising systems around the world, means they are increasingly becoming international information networks. Although akin to the development of manufacturing multinational industrial companies in the 1950s to 1970s, in these financial alliances the question of ownership of remitted returns and, thus, territorial embeddedness, is more complex. In order to maintain its international comparative advantage and competitive position, the LIFFE is also going down this road. Furthermore, the challenge to the London Stock Exchange's market-making order-driven system from the modernisation of continental exchanges and the EC's remote membership regulation is suggestive of a similar process.

Ironically, the deregulation of London as an international financial centre initially re-enforced its territorial embeddedness. However, due to worldwide liberalisation of financial markets and changes in the nature of international financial transactions, there has been a shift in London's regulatory space. This shift threatens a crisis of territorial embeddedness. The response of the Corporation of London to this crisis and to intra-place competition within the capital, is to re-regulate its territory through stimulating further real estate development in the City. A crisis of territory being solved by territory? Territory is a constancy of human experience, but the 'bonfire of vanities' in the world's financial markets in the last decade makes one seriously question global neoclassicism as the objective function of major financial centres such as London.

CHAPTER 7

OFF-SHORES ON-SHORE: NEW REGULATORY SPACES AND REAL HISTORICAL PLACES IN THE LANDSCAPE OF GLOBAL MONEY

Alan C. Hudson

> In creating a global financial market-place the banks altered the geography of the world system. The basic geographical dimensions of space and time were warped to suit the banks operating needs. . . . Nations attempted to control the system through regulation or taxes: tax havens, dots in geographic space but substantial territories in the bankers' world, enabled such restrictions to be by-passed. . . . Time and space in the bankers' world were pliable, moveable, profitable constructions which might or might not correspond with the mundane geography of national territories. (Daly and Logan, 1989: 103)

INTRODUCTION: MONEY FLOWS AND REGULATORY SPACES

This is a story about regulatory spaces and money. Drawing on research conducted in London, New York City, Washington DC, the Bahamas and the Cayman Islands, this chapter explores the links between 'off-shore' and 'on-shore' financial regulatory spaces. In this respect, it addresses one of the latest stages in the historical geography of money, the way in which the development of new monies relates to the emergence of new spatialities of power and social relations, and the development of regulatory spaces. Substantively, the focus is on the development of international banking facilities (IBFs) in the United States, and the relationship of this legally off-shore but physically on-shore space with the off-shore financial centres (OFCs) of the Bahamas and Cayman Islands.

'Regulatory spaces' provide the link between money and the space economy. As particular spatialities of power (Allen, 1997), they are the reason why geography matters in the realm of money. Regulatory spaces, or jurisdictions, can be more or less territorial. Rules may refer to a particular set of activities and/or to a particular geographical territory. As the activities in question slip across regulatory borders, rules which are strongly territorial are increasingly challenged. This is what O'Brien means by 'the end of geography', a situation in which, *'financial market regulators* no longer hold full sway over their regulatory territory: that is, rules no longer apply solely to specific geographical frameworks, such as the nation-state or other typical regulatory jurisdictional territories' (O'Brien, 1992: 1). Leaving aside debates about the newness of globalisation (Hirst and Thompson, 1996) and whether geography can be reduced simply to issues of sovereignty (Agnew, 1994), globalisation can usefully be understood as an increasing mismatch between the spatial extent of activities and the spatial extent of the rules which govern them – a shift from a modern state-centric geo-political economy towards a more postmodern, less territorial world order (Ruggie, 1993; Agnew and Corbridge, 1995). Money and finance occupy a key position in processes and accounts of globalisation. Money flows through space, crossing borders and linking places. As a universal equivalent, money translates value from place to distant place, arguably dissolving the difference between places (Harvey, 1985; Leyshon, 1996; O'Brien, 1992). So, bounded financial spaces always have a precarious existence; their boundaries may be undermined and their fortunes linked through flows of money. In the financial sphere particularly, scales of regulation are temporary fictions. As Dodd argues, 'the conceptual convenience of the distinction between international and national monetary systems should not be allowed to disguise its analytical weakness' (Dodd, 1994: 101).

This chapter begins by looking at the role of money in the shift from a modern to an increasingly post-modern geo-political economy, placing the development of OFCs within the wider context of processes of financial globalisation and the collapse of the Bretton Woods international monetary system in the early 1970s. This is a story about the changing geo-political economy of money; new monies go hand-in-hand with new geographies. Examining the links between on-shore and off-shore, it is demonstrated that the development of OFCs was, in part, a response to increased on-shore restrictions. There is also a description of how US regulatory authorities subsequently sought to regain control of 'their' off-shore dollars. Extraterritoriality, the projection of the United States' legal power beyond its territorial borders, was part of the strategy (Hudson, forthcoming), but this chapter focuses on efforts to develop an attractive on-shore regulatory environment of IBFs to entice the off-shore dollar 'back home'. The chapter considers the initial proposals for IBFs, the predictions which were made as to their impact on the OFCs, and their actual impact, before explaining why IBFs were not as successful as many had expected or hoped they would be.

The explanation for the failure of IBFs, as new regulatory spaces, to attract the off-shore dollar back on-shore, revolves around the fact that money is *still* a social relation, and places are *still* peopled and have histories which matter. As a social relation which links distant creditors and debtors, money and finance – particularly when the value of money is uncertain (Harvey, 1982, talks of 'fictitious capital') – depend on trust, moreover trust which is backed up by legal regulations. In the world of international finance, laws are still overwhelmingly territorial, and so places – bounded territorial regulatory spaces – matter. However, IBFs failed because it is very difficult to create a new regulatory space. These new regulatory spaces were established in real historical places, which, over time, had acquired particular reputations; reputations which, to potential depositors, were not very trustworthy. Money is more than economic; places are too. Flows of money are not solely determined by potential profits; places cannot be created by legislation. Although money flows link spaces, they do not dissolve differences. Places and people are more or less trusted; their past is remembered in their present. To develop Massey's (1984) geological metaphor of the space economy, it may be possible to shift around the topsoil, but the underlying strata continue to shape the characteristics of places.

GEO-POLITICAL ECONOMIES: MODERN AND POST-MODERN

Monies have geographies. Different monies have different geographies. As Leyshon and Thrift put it, 'Each monetary form has its own geography, and the transformation from one monetary form to another has important geographical implications' (1997: 23). As one monetary form is replaced by another, one spatial arrangement of power and social relations is replaced by another. In the early 1970s, this was seen in the transformation from the Bretton Woods international monetary system of fixed (if supposedly flexible) exchange rates to floating rates. The geography of the Bretton Woods system was one of bounded national financial spaces and managed interactions between these spaces. However, as has been well documented (Corbridge, 1994; Helleiner, 1994a; Leyshon, 1992; Strange, 1986 and 1994), the Bretton Woods system was inherently contradictory and doomed to collapse right from the start.

In part, its collapse was due to its geography. Finance was going global, whereas the Bretton Woods regulatory framework remained international, with the US dollar at the centre. Processes of financial globalisation challenged territorial regulation as the organising principle of the modern international political economy. The dynamism of economic activity and money flows was in tension with the stasis and state-centrism of the existing geo-regulatory framework (Leyshon, 1992: 258). The development of new monies divorced from their national spaces – so-called 'stateless monies' – contributed to the collapse of the Bretton Woods geo-political economy. As Leyshon puts it, 'there emerged for the first time an essentially *de-territorialised* economic phenomenon, which possessed a logic and a dynamic completely at odds with the national-centric order of the international regulatory system' (Leyshon, 1992: 260). As a result, 'the post-war model of an international order comprised of a set of interrelated but economically sovereign nation-states was finally exploded by an invigorated . . . increasingly mobile international financial capitalism' (Leyshon, 1992: 261). The geo-political economy of Bretton Woods was shattered by the development of new monies which had their own geographies, monies and geographies which undermined the power of state-territorial regulatory systems. An economic space of flows emerged, apparently divorced from the political space of states (Castells, 1989, 1996 and 1997). The development of stateless monies reconfigured power/space, undermining geographies – spatialities of power and social relations – organised into fixed, mutually exclusive, territorial states (Agnew, 1994). Global finance was, and is, an essentially geographical project (Swyngedouw, 1996).

As the Bretton Woods international monetary system collapsed, new geographies were created. OFCs were a crucial part of this new geography; their development cannot be understood apart from processes of financial globalisation and the production of stateless monies (Hudson, 1996; Martin, 1994; Roberts, 1995). The importance of OFCs derives from their role in unbundling sovereignty – that is, property rights over territory – a practice which articulates the economic space of flows and the political space of states (Burch, 1994; Ruggie, 1993; Hudson, 1996 presents an historical geography of money; see also, Hudson, 1997). The OFCs 'hold down the global' (Amin and Thrift, 1994), providing a gateway which links a seemingly abstract and uncontrollable space of flows with the productive economy and the space of politics. The development of OFCs is a key moment in the transition from a modern to a post-modern geo-political economy in which the organisation of space and power into state-territorial units is increasingly undermined by the mobility of money (Anderson, 1996).

ON-SHORE/OFF-SHORE RELATIONS

As the above discussion suggests, the development of OFCs cannot be understood without looking at on-shore developments, and *vice versa*. On-shore and off-shore financial spaces are linked by flows of money. 'On-shore' and 'off-shore' may define distinct financial spaces, but money flows blur the distinction. From their very beginning OFCs have developed, in part, as a result of on-shore regulation. 'Offshore' means beyond the regulatory reach of the on-shore authority. As Wise argues, 'in the age of instant telecommunications, insularity is not determined by geography. Today, offshore banking centers are not necessarily physical islands set off by the oceans; rather, they are islands surrounded by a sea of regulation' (Wise, 1982: 300). For the Caribbean OFCs, for example, the United States is the most significant on-shore regulatory authority. The initial development of the Bahamas and the Cayman Islands' OFCs was very much tied up with US efforts to defend the Bretton Woods geo-political economy, a state-territorial regulatory framework based on the dollar as the inter-state numeraire.

As Triffin pointed out in 1960, the Bretton Woods international monetary system was bedevilled by a dilemma: how could the dual goals of lubricating increasing volumes of international trade on the one hand, and maintaining confidence in the dollar as the international measure of value on the other, be achieved when the dollar was backed by a relatively inelastic stock of gold? This dilemma was heightened by the increasing internationalisation of business, dollar investments abroad, US expenditures on the Vietnam War, and the resultant growth of dollar holdings outside the regulatory reach of the United States.[1] The United States faced the related problems of an increasing balance of payments deficit[2] and doubts about the stability of the dollar. To finance the balance of payments deficit, more dollars were printed; this made holders of dollars more doubtful that their holdings could and would be redeemed for gold. These trends threatened to undermine the role of the dollar and the Bretton Woods international monetary system. In an effort to tackle these problems and to defend the Bretton Woods system, the United States introduced a series of capital controls and financial regulations. By restricting the outflow of dollars from its territorial space economy, the United States hoped to regain control of its currency, address its balance of payments deficit and restore confidence in the dollar.

The first measure considered by the Kennedy administration in 1961 and 1962 was a tax reform. This was intended to strengthen US trade and goods exports at the expense of capital export growth by eliminating foreign tax credits. However, it was strongly opposed by US multinationals and was never implemented. The second measure, enacted in 1963, was the Interest Equalisation Tax (IET). The IET acted as a tariff, influencing the supply and demand of capital indirectly through the market by increasing the costs of new US issues of foreign equities in an effort to minimise capital outflows. The IET was intended as a temporary measure, but in actuality was renewed every two years until 1973, with vigorous opposition from the transnational banks at each renewal. In his pioneering but neglected book *Dollars and Borders*, Hawley explains that:

> The IET stimulated the initial rapid growth of the Eurocurrency system in 1963, promoting the internationalisation of finance. In so doing the IET aided in denationalizing the Eurocurrency system by placing it beyond the effective control of national governments and international agencies, ultimately creating a financial structure which was instrumental in the downfall of the dollar in 1971. (1986: 62)

Hawley goes on to describe the capital controls programme which was instituted on a voluntary basis by President Johnson in 1964 and made mandatory in 1968. The capital controls programme aimed to limit US foreign direct investments, US deposits in foreign banks, and the holdings of foreign assets by US transnational banks and the largest US transnational corporations. These capital controls shaped the development of international finance from the mid 1960s until they were cancelled in 1974, in belated realisation that the regulations had been circumvented. As one commentator wryly observed, 'banks did not invent the Euro market. Governments created it by seeking to control the natural flow of money' (Aliber, 1979: 19). Johns (1983) explains further how 'national friction structures and distortions' in the US regulatory environment stimulated the development of the Euro markets and off-shore finance. Prohibitions on inter-state banking (as stipulated under the McFadden Act of 1927), the divide between commercial and investment banking (Glass-Steagall Act of 1933), and the existence of interest rate ceilings (Regulation Q) and reserve requirements (Regulation D) hindered the on-shore competitiveness of major US banks and eventually pushed them off-shore. Thus, the Euro markets, dollar-denominated off-shore business based chiefly in London, developed rapidly, and were given a further boost when OPEC's (the Organisation of Petroleum Exporting Countries) inflated petrodollar earnings were deposited in, and recycled through, the Euro markets after the 1973–4 oil shock (Helleiner, 1994b; Martin, 1994).

Some of the smaller US banks, faced with the high infrastructural costs of a London base realised that the Caribbean OFCs offered a cheaper and equally attractive regulatory environment – free of exchange controls, reserve requirements and interest rate ceilings, and in the same time zone as New York – and moved their Euro market operations to these emerging off-shore centres. In addition, these centres provided a secure, politically stable environment, with increasingly good communications facilities. The number of overseas branches of US banks increased from 180 in 1965 to 732 in 1975, the Caribbean component increasing from five to 164 branches (Johns, 1983: 29). The development of the Bahamas and the Cayman Islands' OFCs has always been influenced by regulatory practices in the United States. In fact, as Edwards laments, 'the remarkable development of off-shore dollar banking is at bottom a history of regulatory myopia, together with a good bit of regulatory mismanagement' (Edwards, 1981: 6).

As the dollars flowed off-shore, the United States lost control of its currency and ultimately had to abandon its commitment to the Bretton Woods international monetary system, cutting the promised link between the dollar and gold, or 'closing the gold window' (Gowa, 1983). However, the United States has never abandoned its desire to control or regulate its dollars and/or the off-shore spaces through which its dollars flow. The dollar's escape from the regulatory and supervisory clutches of the US authorities, although initially for reasons of profitability and competitiveness, led to further problems for the United States, particularly as the activities hosted by the OFCs were largely hidden by strong secrecy laws. The US state authorities, particularly the Treasury and Justice Departments, have long been opposed to the development of the Bahamas and Cayman Islands' OFCs.[3] The OFCs have undoubtedly facilitated tax evasion and avoidance by US citizens, and the laundering of money from narcotics trafficking and organised crime. As one commentator suggested:

> The Bahamas [and other OFCs] must do things which are not allowed in the US because to do things which are allowed in the US is non-competitive, since in every instance the US does it better than the Bahamas do. The Bahamas are therefore compelled in banking and trust operations to appeal to unallowable activities and by inference to appeal to activities disallowed in the US. (Blum, 1984: 144–5)

In response, the United States has tried to limit the role of OFCs, to regain control of its dollars, and to boost the competitiveness of US on-shore banking. The US strategy has had two parts: first, trying to entice the dollars back on-shore; and, second, trying to regulate the off-shore territories. As Wise puts it, when regulations are undermined 'policy-makers are presented with just two options. Either they must seek to mitigate regulations in the direction of conditions existing in the external market, or, conversely, they must seek to gain control over the external market' (Wise, 1982: 312). US agencies have tried, with some success, to regulate the off-shore territories by reshaping the regulatory landscape and extending US legal power extra-territorially into the off-shore jurisdictions (Hudson, forthcoming). However, the focus here is on US efforts to entice the off-shore dollar back on-shore.

OFF-SHORES ON-SHORE: INTERNATIONAL BANKING FACILITIES

DEVELOPING OFF-SHORES ON-SHORE

Given that the major attraction of the Bahamas and Cayman Islands' OFCs is their relatively 'unregulated' environment, and that financial capital is potentially highly mobile, the development of OFCs is necessarily intertwined with developments elsewhere in the financial regulatory landscape. One might expect on-shore deregulation to erode the competitive advantage of off-shore jurisdictions. As the Governor of the Central Bank of the Bahamas commented, 'technological advances together with global deregulation and liberalisation of financial markets have undoubtedly intensified competition and may well reshape the contours of off-shore financial activities permanently' (Smith, 1990). The OFCs' positions as places in a relational regulatory landscape are modified by regulatory changes elsewhere.

Since the mid 1970s and the 'Mayday' deregulation of the New York Stock Exchange, a deregulatory trend has swept across international financial markets leading to, for instance, London's 'Big Bang' of 1986. In addition, in 1986 a Japanese off-shore market was established, offering a liberal regulatory environment based in Tokyo. This facility proved to be attractive to international financial business, attracting US$400 billion of funds in its first two years (Johns and Le Marchant, 1993: 77). However, in terms of impact on the Caribbean OFCs, a similar move to establish IBFs in the United States was of greater importance, a move which was described by the *Financial Times* as a 'carrot' to entice off-shore business to US shores (*Financial Times*, 1983b). As Wise suggests, 'the United States threw a lasso around the Bahamas and the Cayman Islands, in order to pull the operations transacted there back to the American shore' (1982: 300).

US-based IBFs came into existence at the beginning of December 1981 and permitted the establishment of 'banking entities' (in reality, another column in a spreadsheet rather than a physical bank) based in the United States, which would be subject to less stringent regulations than international banks in the United States were used to. Specifically, there would be no reserve requirements (in contrast to the 3% imposed by Regulation D), no interest rate ceilings, and banks would be exempt from the 48-hour 'notice-of-withdrawal' requirement. Foreign banks and official institutions were permitted to place 'overnight funds' in IBFs to take advantage of short-term interest-rate differentials. As two officials from the Federal Reserve Board recalled, 'the purpose was to allow these banking offices to conduct a deposit and loan business with foreign residents, including foreign banks, without being subject to reserve requirements or to the interest-rate ceilings then in effect' (Key and Terrell, 1988). IBFs offered an escape from some of the regulations that had pushed banking off-shore during the 1960s and 1970s. In order to ensure that the IBFs remained an international wholesale banking market, individual and small-scale clients were discouraged from using them; a minimum withdrawal/deposit limit of US$100 000 was set; individual clients had to give 48 hours' notice of withdrawals; and, most significantly, US citizens were not permitted to use them.

The idea of facilitating off-shore banking on-shore, complicating the relationship between territorial states and their regulatory jurisdictions, had in fact been around for some time before the establishment of IBFs in 1981. A 'foreign window' for international banks based in New York had been proposed by one governor of the Federal Reserve Board in 1969 as a way of avoiding the restrictions imposed by the capital controls programme, but this was rejected by the Federal Reserve

Board due to concern about the potential effect on monetary policy (Key and Terrell, 1988; Johns, 1983). In 1977, the Chairman of Citibank, Walter Wriston, revived the idea of IBFs, and, in 1978, the idea received the support of the New York Clearing House Association and the New York state and city authorities which agreed to free international banking from their taxes if the Federal Reserve Board would approve IBFs. Approval was given in 1979, and detailed legislation was drawn up and passed in June 1981.

This brief history of the IBFs proposal hints at the complexity of negotiations, actors and motives that led to their establishment. Prior to 1981, no agreement could be reached between the interested parties – the Federal Reserve Board, international banks and city and state authorities – but eventually all parties felt that they could get something out of IBFs. Hawley describes the combination of motives behind the IBFs, saying that 'while transnational banks wanted to use the IBF as a wedge for deregulation, Federal Reserve officials saw it as a way to make the best out of a bad Euro currency situation' (1986: 139). An official of the Federal Reserve Board, when asked why IBFs were set up, responded:

> Briefly, the large US money centre banks had been down here for five years previous asking us to do it. They had this grand vision that London was going to migrate to New York, and all the jobs. We didn't quite see it that way and kind of dragged our heels on it. We thought, you know, we'd lose control of the monetary aggregates if you had all this reserve free banking going on in the US. Ultimately they wore us down . . . New York state passed some tax legislation, contingent on us approving IBFs. To give the banks relief they would basically say that IBFs are not a part of New York's tax base. And that pushed us a little harder. You know, if New York state was going to go that far it was hard for us to stand in the way. (Simons, Personal Interview, Washington DC, 1994)

The US banks, and particularly the New York-based ones, supported the IBFs proposal as they wanted to maintain their competitiveness and that of New York as a financial centre. They wanted to be able to conduct international banking business from the United States rather than having to go overseas. The motive of the New York state and city authorities in supporting the IBFs proposal was to generate employment. An interview with one of the architects of the IBFs illustrates this:

> In New York state it was a fairly simple proposition. Who was behind it? The State was behind it because the burdens of operating in New York state, the taxes, etc., were very high compared to, say, Chicago. By creating an IBF you lower the tax structure, and all kinds of things happen. So in this case it was a desire to keep New York the centre of the financial community in this time zone, and to make the cost structure as reasonable as possible for the participating banks. That's really what happened. There were lots of other reasons, but that's what it came down to. It was thought that there would be benefits . . . employment, keeping the banks here instead of moving off-shore. You see, for example, if you're a foreign bank here, with the taxes and all the other costs that are involved, you might say 'whoa, I don't do that much business, I'll use the Cayman Islands or go to the Bahamas.' So it's employment, taxes, and so forth. There are a lot of benefits. (Hughes, Personal Interview, New York, 1994)

A banking regulator at the Federal Reserve Board also explained about the motives of New York state in supporting the IBFs proposal. In his view, 'there was first this notion that you could attract this business here on-shore. The Federal authorities were basically ambivalent although supportive of the notion of these IBFs. The effort was really driven by the states, especially the state of New York, which believed that this would lead to increased employment, increased business' (Lane, Personal Interview, Washington DC, 1994). Other key agents have argued that the Federal Reserve Board eventually supported the proposal as a way to enhance their regulatory powers. Thus, as a representative of the American Banking Association put it:

> The feeling was that US banks were going outside the US, were escaping oversight and regulation, so there was concern by the bank regulators that maybe there were things going on that were being pushed outside the US, out of the view of their safety and soundness attempts, that it was better to loosen the regulations in the US, on a restricted entity basis, so you can keep a closer watch on them. So for that it had to be the regulators that were driving these moves. (Thompson, Personal Interview, Washington DC, 1994)

A further regulatory motive was assigned to Federal Reserve support for IBFs, with some officials suggesting that they were intended to sort out the off-shore wheat from the chaff, as legitimate business would now have no reason to use the OFCs. A Federal Reserve official explained that:

> We installed in the US an international banking programme about five, seven, maybe ten years ago, which was designed to provide the benefits of an off-shore centre as it relates to taxes, reserve requirements, and depository insurance relief, under the misguided belief that we could attract all this business right here in the US. It would then differentiate between those conducting legitimate, loosely, versus illegitimate business. (Lane, Personal Interview, Washington DC, 1994)

Hawley suggests that the Federal Reserve Board's shift of position to supporting the IBF proposal in the late 1970s was linked to negotiations with the UK about the international regulation of banking: the Federal Reserve Board hoped to use the IBF proposal to pressure the UK into accepting internationally coordinated banking supervision. The development of IBFs was certainly a complex process, and a somewhat bizarre situation. One commentator asked rhetorically, 'How do we find ourselves in the extraordinary position of having to create special banking facilities to repatriate to the US a gigantic financial market whose principal commodity is none other than our own currency?' (Edwards, 1981: 6). The answer to this question lies in the de-linking of US dollars from US territorial regulation since the development of the Euro markets.

THE END OF OFF-SHORE?

Many commentators predicted that the introduction of IBFs would result in the return of much of the Euro dollar business that took place in London and the

Caribbean back to the United States. For example, writing at the beginning of the 1980s, Ashby estimated that by 1990 London's share of the Euro dollar market would decline from 32% to 20%, that of the Bahamas/Cayman Islands would fall from 11% to 2%, and New York's share would increase from 0% to 18% (Ashby, 1981; see also Johns, 1983: 235). Such quantitative predictions of the decline of off-shore centres were complemented by the gleeful hopes of many US commentators. Ashby suggested that 'the main effect of the introduction of IBFs . . . will be to dull the shine on those brass plates, as US banks will shift their Euro currency operations back home' (Ashby, 1981: 97), while Ireland remarked that 'the US Federal Reserve Board's decision to grant permission for the establishment of off-shore banking facilities in New York has sent a small frisson through those Caribbean central banks which currently host the Euro currency operations of US banks' (1981: 51). Another commentator (obviously not accepting the argument that all markets are regulated institutions), argued prematurely that 'this result is not surprising since the Caribbean markets are more the result of US regulation than the result of market forces' (Campbell, 1982: 537). Whereas many US commentators predicted and hoped that IBFs would signal the end of OFCs, financiers in the Bahamas, although worried, did not entirely accept such views. The Governor of the Central Bank of the Bahamas, William Allen, recognised that 'New York clearly poses a threat. But at present we are more worried by Miami. In any event a big hole could be knocked in our off-shore banking business as we know it today. And we are well aware of the need to respond to a changing situation' (*Financial Times*, 1983a and 1983b). The *Nassau Guardian* reported Allen's comment that 'the position that the international banking facility spells doom for off-shore activity in the Bahamas appears, however, not to be substantiated'. Backing up this assertion, it explained, 'It seems therefore that foreign banks operating in the Bahamas are hardly likely to be keen on moving their operations from their Bahamian locations to US offices where they would deny themselves the advantages and benefits which motivated them to establish operations in the Bahamas in the first place' (*Nassau Guardian*, 1981).

NOT THE END OF OFF-SHORE

Many IBFs were established in a very short time from the first day (3 December 1981) when they were permitted. By the beginning of September 1982, some 395 IBFs had been established in the United States, of which 176 were in New York (Johns and Le Marchant, 1993: 76). The initial rapid growth of IBFs, with assets of US$63 billion by the end of the first month, was due to the repatriation of funds from London, Luxembourg and Nassau (Walmsley, 1983). Walmsley estimates that US banks' funds deposited in London and Caribbean branches fell, respectively, by 11% and 40% between November 1981 and July 1982 (Walmsley, 1983: 85). However, this early rapid growth did not continue. Figure 1 compares the amount of international banking activity (external assets) taking place in IBFs from 1981 to 1991 with that hosted by the Bahamas and Cayman Islands from 1977–91. By December 1987, 540 banking institutions had established IBFs, with external

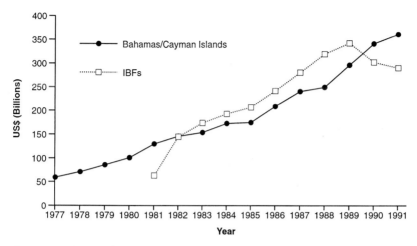

Figure 1 Volumes of Banking Activity: IBFs and Bahamas/Cayman Islands (*External Assets*). Data source: Bank for International Settlements, *International Banking Statistics*

assets of US$281 billion. New York IBFs accounted for three-quarters of IBF activity. This compared with London's US$428 billion and the combined Bahamas/Cayman Islands' external assets of US$241 billion (Key and Terrell, 1988; Bank for International Settlements, 1993). In contrast to Ashby's prediction of 18% of Euro dollar business being based in New York by the end of the decade, by 1988 IBFs hosted only 7% of total international banking, with New York IBFs hosting less than 6% (Key and Terrell, 1988). From 1989 to 1991 the volume of IBF external assets actually decreased from a peak of US$343 billion to $291 billion, whilst activity in the Bahamas and Cayman Islands increased from US$297 billion to $362 billion.

The IBFs neither persuaded the London Euro dollar market to migrate to New York nor spelled the end for traditional off-shore centres. However, there was certainly some loss of business from the Bahamas and Cayman Islands. Representatives of the Bahamas off-shore financial community accepted that 'unquestionably US banks, and possibly some others, have transferred part of their external positions from the books of their Nassau branches to the books of their IBFs, preferring central administration of assets and liabilities from a base such as New York' (Central Bank of the Bahamas and the Association of International Banks and Trust Companies, 1986: 10). The transfer of business by some of the major US banks with Euro dollar booking centres in the Caribbean was described by an ex-governor of the Central Bank of the Bahamas as a 'political' move. He explained that:

> One of the banks that played a very important role, an important political role, in the development of IBFs, was Citibank. Citibank was in the Bahamas. Once the legislation was approved, Citibank had almost to make a political response to it, to move to New York from the Bahamas. It did this and as a result some of the American banks pulled

their off-shore operations back to the US. Certainly the New York banks did that to a great extent, and so obviously the footings changed. But the Swiss banks and the other non-American financial operations, they didn't feel the need to respond to the IBFs. (Talbot, Personal Interview, Bahamas, 1994)

The impact of the IBFs on the OFCs was far from simple; in fact, as US residents were not permitted to use them, it resulted in a rather strange situation. As a Federal Reserve Board publication noted: 'the current regulatory situation has produced a paradox: non-US residents are now encouraged to conduct their banking transactions in the United States, while US residents have incentives to book their transactions, particularly their deposit accounts, off-shore' (Terrell and Mills, 1983: 12). Therefore, even though many US banks established IBFs they tended to retain their Caribbean entities too. As a central banker in London put it, 'lots of IBFs were set up, but few banks closed their Caribbean presence. In many cases IBFs and the Caribbean operations were both operated from New York anyway, by the same people as two books, so the entities were used selectively depending upon the specific case' (Gilling, Personal Interview, London, 1993). Without exception, bankers in London, the United States and the Caribbean centres acknowledged that the impact of the IBFs had been less than expected. Key and Terrell conclude that 'IBFs have not turned out to be the dramatic innovation that some had predicted, and that IBFs simply provide another centre for booking transactions with foreign residents in a regulatory environment broadly similar to that of the Euro market' (Key and Terrell, 1988).

Why Not the End of Off-shore?: The Path-dependence of Place

The IBFs failed for a variety of reasons, including economic, political, social and historical reasons. Economically, in terms of their potential profitability and flexibility, they were at a competitive disadvantage in relation to the OFCs. A Federal Reserve report acknowledged that the IBFs still impose more restrictions on international banking than other Euro market centres do. These restrictions include the fact that IBFs cannot do business with US residents; the minimum maturity period of 48 hours for non-bank foreign residents; the minimum transaction of US$100 000 for non-bank customers; and the fact that IBFs cannot issue negotiable instruments (Key and Terrell, 1988; see also *Financial Times*, 1984). As one regulator in the United States explained, 'banks can do more things and they have more flexibility in the Bahamas' (Evans, Personal Interview, Washington DC, 1994).

A second set of reasons for the IBFs' failure related to the fact that potential investors did not see US 'country risk' as particularly advantageous. One of the supposed key attractions of IBFs was that they would avoid restrictive on-shore regulations (which we have seen to be only partly true), and yet offer US country risk, a risk which was assumed to be lower than that experienced in the Caribbean

OFCs. However, a Federal Reserve Board report explained that in an age of consolidated banking supervision 'sovereign risk' has lost much of its meaning:

> The view that depositors would perceive clear advantages in the sovereign risk associated with deposits subject to US law does not seem justified. Sophisticated international depositors do not appear to perceive a significant difference in sovereign risk between deposits at branches of a US bank located in other major international financial centers and deposits at that bank's IBF in the US; in both cases the deposits are backed by the US bank, which is supervised on a worldwide consolidated basis by US bank regulatory authorities. (Key and Terrell, 1988: 28)

However, the most important reasons for the failure of the IBFs to bring the off-shore dollar back home were social and historical. A particular issue was concern about the US authorities' access to account information. A representative of the Central Bank of the Bahamas explained that 'the idea behind the IBFs was to bring on-shore the off-shore dollar. It has not worked. It has not worked for a very simple reason. The banks don't want full disclosure. The banks don't care about the money being back on-shore. The Federal Government want it back on-shore. They want to be able to control that money' (Johnston, Personal Interview, Bahamas, 1994). As Key and Terrell put it, 'some non-bank customers may want to keep their accounts outside the United States for reasons of secrecy because they view such accounts as subject to less scrutiny by US and other authorities than an account at an IBF' (Key and Terrell, 1988: 21). More colourfully, Miller contrasts the off-shore centres with the IBFs: 'In the Caymans, the banks work for you. In the US, they work for the IRS [Internal Revenue Service], your ex-wife, and whoever else wants to know your financial position' (Miller, 1981: 41). Off-shore money has always been about secrecy and confidentiality as well as profitability.

The impression gained from interviews in the Bahamas and Cayman Islands was that clients, particularly those from Latin America who are familiar with the heavy-handed approach of the United States, were very wary of placing funds in the United States, even if for regulatory purposes they were supposedly off-shore. A Bahamian lawyer describes why a client might not wish to use an IBF:

> If people are looking at moving away from their regulatory authorities they don't go in the same country to set up entities. If you're within their borders you're still subject to their control, their disclosure, and to their ability to penetrate the system. You're literally right in their yard. So those who are still looking to have funds which are coming from international sources, not be subject to possible disclosure or knowledge of their [US] authorities, will not use the IBFs. They will use the OFCs or other countries outside of the US. (Young, Personal Interview, Bahamas, 1994)

Although legislation is necessary in the creation of an attractive space for international finance, it is not sufficient. Legislation could not change the fact that IBFs were still in the United States. Legislation could not erase the memory of previous efforts by the United States to access confidential account information (Hudson, forthcoming). Although the IBFs were supposed to provide 'off-shore' facilities from New York and other US cities, a regulator in Cayman clearly explained the fear that other US regulatory and enforcement authorities would gain access to account information:

> It didn't work because people don't have confidence in the US system in being able to separate out different zones. They don't have confidence in the fact that they can have an IBF that can have information in it that can't go to other sections. If the Department of Justice has something then of course the IRS has it, etc. Unfortunately, the US doesn't have that good a record with being able to streamline and isolate their different departments. (Fry, Personal Interview, Cayman Islands, 1994)

A representative of the Central Bank of the Bahamas explained another concern, namely the potential reversibility of the IBFs' legislation: 'Some of them would open an IBF, but they kept their same operation off-shore because if one government brought in the legislation, another government could take it out, and that has been the history of banking legislation' (Johnson, Personal Interview, Bahamas, 1994). Legislation could clearly be altered in the OFCs too, but the point is that the legislative history of the off-shore centres does not include such swings in policy. As Yassukovich suggested, 'it is perhaps too early to assess the degree of commitment of the US authorities for the IBF concept. Abrupt regulatory changes in the United States are not unknown' (Yassukovich, 1981: 253). The history and reputation of the United States and its regulatory authorities meant that banks and customers lacked confidence in the IBFs. Situated within a jurisdiction remembered for regulatory changes and an inability to resist looking into the financial affairs of people and institutions, IBFs and their regulatory authorities were not trusted. As Wise suggested:

> although the US is recognised as a low political risk, it may be regarded as a significant regulatory risk. The past capital and credit restraint programs and more recently the freezing of Iranian assets held in US banks are examples of such regulatory actions. This history may be the most significant determinant of whether international banking facilities are able to attract the degree of the external market that many proponents claimed. (Wise, 1982: 328–9)

A regulator at the Federal Reserve Board in Washington DC got to the heart of the matter, explaining that IBFs are purely fictional entities created by legislation:

> Well, banks had been operating the Nassau books, or the Cayman Islands books, or the Netherlands Antilles books on premises for years. I mean they already had a structure in place to do this. All you offered them is a different title on top of the spreadsheet, computer run, listing of customer business. So for these banks they didn't need a whole cadre of people to come in and do something they were already doing. (Lane, Personal Interview, Washington DC, 1994)

A representative of the American Bankers Association described the IBFs as 'no big deal' and, invoking a path-dependence type argument, argued that 'these bases have been well established for international monetary flows and, perhaps if we'd always had IBFs in the US they wouldn't have started in these OFCs, but since we haven't they've been very functional' (Thompson, Personal Interview, Washington

DC, 1994). The (financial) genie was pushed out of the (US regulatory) bottle in the 1960s and could not now be persuaded to return.

Conclusions: Money Rules or Placing Trust?

The emergence of IBFs in the United States played an important role in the development of the Caribbean OFCs, spurred on processes of financial globalisation and liberalisation, and contributed to the construction of a post-modern geo-political economy. New monies and new geographies were created together. Money flows, linking the fortunes of on-shore with off-shore. Jurisdictions may seek to attract these flows by constructing attractive regulatory environments. For some commentators, the resultant processes of place competition lead to the 'end of geography' (O'Brien, 1992). Even for Harvey, whose Marxist political position is in stark contrast to O'Brien's free-marketeering preferences, 'commodity exchange and monetisation challenge, subdue and ultimately eliminate the absolute qualities of *place*, and substitute relative and contingent definitions of places with the circulation of goods and money across the surface of the globe' (Harvey, 1985: 11). Harvey and O'Brien both suffer from an excess of economic reasoning and restrictive conceptualisations of 'places'. Although the author finds Harvey's analysis of uneven capitalist globalisation more persuasive and useful than O'Brien's global homogenisation thesis, he would argue that although money flows link the fortunes of places, they do not dissolve their differences. Places are not purely economic nodes[4] and, as long as they are socialised entities, never will be (on the social and cultural determinants of international financial centres, see also Thrift, 1994; Leyshon and Thrift, 1997). The history of places is not easily transferable, and their more or less trustworthy reputations cannot be legislated away.

The development of IBFs admirably illustrates this, both in the attempt and the failure to bring the off-shore dollar on-shore. The IBF phenomenon shows that places can be partially constructed through legislation. However, the failure is more interesting, illustrating that money and the places between which it flows are social and historical, as well as economic constructions. The IBFs failed for economic, political, social and historical reasons.[5] Economically, the US regulatory authorities – clinging to a disappearing dream of sovereignty – were unwilling to construct the IBFs as truly off-shore centres. Financial activity in these on-shore off-shores remained more restricted and less potentially profitable than in the Caribbean centres, which, in contrast, made fruitful use of their remaining legal sovereignty. As Dodd puts it, 'Offshore transactors rely on the strength and consistency of the monetary and fiscal sovereignty of individual states. It is from differences between individual regulatory environments – which are based on the sovereign right of each state to legislate independently – that commercial incentives are derived' (Dodd, 1994: 100). In addition, in a globalising political economy with globally consolidated banking supervision, US 'sovereign risk' provided little incentive for off-shore dollars to return to the United States. However, it is the social and historical reasons for the failure of IBFs which are the most illuminating. Off-shore money has always been about secrecy and confidentiality as well as profitability. Many banks and their wary customers did not trust the US regulatory authorities, fearing that they would continue – as they had historically – to pry into their financial affairs and to pass this information around government departments. In addition, bearing in mind the historical record, the banks and customers did not trust the United States to retain the IBFs over time; as a legislative decision, it could be reversed. Finally, and again because of historical path-dependence, once banks and their customers had grown to trust the Caribbean OFCs, they were reluctant to switch their allegiance to the IBFs.

The failure of IBFs illustrates that financial space is made up of social and historical places rather than abstract timeless nodes in an economic network. Money flows and the development of financial spaces are only partially explained by economics. Money is more than economic; flows of money are not solely determined by potential profits. As Dodd puts it, 'it is not at all clear that international monetary networks mark the emergence, teleologically as it were, of a market stretching across geo-political boundaries whose operation approximates more and more closely to a perfection derived from economic reasoning' (1994: 102). In suggesting that, 'as a universal equivalent, money becomes a form of communication about economic value, which can link social actors separated by vast tracts of space and time' Leyshon (1996: 65) marginalises, but does not erase, the sociality of money. Money remains a social relation; the communicating actors are social. Their reasons and actions are communicative, rather than purely economic and instrumental. For their communication to succeed, there must be trust. Places and peoples are more or less trusted; their past is remembered in their present. Money is a socio-historical geography.

Part IV

Money and the Local Economy

CHAPTER 8

FINANCING ENTREPRENEURSHIP: VENTURE CAPITAL AND REGIONAL DEVELOPMENT

Colin M. Mason and Richard T. Harrison

INTRODUCTION

In general, entrepreneurs follow an established 'pecking order' in their preferences for, and use of, finance to start and grow a business (Meyers and Majluf, 1984). Where possible, they prefer internally, rather than externally, generated finance, and so will tend to rely on personal finance and retained earnings generated by the business. However, with the exception of small-scale start-ups and low-growth businesses, these internal sources are often insufficient to fund the development of the enterprise, necessitating reliance on external sources. In essence, the choice here is between debt and equity finance. With debt finance, the entrepreneur does not have to give up any ownership of the company, but does have to make regular payments of interest and principal to the lender. Equity finance, on the other hand, does require the entrepreneur to give up partial ownership. However, the entrepreneur has considerable discretion over whether the investment is repaid and whether dividend payments are made. Hence, in theory at least, the entrepreneur faces a trade-off between paying interest or giving up some ownership and control (Bygrave, 1997a). Debt finance will be preferred to equity finance because entrepreneurs have, for the most part, a pressing desire to maintain control and ownership of the business. If equity finance is sought, finance from family and friends will be preferred over finance from institutional sources (e.g. venture capital or public equity markets such as the North American Securities Dealers Automated Quotation and the London Stock Exchange).

There are circumstances, however, in which these preferences cannot be met. For example, at the start-up stage, where there is neither track record nor collateral to support bank lending on the scale required, a combination of internal finance sources, bootstrapping[1] (Winborg and Landström, 1997; Bhide, 1992) and equity finance from external sources will be required. Equally, for companies seeking to exploit significant growth opportunities, the financial requirements of the business may rapidly exceed its capability of generating funds internally and of attracting additional debt finance on account of the limited availability of business collateral (Binks and Ennew, 1996).

In such circumstances, the ability of these businesses to obtain equity finance is likely to determine whether they are able to start, grow and survive. Of course, there are businesses which have little or no need for external sources of finance.

Indeed, some companies have grown to a significant size without raising external finance at some stage (Bhide, 1992; Bygrave, 1997a). However, they are exceptions. Studies of technology-based firms have established that there is a strong link between the success of the business and the amount of initial financing (Roberts, 1991a). Recourse to the stock market is not feasible for all but a handful of small, growing firms on account of the high administrative and pricing costs involved in raising new equity capital and the compliance costs (legal and reporting requirements) of maintaining a public quotation, which both discriminate against smaller issues. Therefore, *venture capital* has emerged to bridge this financing gap for small, growing firms which are unable to finance growth (and less often their start-up) from internally generated sources of finance and debt finance and are too small to access public equity markets.

There are two main sources of venture capital. The first, and oldest, source is the informal venture capital market, comprising wealthy private individuals, or 'business angels', the majority of whom are self-made with substantial business and entrepreneurial experience, and are prepared to invest either on their own or with others directly in small businesses (Mason and Harrison, 1994 and 1997a). Their investments are typically fairly small (less than £50 000/US$100 000), and their role is most significant at the conception ('seed'), start-up and, to a lesser extent, early growth stages. Their involvement quickly diminishes in later financing rounds and where investment requirements are large (Freear and Wetzel, 1990; Freear, Sohl and Wetzel, 1995b). The second source is the institutional venture capital market. Professional investment companies raise finance from financial institutions (e.g. banks, insurance companies and pension funds) to invest in businesses with high growth prospects. Their role is to identify investment opportunities with the potential for high returns, structure the transaction, add value to the post-investment relationship and secure an exit for the investment. Across Europe there are some 500 specialist venture capital companies (European Venture Capital Association, 1995), of which just over 100 are in the UK. The United States has around 600 venture capital firms (Timmons and Bygrave, 1997).

There are two types of venture capital company. First, 'independent' venture capital firms create one or more funds of a specified size, duration and investment focus, and then seek financial institutions and other investors (e.g. companies, wealthy families, universities and government) to invest in the fund. The venture capital firm gets its return in two forms, partly as a fee for management services in the investment process, and partly as a share of the capital gains realised by the fund. Larger venture capital firms may manage three to five funds at any one time. Second, 'captive' funds are the in-house venture capital subsidiaries of financial institutions (e.g. banks). The attraction of venture capital for institutional investors is the potential for superior returns compared to other forms of investment (e.g. stocks, bonds and gilts). However, in terms of their overall investment portfolio, venture capital is a marginal category, comprising less than 1% of total annual institutional investment in the UK (Murray, 1996). This limited involvement may explain why most institutions derive their exposure to venture capital investments by investing in independent venture capital funds rather than establishing a specialist in-house venture capital subsidiary (Murray, 1996).

The aim of this chapter is to demonstrate that the availability of venture capital within developed countries is geographically uneven and that this in turn has adverse consequences for the economic prospects of venture capital-deficient regions. While data availability means that much of the discussion will be concerned with the institutional (or formal) venture capital industry, the chapter will also discuss the important but much less visible role of the informal venture capital market in financing entrepreneurship and regional development. The chapter begins with some essential background information. In the following three sections the main features of venture capital are discussed, the growth and evolution of venture capital in the United States and Europe is reviewed, and evidence is presented to demonstrate its economic significance for national and regional economic development. The geographical dimensions of venture capital activity are then examined. In view of the link between venture capital and economic development, governments have responded with various initiatives which attempt to stimulate the supply of venture capital in regions where it is underdeveloped or absent. The concluding section offers a critical review of these policy responses.

THE NATURE OF VENTURE CAPITAL INVESTMENT

Venture capital is a distinctive form of industrial finance which is part of a more broadly based private equity finance market which includes both organised and informal (or business angel) venture capital markets and pools of capital dedicated to non-venture capital private equity investments (Brophy, 1997). Venture capital can be defined as the provision of finance by professional investors to businesses that are *not quoted* on a stock market and which have the potential to grow rapidly and become significant businesses in international markets. Venture capital is *equity-oriented*. Although venture capitalists may use a number of different financing instruments, the majority of their investments are either pure equity or in a form that can be converted into equity under agreed contractual conditions. The objective is to achieve a *high return* on the investment in the form of *capital gain* through an exit, achieved by the sale of the equity stake rather than by dividend income. Exit is normally achieved either through an initial public offering (IPO), involving the flotation of the company on a stock market where its shares can be traded freely, or through a trade sale in which the venture capital fund, normally along with all of the other shareholders in the company, sell out to another company. Venture capitalists normally expect to make their exit between three and seven years after making the investment. Therefore, venture capital is a *high risk* investment. Although the investor shares in the success of the business, he or she will lose the investment if the business fails. As equity finance is subordinated to other forms of finance (e.g. debt finance provided by the banks), shareholders are at the back of the queue in the event of the failure of the business. Furthermore, once an investment is made, it will be illiquid for several years and cannot be realised unless and until an exit is achieved. Finally, venture capitalists are normally minority shareholders, hence although they will have seats on the board of directors, they are unlikely to have outright voting control.

In order to compensate for these risks, institutional venture capitalists are highly selective in the types of businesses in which they invest. Typically, less than 1% of proposals are eventually financed (Fried and Hisrich, 1994; Sweeting, 1991; Tyebjee and Bruno, 1984). In their evaluation of investment opportunities, venture capital funds place particular emphasis on the competence of the management team, the growth potential and profitability of the industry and the attributes of the product/service (Dixon, 1991; Hall and Hofer, 1993; Muzyka, Birley and Leleux, 1996; Roberts, 1991b). Venture capital funds which invest in early-stage businesses give particular emphasis in their investment appraisal to unique proprietary products with high growth potential, whereas later-stage investors demand a market-proven product. However, all venture capitalists, regardless of their investment focus, demand high quality management (Elango, Fried, Hisrich and Polonchek, 1995) – 'there is no question that, irrespective of the horse (product), horse race (market), or odds (financial criteria), it is the jockey (entrepreneur) who fundamentally determines whether the venture capitalist will place the bet [invest] at all' (MacMillan, Siegal and Subba Narasimba, 1985: 119). In terms of financial return, venture capitalists seek companies that can provide an internal rate of return^2 of at least 30% in the case of established companies, rising to 60% or more for seed and start-up investments (Murray and Lott, 1995). This is a much higher return than is available from investing in quoted companies and other forms of 'safer' investment. Put another way, venture capital funds aim for a five to tenfold return on their investment in about five to seven years.^{3} Such a high return is necessary to compensate not just for the higher risks and illiquidity of investing in unquoted companies, but also for losses from unsuccessful investments. In a typical successful venture capital portfolio, of every 10 investments just two will meet or exceed the target rate of return, another two will be total failures and six will comprise the 'living dead', that is investments which have failed to meet expectations and are unlikely to produce a profitable exit, but nonetheless have a stable existence (Bygrave, 1997b). Thus, only firms able to demonstrate the probability of achieving exceptional returns are candidates for venture capital.

A slightly different set of considerations arises in the case of management buy-outs (MBOs) and management buy-ins (MBIs). MBOs involve the purchase of a company or, more usually, a division or subsidiary, from its existing owners by its incumbent management team (Wright, Thompson, Chiplin and Robbie, 1991). MBIs are a similar form of transaction, but involve the purchase of an existing subsidiary, division or company by an incoming management team (Robbie and Wright, 1996). Typically, the purchase price is met by a combination of the personal financial resources of the management team, bank debt and venture capital. Opportunities for MBOs/MBIs normally occur in four situations: first, when a company re-focuses on its core businesses by disposing of unrelated activities (often correcting the problems of past unsuccessful diversifications) in order to maximise firm value or is undertaking a forced reorganisation in response to financial difficulties; second, where family businesses face succession problems; third, arising from receiverships; and fourth, as a means of privatising government-owned companies (Wright and Robbie, 1997).

It is not essential for the MBO/MBI target to show substantial growth prospects. Instead, the key consideration is cash flow generation. The opportunity for significant financial return is based on the initial pricing and financial structure of the deal, which involves substantial senior and unsecured (mezzanine) debt. In addition, the performance of the business is expected to improve as a result of the new freedom of the management team from the controls imposed by the parent company to develop their own strategy and their additional motivation from having an equity stake in the business (and often substantial personal financial assets at risk). Other attractions of MBO/MBI transactions for venture capital firms are that the management team is clearly identified, the market position is understood, the potential unrealised value in the business can be clearly identified and analysed, and the time horizon to exit from the deal (two to three years typically) is shorter than for early-stage deals. The major risk in an MBO/MBI is that the business could fail as a result of over-gearing,[4] which would arise if cash flow generation is insufficient to pay off the debt in the business (Murray, 1996).

Venture capitalists also take a variety of other steps to minimise their risk. First, they rely on known and trusted networks of professional intermediaries, such as accountants, solicitors and consultants, for referrals on companies seeking finance (Fiet, 1995; Murray, 1994). A firm that seeks venture capital requires not merely a technically competent business plan, but also to secure social endorsement from the key players in the venture capitalist's network (Steier and Greenwood, 1995). Second, some investments are syndicated with other venture capital funds to spread financial risk and allow the sharing of expertise and information between co-investors (Lerner, 1994). Third, the investment may be structured as a series of rounds of finance, each round being contingent on the enterprise meeting certain milestones (Gompers, 1995). This gives the venture capitalist the option to abandon, re-value or increase the capital committed to the project as new information becomes available (Salhman, 1988). Fourth, venture capitalists structure the investment agreement in various ways (such as entitlements to profits and equity, and obligations on management) to protect their investment in the event of both negative and positive future events (Sahlman, 1990). Finally, venture capital firms normally have some form of 'hands on' involvement in the management of the businesses in which they invest, both to minimise downside risk but also to add value to their investments (Bygrave and Timmons, 1992; Fried and Hisrich, 1995; Gorman and Sahlman, 1989; MacMillan, Kulow and Khoylian, 1989). This involvement includes strategic, social/supportive and networking roles (Bygrave and Timmons, 1992), but rarely technical and production matters. Hands-on involvement is most common in the case of venture capital funds which specialise in early-stage investments (Sapienza, 1992).

Much of the foregoing discussion also applies to the informal venture capital market, but there are some key differences of focus and emphasis. Business angels, like their institutional venture capital counterparts, are strongly, though not exclusively, equity-oriented, but are more likely to rely on simple shareholder agreements and covenants, and are less likely to specify *ex ante* the holding period or exit route (Mason and Harrison, 1994; Mason and Harrison, 1996a). As informal investors have less power to replace or modify the management team, issues of management capability and trustworthiness, together with the investor's product and market knowledge, become most significant in the investment decision-making process (Mason and Harrison, 1996b; Mason and Rogers, 1997). As in the formal venture capital market, informal equity investors rely heavily on personal social and business networks for referrals on companies seeking finance, supplemented by the growing number of business introduction services which have been established to improve the flow of information, and hence capital, in the marketplace (Harrison and Mason, 1996; Mason and Harrison, 1997b). Once business angels invest in a business, they are likely to play a value-adding hands-on role in its management (Harrison and Mason, 1992; Mason and Harrison, 1996a). There is some evidence to suggest that there is a complementarity between the informal and institutional venture capital markets: business angels can provide the start-up and early-stage capital in relatively small tranches that provide opportunity for a business to establish and develop to the point at which institutional venture capitalists provide subsequent, larger rounds of development capital (Freear and Wetzel, 1990). Accordingly, informal venture capital has been identified as a key element in financing entrepreneurial ventures (Advisory Council on Science and Technology, 1990; Bank of England, 1996).

THE GENESIS AND GROWTH OF VENTURE CAPITAL

Informal venture capital is a long-established, although largely undocumented, source of business finance. For example, at the turn of the twentieth century Henry Ford raised the initial finance for his motor company from a group of five business angels (Gaston, 1989). The United States has the most extensive and well-developed informal venture capital market (Freear, Sohl and Wetzel, 1995b), but there is evidence that, although not as well developed as in the United States, the informal venture capital market also plays a significant – and growing – role in business financing in Western Europe (Mason and Harrison, 1994; Landström, 1993; Lumme, Mason and Suomi, 1998), Canada (Short and Riding, 1989) and Australia (Hindle and Wenban, forthcoming).

Institutional venture capital originated in the United States with the formation of American Research and Development in 1946. The Small Business Investment Company Act of 1958 led to the creation of small business investment companies (SBICs), privately organised corporations or partnerships licensed by the Small Business Administration and given tax advantages and access to government loans at favourable rates (O'Shea, 1995). However, rapid growth has only occurred from the early 1980s, following reductions in the rate of capital gains tax and new rules which explicitly allowed pension funds to invest in venture capital. These changes, particularly the latter, resulted in a massive flow of investment by financial institutions into venture capital funds (Bygrave and Timmons, 1992; Gompers, 1994), reinforced by a growth in demand for venture capital from emerging technology-based businesses.

The development of the institutional venture capital industry during the 1980s and 1990s has been cyclical (Figures 1 and 2). In part at least, the flow of money

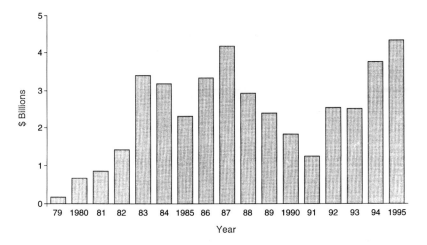

Figure 1 Annual Commitments to Independent Private Venture Capital Funds in the United States, 1979–1995. Note: Independent Funds Only (Excludes SBICs). Source: Bygrave and Timmons (1996)

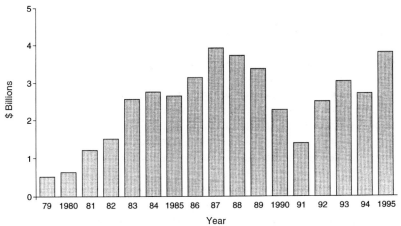

Figure 2 Annual Amounts Invested by US Venture Capital Funds, 1979–1995. Note: Includes Independent and Captive Funds. Source: Bygrave and Timmons (1996)

into venture capital is related to the state of the public equity markets, and in particular the market for initial public offerings (IPOs). As venture capital funds, at least in the United States, generate most of their capital gains from IPOs, a strong IPO market offers higher rates of return to venture capital funds, and hence to their investors (Timmons and Bygrave, 1997). If institutional investors are realising a profit from their investments in venture capital funds, they will be encouraged to reinvest and commit additional funds to this part of their portfolio in the following years. The revival of the US venture capital industry from its cyclical downturn in the late 1980s and early 1990s can be related to a

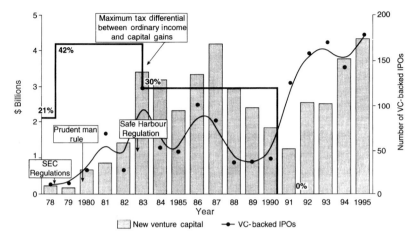

Figure 3 *Influences on the Flow of New Finance to Independent Venture Capital Funds in the United States, 1978–1995.* Source: Bygrave and Timmons (1996)

combination of a robust IPO market and demand for finance from a new technological wave of information technology businesses (Timmons and Bygrave, 1997), which have encouraged institutional and other investors to allocate an increasing proportion of their assets to venture capital. Figure 3 illustrates the various influences on the flow of institutional finance to independent venture capital funds in the United States.

During the 1980s, institutional venture capital emerged and expanded rapidly in the UK, and subsequently in the rest of Western Europe, as well as in Canada and Australia. Here again, growth has been cyclical. For example, the European venture capital industry experienced a decline in fund raising between 1989 and 1993 and a subsequent levelling-off in investment activity, a shake-out of funds, and a resumption in expansion since 1993 (Figures 4 and 5). More recently, venture capital has also taken root in other regions, notably the Far East, so that by the late 1990s it could be said that venture capital has become a truly global phenomenon. However, there are important differences between countries in the form that institutional venture capital takes.

Bygrave and Timmons (1992) make a distinction between two forms of venture capital. 'Classic' venture capital funds, which raise their finance from patient investors (for example, wealthy families and individuals), and are managed by investors with entrepreneurial experience and industry knowledge, provide *new* equity finance to investee businesses at their early stage of development. 'Merchant' venture capital (or business development capital) funds, which raise their finance from institutional investors who have short-term investment horizons, are typically managed by MBA-trained professionals with backgrounds in investment banking or other financial or consulting organisations, but with little company-building experience. These skills are more appropriate to MBO/MBI investments and to investments in established companies. Therefore, merchant venture capital is predominantly involved in the refinancing of existing businesses via equity *replacement*.

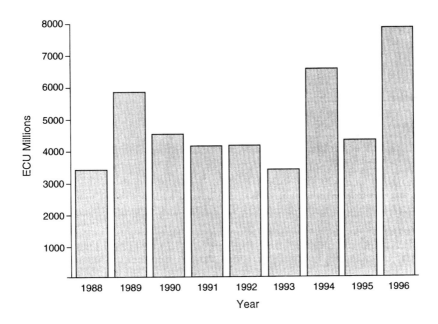

Figure 4 New Finance Raised by European Venture Capital Funds, 1988–1996. Source: EVCA

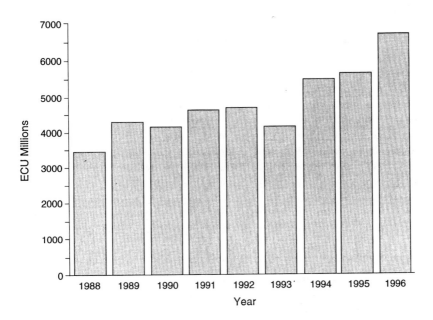

Figure 5 Annual Amounts Invested by European Venture Capital Funds, 1988–1996. Source: EVCA

The growing dominance of investments by financial institutions in venture capital funds in the United States since the early 1980s has resulted in a significant expansion of merchant venture capital. As a consequence, although classic venture capital is still significant, it now accounts for a minority of US institutional venture capital activity (Figure 6(a)). In 1995 it accounted for 42% of all investments and one-third of the dollar amount invested. There has, however, been a slight increase in seed, start-up and early-stage investing since the mid 1990s (Timmons and Bygrave, 1997; VentureOne, 1997) as institutional and other investors have increased their commitments to funds specialising in classic venture capital. In Europe, merchant venture capital is the major form of institutional venture capital, with classic venture capital in long-term decline since the mid 1980s (Figure 6(b)).

There are three reasons why classic venture capital is on such a limited scale in Europe. First, as Bygrave and Timmons (1992: 92) comment, 'in Europe, ... the presence of banks as dominant institutional backers has had a direct impact on the shape of strategy and practice in venture capital investing'. Institutional investors, including longer-term investors such as pension funds and insurance companies, are under pressure to demonstrate better than average short-term results. This pressure for a quick return has extended to their investments in venture capital funds, even though such investments are an insignificant part of their overall investment portfolio, resulting in a mismatch between the long-term discipline of classic venture capital and the shorter-term investment horizons of institutional suppliers of finance. Funds which specialise in MBOs have been favoured because, as noted earlier, investments in such businesses involve less risk as they are backing established management and established products in mature markets. Second, the preference of institutional investors to invest in funds specialising in MBOs/MBIs can also be explained as a rational response to the much higher returns from such funds (O'Shea, 1995; British Venture Capital Association, 1996).[5] This is in marked contrast to the United States where early-stage funds have yielded higher returns than later-stage funds (Bygrave and Timmons, 1992). Third, demand-side constraints also play a role; for example, it is argued that Europe lacks start-ups with the potential to grow rapidly because management teams tend to be inexperienced and more risk averse (Confederation of British Industry, 1997; European Venture Capital Association, 1995).

ECONOMIC SIGNIFICANCE OF VENTURE CAPITAL

The venture capital industry is small compared to other financial markets. The amount of capital which US venture capitalists manage is less than the total assets of one of the larger US banks, and the annual amount invested during the 1990s (US$3–4 billion) is less than the sum that changes hands on an average morning on the New York Stock Exchange (Wetzel, 1994). In the UK, the annual amount invested by the institutional venture capital industry (£3.2 billion in 1996) is small in relation to the total amount of bank borrowing by small businesses (£34.1 billion in 1997) (Bank of England, 1997). The informal venture capital market is

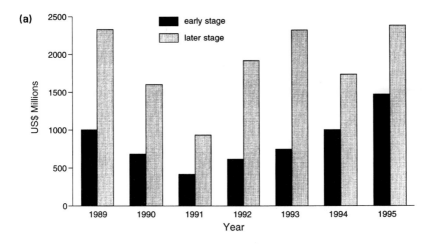

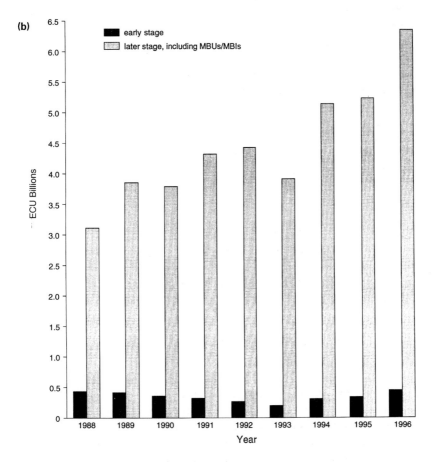

Figure 6 Amount Invested by Stage (a) US Venture Capital Funds; (b) European Venture Capital Funds. Sources: (a) Bygrave and Timmons, 1996; (b) EVCA

larger, although difficult to measure. Wetzel (1994) has estimated that in the United States about 250 000 business angels invest US$10–20 billion annually in over 30 000 companies, while Mason and Harrison (1997c) estimate that there are 18 000 active and potential business angels in the UK who invest around £500 million in about 3500 businesses. These amounts are still small in terms of overall business financing.

Nevertheless, the economic impact of venture capital is significant. Companies that have been backed by institutional venture capital funds add value through technological and market innovations, while their economic contributions have included significant job creation, new research and development (R&D) expenditures, export sales and the payment of taxes (Bygrave and Timmons, 1992). The European Venture Capital Association describes venture capital-backed companies as 'engines for our economies' (1993: 4). In the UK, between 1990 and 1995, venture capital-backed companies increased employment by 15%, while over the same period total employment grew by less than 1% and employment in the largest companies (the FT-SE 100) declined by 1%. Venture capital-backed companies increased sales on average by 34% per annum, five times faster than FT-SE 100 companies, expanded exports by 29% per annum and raised investment by 28% per annum compared with GDP growth of just 5% (Coopers and Lybrand, 1996). In the United States, between 1991 and 1995 venture capital-backed companies grew by 34% per annum in employment terms compared with a 4% annual reduction in employment in Fortune 500 companies. Sales revenue also grew by 38% per annum over the same period, compared with 3.5% in the Fortune 500 companies. More specifically, an analysis of the fastest growing companies in the United States indicates that venture capital-backed companies increased their revenues by 37% in 1995, compared with 23% for non-venture capital-backed companies in the fast growth category, and employed 114 workers in the United States on average compared with 60 in non-venture capital-backed companies (Coopers and Lybrand/VentureOne, 1996). However, as the next section makes clear, venture capital investments are unevenly distributed across space, and so these benefits are concentrated in particular regions.

THE LOCATIONAL DISTRIBUTION OF VENTURE CAPITAL INVESTMENTS

SPATIAL PATTERNS OF INSTITUTIONAL VENTURE CAPITAL INVESTMENT ACTIVITY

In the UK, venture capital activity is disproportionately concentrated in the southeast region, the most prosperous part of the country. From 1992 to 1996, the South East accounted for 36% of investments and 42% of the amount invested. The next largest regional concentrations are in Scotland (in terms of number of investments rather than the amount invested), the North West and the West Midlands (Table 1, Figure 7). However, in order to control for the varying size of

Table 1 The Regional Distribution of Venture Capital Investments in the UK, 1992–1996

Region	Amount invested			Number of investments			Average size of investment
	£ million	%	LQ	Number	%	LQ	£000
South East	3853	42.4	1.21	1963	36.3	1.04	1962.8
South West	516	5.7	0.61	314	5.8	0.62	1643.3
East Anglia	247	8.7	0.68	208	3.8	0.96	1187.5
West Midlands	966	10.6	1.25	474	8.8	1.03	2038.0
East Midlands	752	8.3	1.20	324	6.0	0.87	2321.0
Yorkshire/Humberside	501	5.5	0.73	430	7.9	1.06	1165.1
North West	1034	11.4	1.26	514	9.5	1.06	2011.7
North	331	3.6	0.96	233	4.3	1.13	1420.6
Scotland	601	6.6	0.87	635	11.7	1.54	946.5
Wales	192	2.1	0.42	182	3.4	0.67	1054.9
Northern Ireland	103	1.1	0.32	127	2.3	0.67	811.0
Total	9096			5410			1681.3

Source: Calculated from British Venture Capital Association statistics

region, it is more meaningful to relate each region's share of venture capital investments to its share of the total stock of VAT-registered businesses by means of location quotients.[6] Because of differences in the average size of investments in each region (Table 1), the picture differs somewhat depending on whether number of investments or amount invested is considered. In terms of the amount invested, four regions have attracted more than their 'expected' share of venture capital investments in 1992–6, namely the North West, the West Midlands, the East Midlands and the South East (all with LQs > 1.2). At the other extreme, Northern Ireland, Wales (LQs < 0.5), the South West and East Anglia (LQs < 0.7) are venture capital-deficient regions. In terms of numbers of investments, the majority of regions have location quotients of between 0.8 and 1.2, indicating a less uneven distribution relative to each region's share of the business stock than in the case of the amount invested. Just four regions stand out. Scotland has attracted considerably more than its expected share of investments, whereas Northern Ireland, Wales and the South West again stand out as attracting less than their 'expected' shares of venture capital investments (Table 1).

The regional distribution of venture capital activity in the UK has become less uneven over time (Figure 7). From 1987 to 1991 the South East attracted 53% of venture capital investments by value, 10 percentage points higher than its share in 1992–6. Even more significantly, the South East and Scotland were the only regions to have attracted more than their fair share of venture capital investments in this period. Every other region had a lower than expected share of venture capital investments (Martin, 1989 and 1992; Mason and Harrison, 1991). During the 1990s, there has been a major expansion in venture capital investment activity in the East Midlands, the West Midlands and the North West, the effect of which has been to reduce, although not eliminate, the over-concentration of venture capital activity in

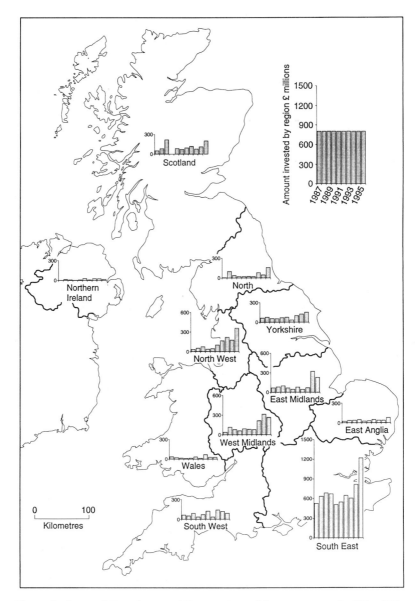

Figure 7 *Regional Distribution of Venture Capital Investment in the UK, 1987–1996.* Source: BVCA

the South East. However, the position of the venture capital-deficit regions – Northern Ireland, Wales, the South West and East Anglia – has not improved.

Further insight into the geography of venture capital investment activity in the UK is gained by differentiating between merchant venture capital investments and classic venture capital investments. MBO/MBI investments now account for almost three-quarters of the amount invested by venture capital firms in the UK. Venture

Table 2 *The Regional Distribution of Venture Capital Investments: MBO/MBI and Other Investments, 1994–1996*

Region	MBOs/MBIs			Other*			MBO/MBI as % of total
	£ million	%	LQ	£ million	%	LQ	
South East	1802	38.3	1.06	851	44.6	1.23	67.9
South West and Wales	309	6.6	0.47	186	9.7	0.69	62.4
Midlands and East Anglia	1240	26.4	1.37	344	18.0	0.93	78.2
North**	1119	24.6	1.06	342	17.9	0.77	77.2
Scotland	199	4.1	0.57	187	9.8	1.32	51.6
Total	4704	100		1910	100		71.1

Notes: * This mainly comprises investments in start-ups, early stage and expansions
** North West, Yorkshire/Humberside, North and Northern Ireland

Source: British Venture Capital Association

capital activity in the Midlands and North is dominated by MBO/MBI investments, whereas classic venture capital activity is relatively more significant in Scotland and the South West/Wales (Table 2). Given the much larger size of MBOs/MBIs, this disaggregation helps to explain regional differences in the average size of investments. In particular, it explains why Scotland has less than its fair share of venture capital investments in terms of the amount invested, whilst having a substantially greater than expected share of the total number of venture capital investments. It can also be concluded from this evidence that the growth in venture capital investment in the Midlands and northern regions of the UK during the 1990s is the outcome of increased MBO/MBI activity (Wright, Robbie and Ward, 1994). However, classic venture capital continues to be disproportionately concentrated in the South East and Scotland.[7]

Venture capital investment in the United States has exhibited greater spatial variations. Historically, the most significant concentrations have been in Silicon Valley and New England, centred on the Greater Boston area (Florida and Kenney, 1988a). As Figure 8 shows,[8] this pattern continues. However, over 60% of US venture capital was invested outside these traditional venture capital regions, with Texas, New York Metro and the Mid-west regions showing strongly. There is also evidence, at a third-tier regional level of investment activity, of intermittently successful regions for venture capital investment, such as in the South West. However, standardising venture capital investments in dollar terms by share of the 1995 civilian labour force, the relative dominance of Silicon Valley, San Diego, New England, Colorado and, to a lesser extent, New York Metro, Philadelphia Metro and Texas becomes clear (Table 3). By contrast, although the Mid-West, the DC/Metroplex area and the North West are significant regional clusters of venture capital in absolute terms, they account for a lower share of the total than would be expected on the basis of the overall size of these regions. These figures strongly suggest that, in addition to inter-state or inter-regional variations, intra-state variations (most clearly seen in California where the performance of the Sacramento and Los Angeles areas is significantly poorer than that of Silicon

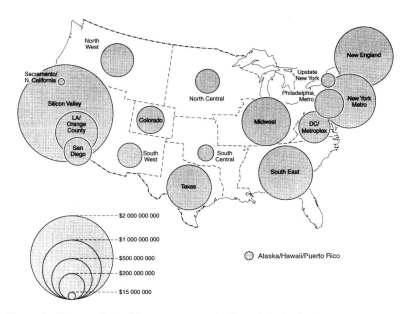

Figure 8 *Venture Capital Investments in the United States by Region, 1997 (Quarters 1–3).* Source: Price Waterhouse (1997)

Table 3 *The Regional Distribution of US Venture Capital Investments by Value, 1996 and 1997*

Region*	% of total venture capital investment		Location quotient***	
	1996	1997**	1996	1997**
Alaska/Hawaii/Puerto Rico	0.3	0.3	0.45	0.39
Colorado	3.6	2.4	2.28	1.54
DC/Metroplex	3.2	3.4	0.86	0.61
LA/Orange County	4.7	5.9	0.85	1.06
Mid-West	9.4	7.7	0.44	0.36
New England	13.4	11.1	2.77	2.30
New York Metro	7.0	9.2	1.01	1.33
North Central	1.8	1.9	0.28	0.29
North West	4.2	3.7	0.97	0.84
Philadelphia Metro	3.5	2.7	1.52	1.17
Sacramento/Northern California	0.2	0.2	0.23	0.17
San Diego	3.3	2.9	3.55	3.15
Silicon Valley	24.1	29.6	5.59	6.87
South Central	1.4	0.8	0.31	0.18
South West	11.4	9.2	0.69	0.56
South East	1.5	1.9	0.42	0.42
Texas	7.0	6.5	0.97	0.89
Upstate New York	0.1	0.6	0.05	0.28

Notes: * Regions are defined in the Appendix of this chapter
** Data for quarters 1–3 (January to September) only
*** Relative to the distribution of the 1995 civilian labour force

Source: Price Waterhouse National Venture Capital Survey

Valley and San Diego) and a metropolitan effect are of particular importance in understanding and describing the geography of venture capital.

Explaining the Uneven Nature of Venture Capital Investments

The uneven distribution of venture capital investments reflects both demand- and supply-side factors. On the demand side, the key influence is the geographical distribution of investment opportunities that are likely to meet the criteria of venture capital funds. In the UK, the demand side comprises two distinct elements. First, investment opportunities for classic venture capital funds are disproportionately concentrated in the South East, which has both the largest volume of new firms and the highest rate of new firm formation (Keeble and Walker, 1994) and, more significantly, which contains the highest proportion of fast-growing SMEs (Keeble, 1997) and the greatest concentration of new technology-based firms (Keeble, 1994). Moreover, fast-growth innovative businesses in the South East are more likely to have raised external equity finance (Thwaites and Wynarczyk, 1996). This suggests that spatial variations in the demand for venture capital arise, in part, from differences in attitudes towards ownership and control and financing preferences across the UK. Spatial variations in the demand for venture capital may also be associated with differences in the quality of professional advisers (e.g. accountants and lawyers) available to entrepreneurs in different locations (Mason and Harrison, 1991; Hustedde and Pulver, 1992).

Second, opportunities for MBOs and MBIs are related to the pre-existing spatial division of labour within large companies. Divisions and subsidiaries which have a fairly complete range of management functions, including R&D and sales and marketing, provide the greatest potential for MBOs/MBIs. The South East has consistently been the principal location for MBOs/MBIs, although its share of total MBO/MBI activity is broadly in line with its 'expected' share. There has been an increase in MBO/MBI activity in other regions during the 1990s, notably the North West, the East Midlands and West Midlands (Centre for Management Buyout Research, 1996), which corroborates earlier observations concerning the reduction in regional inequality in venture capital investments over this period. However, there is less potential for MBOs/MBIs in peripheral regions, such as the North and Wales, because these regional economies are dominated by production-oriented branch plants which have a truncated range of management functions and are integrated into intra-company flows of semi-finished goods (Phelps, 1993).

In the United States, the geography of venture capital investments is closely related to the distribution of high technology industry (Florida, Smith and Sechoka, 1991; Florida and Smith, 1993), and specifically to high technology agglomerations characterised by high rates of formation of innovative new businesses (Saxenian, 1994; Lyons, 1995; Roberts, 1991a; Mitton, 1991). In 1996, 69% of investments and 62% of dollars invested were in high technology;[9] these proportions rise to over 75% in Silicon Valley, San Diego, the North West and (for dollars only) New York Metro (Price Waterhouse, 1997).

Table 4 The Location of Venture Capital Firms in the UK

Region	Head-offices	Including regional/branch offices	
		No.	%
London	63	70	41
Rest of the South East	6	11	6
East Anglia	3	4	2
South West	2	5	3
East Midlands	2	5	3
West Midlands	3	10	6
North West	1	15	9
Yorkshire/Humberside	5	16	9
North	2	6	4
Wales	1	4	2
Scotland	9	20	12
Northern Ireland	4	4	2
Total	101	170	100

Source: British Venture Capital Association (1997b)

The key supply-side influence on the uneven distribution of venture capital investments is the concentration of venture capital firms in a small number of locations. In the UK, nearly two-thirds of venture capital firms are based in London (Table 4). This concentration arises from London's long-established position as the pre-eminent financial centre in the UK, containing various financial and related institutions and a diverse range of investment expertise and business consultancy services. Thus, it was inevitable that the venture capital industry, a further specialist division of the finance industry, should develop there for reasons connected with the localising effects of the spin-off process, the availability of professional labour and the existence of agglomeration economies (Martin, 1989). A significant number of the venture capital offices outside the South East were either public sector creations (although these are now mostly in private ownership), or have been established to invest local authority, government or EU funds to promote local economic development. However, there has been a growth in the number of branch offices of venture capital companies in the regions because London-based firms have recognised the need to obtain direct access to local intermediary networks (Wright, Robbie and Ward, 1994). These regional offices mainly have an MBO/MBI and development capital investment focus (Gibbs, 1991).

The situation in the United States is somewhat more complicated. At state level, the major concentrations of venture capital offices are in California, New York and Massachusetts, whereas at Metropolitan Statistical Area (MSA) level the biggest concentrations are in New York, Boston and San Francisco, with somewhat smaller concentrations in Chicago, San José, Dallas and Houston (Florida and Smith, 1993). California's dominance is even greater in terms of capital under management. There is an important distinction between venture capital firms in New York and Chicago, which tend to be captive funds tied to large financial institutions, and those elsewhere which are predominantly independent firms (Florida and Kenney, 1988a).

The location of venture capital funds also influences the geographical distribution of investments. Distance impedes venture capitalists' access to information regarding both investment opportunities and the performance of investee businesses (Green, 1991). The earlier discussion of the venture capital investment process noted that venture capital firms rely on information-sharing with other venture capital firms, accountants, banks, lawyers, consultants, universities and a variety of other organisations in order to identify investment opportunities, organise investments and mobilise resources for its investee companies. As a result of the intensive nature of this information flow, it tends to be personalised, informal and, therefore, localised (Florida and Kenney, 1988a). This then biases investments by venture capital firms in favour of companies located relatively close to their office. The earlier discussion also highlighted the hands-on investment style of most venture capital firms in order to limit risk and add value to their investments. This involves venture capital firms in close contact with investee companies in order to contribute technical skills, operating experience and networks. As Saxenian has observed in Silicon Valley, 'geographical proximity help[s] build and sustain these relationships' (1994: 39). Venture capital funds are discouraged by the additional costs and greater difficulties of interacting on a face-to-face basis with entrepreneurs who are not located within easy travelling distance.

However, the extent to which distance between the venture capital firm and the location of the investee business acts as a constraint on investment activity depends on the investment focus of the venture capital firm. Classic venture capital funds are more involved with their investee companies in both a monitoring and advisory capacity because the management team in most early stage ventures is inexperienced and there is greater uncertainty associated with the technology, markets and competitive situation (Gupta and Sapienza, 1992). Venture capital firms have less need to closely monitor their MBO/MBI investments, or to become actively involved, because these firms have an established management team, established products or services and established markets. Elango, Fried, Hisrich and Polonchek (1995) therefore argue that whereas the market for early-stage venture capital investments, and also smaller later-stage investments, are localised, the market for large later-stage investments, including MBOs and MBIs, is national, with distance playing little or no role in the investment process. Yet even for this type of investment, distance may hinder the identification of investment opportunities. Murray and Wright (1996) note that London-based captive MBO funds have opened regional branch offices to develop local networks of professional intermediaries in order to better identify local investment opportunities.

It was noted above that syndication of investments is a way in which venture capitalists can pool expertise and share risks. Typically, syndicated investments involve a single lead investor which undertakes the monitoring and provides any assistance that is required by the investee company, with the other members of the syndicate playing more passive roles. Syndication plays a crucial role in the geography of venture capital activity because it loosens, although it does not eliminate, the spatial constraints on venture capital investing (Florida and Kenney,

1988a). Passive members of a syndicate do not need to have spatial proximity to investee companies: only the lead investor needs to be within easy travelling distance because of its direct involvement with the investee business. Florida and Kenney (1988a) note that syndication enables venture capital firms based in the major financial centres of New York and Chicago to invest in technology-based firms in Massachusetts and California, and also results in California being a net importer of venture capital. However, there is also considerable syndication between venture capital firms in the same state.

There is strong circumstantial support for the supply-side explanation in the case of the UK. Scotland, which has the largest number of venture capital offices of any region outside the South East, has attracted significantly more than its fair share of venture capital investments. The South West, Wales and Northern Ireland have each attracted less than their expected shares of venture capital investments. They are also the regions with the fewest venture capital offices (see Table 4). The increase in the number of venture capital funds in provincial centres and the establishment of regional branch offices by some London-based funds since the late 1980s (Mason and Harrison, 1991; Wright, Robbie and Ward, 1994) is also consistent with the reduction in regional inequality in the distribution of venture capital investments in the UK. These regional offices are often the lead investor in syndicates (Gibbs, 1991). Nevertheless, the role of the demand side must not be totally dismissed. The UK's largest venture capital fund, 3i plc, has a network of offices throughout the country, and although it has a reasonably wide regional distribution of investments in terms of number of deals, the South East/East Anglia still dominates in terms of the amount invested (BCR, 1995).

In the case of the United States, on the other hand, venture capital investment does not appear to be related to the distribution of venture capital supply (at the MSA level), indicating that investments are not determined by the location of funds (Florida and Smith, 1993). This may reflect the operation of the syndication process which, as noted above, loosens the spatial constraint on investing. However, this finding might also be influenced by the spatial scale of analysis: the fact that Silicon Valley comprises several MSAs means that investments by venture capitalists based in San Francisco in firms located in San José, Santa Clara and Santa Cruz MSAs would not be regarded as local. Supporting the supply-side explanation, Friedman (1995) finds that the presence of venture capital firms is one of the characteristics associated with the geographical distribution of small, fast-growth firms.

The Geography of the Informal Venture Capital Market

Little is known about the geographical distribution of investments made by business angels. According to Gaston, referring to the United States, business angels 'live virtually everywhere' (1989: 4). However, in view of their entrepreneurial backgrounds, they are likely to be most numerous in regions with a large SME sector, a high proportion of fast growth firms, a tradition of successful companies going public through IPOs or being acquired (thereby releasing capital to the owner-managers), and a concentration of wealth and income. To the extent to which these characteristics tend to be defining characteristics of economic core regions of a national economy, business angels are likely to be concentrated in the same regions as venture capital activity. UK evidence indicates that business angels are found throughout the country (Mason and Harrison, 1994), and although they are over-represented in the southern areas (and have invested more than business angels elsewhere in the country) they are nevertheless significant in all regional economies (KPMG Management Consulting, 1992; Harrison, McIntyre and Mason, 1997).

More importantly from a regional development point of view, however, investments by business angels are characterised by a high level of geographic localisation, with a majority of all such investments made within 50 to 100 miles (one to two hours' driving time) of the investor's home (Mason and Harrison, 1994; Landström, 1993, Freear, Sohl and Wetzel, 1994). This is a reflection of three factors: first, to an even greater extent than institutional venture capital, informal investment is largely a hands-on process involving significant on-going involvement in the business; second, reliance primarily on personal social and business networks for information on potential deals results in business angels having superior information about investment opportunities close to home; third, most business angels are reluctant or unable to make the time commitment to visit potential investee businesses in distant locations. As such, business angels represent an important mechanism for the retention and recycling of entrepreneurial capital within regions. Furthermore, notwithstanding the likelihood of regional variations in the distribution of business angels and in informal investment activity, this source of investment capital probably supports a larger volume of classic-type investment activity than institutional venture capital funds in most, if not all, regions (Mason and Harrison, 1995).

IMPLICATIONS OF THE UNEVEN GEOGRAPHICAL DISTRIBUTION OF VENTURE CAPITAL INVESTMENTS

The implication of the previous section is that a local venture capital industry is essential for regional development. Indeed, Bygrave and Timmons (1992) suggest that it is unlikely that a country or region can be competitive in the commercial exploitation of innovative processes, products or services without a strong local venture capital industry. The presence of a local venture capital industry specialising in 'classic' venture capital is vital if a region is to enjoy a high level of entrepreneurial start up and growth. Access to capital is a major constraint on the start-up and growth of firms (Advisory Council on Science and Technology, 1990; Roberts, 1991a; Storey, Watson and Wynarczyk, 1989), so that the differential spatial access to the availability of risk capital may be expected to lead to spatial variations in entrepreneurial activity. As such, the supply of venture capital will itself contribute to increased demand for venture capital arising from new business formation and growth. Furthermore, the importance of venture capital is not simply as a source of finance. Florida and Kenney emphasise that venture capital firms are an integral part of a well-developed local technology infrastructure:

> venture capitalists enhance such environments by acting as both catalyst and capitalist, providing the resources and the contacts to facilitate new business start-ups, spin-offs and expansions. Because they sit at the centre of extended networks linking financiers, entrepreneurs, corporate executives, head-hunters and consultants, venture capitalists have a propulsive effect on business formation. (1988a: 43–4)

Therefore, venture capital firms should be seen as an important element in the 'institutional thickness' of regions, which is a central factor in underpinning and embedding local economic activity (Amin and Thrift, 1994). A local venture capital industry is probably not essential for a region to be able to attract large later stage and MBO/MBI financing because venture capital firms are able to spread the fixed transaction costs of travel and travel time over a larger investment. Moreover, such deals are less dependent than early-stage investments on localised information networks (Elango, Fried, Hisrich and Polonchek, 1995) and the venture capital fund is likely to be less involved in providing assistance to the management team.

The uneven geography of venture capital activity is also significant for another reason. As noted earlier, venture capital firms raise their investment capital from financial institutions, notably pension funds, insurance companies and banks which, in turn, are the repository of the savings of individuals throughout the country in the form of occupational pension contributions, insurance premiums, investment trusts/mutual funds and bank savings accounts. The effect of the uneven spatial distribution of venture capital investments is to drain venture capital-deficient regions – typically peripheral, economically depressed regions – of savings which are invested in a small number of economically buoyant geographical locations (Mason and Harrison, 1989).[10] It is noteworthy in this context that the successful experiment in community economic development in the Mondragon region of Spain is based on the establishment of a local financial institution which retains and recycles local savings for investment in new enterprises, and demonstrates that community economic development must be based on local capital entrapment (MacLeod, 1997).

POLICY RESPONSES TO REGIONAL VENTURE CAPITAL GAPS

National and local/state governments have responded to the existence of regional 'gaps' in the supply of venture capital with initiatives designed to stimulate the creation of locally managed pools of classic venture capital in venture capital-deficient regions. These initiatives have taken two forms: direct and indirect measures (Organisation for Economic Cooperation and Development, 1997a). The first approach has been to encourage the creation of private sector funds through initiatives which increase the reward for investors (e.g. tax incentives), reduce the risk for investors (e.g. equity guarantee schemes) or reduce appraisal or operating costs (e.g. subsidies), thus increasing portfolio return. The second approach is for the government to become directly involved as an equity investor, by investing in private sector venture capital funds, providing loans to private venture capital companies at a preferential rate of interest, or by establishing its own venture capital fund. In the United States, the SBIC programme enables privately organised and managed investment firms to use their own capital to leverage additional funds borrowed at favourable rates through the Federal Government. Various state governments have also established their own venture capital funds (Florida and Smith, 1990). UK local authorities established local venture capital funds during the 1980s, but this activity was subsequently prohibited by legislation, although most of these funds have continued under private ownership. The development agencies in Scotland and Wales also have their own venture capital divisions.

However, there are many criticisms of such approaches (Florida and Kenney, 1988a and 1988b; Florida and Smith, 1990; Standeven, 1993). First, the effect of these programmes may simply be to replicate and reinforce existing spatial biases in the venture capital industry (Mason and Harrison, 1989). This can be illustrated by the SBIC programme in the United States. Certainly, SBIC investment is less concentrated spatially than venture capital: the top 15 states account for 89% of venture capital (and the top five for 63%), but for only 69% of SBIC investment (31% in the top five venture capital locations). Furthermore, while only six of these states record more than their expected share of venture capital investment, nine also record higher than expected shares of SBIC investment (Table 5). Thus, the SBIC programme has broadened access to venture capital on a spatial basis. Nevertheless, three of the top five states for private venture capital (and nine of the top 15) have attracted more than their expected share of SBIC investment, and there remain a significant number of states which have very low levels of both private venture capital and SBIC investment.

Second, many commentators argue that venture capital funds are constrained by the restricted supply of viable, high-potential businesses. This constraint is greatest in economically lagging regions. In this context, an increase in the supply of venture capital as a result of direct or indirect government programmes will create distortions in the market which over the long term could drive out private sector venture capital funds. Where private sector venture capital funds are in competition with government venture capital funds or subsidised private sector funds for the same set of investment opportunities, they will be at a significant disadvantage both in terms of their higher costs and higher rate of return requirements to satisfy their investing institutions. Therefore, government funds and subsidised funds will be able to offer better terms to potential investee businesses. One possible consequence is that private funds will withdraw from the market. Alternatively, if private venture capital companies choose to compete with government and subsidised funds, the effect will be to drive down their rates of return which, over the long term, will discourage institutions and private individuals from providing further investment funds. In this situation, private sector venture capital firms will be forced to withdraw from the market because they are unable to raise new funds (Riding and Orser, 1997).

Table 5 State Level Distribution of Private Venture Capital and SBIC Investment in the United States

State*	Private venture capital invested, 1996		SBIC investment, 1992–1996 inclusive		Location quotient	
	US$ million	%	US$ million	%	Private venture capital	SBIC
California	3749.3	37.4	632.0	13.9	3.21	1.19
Massachusetts	1209.0	12.1	199.4	4.4	5.06	1.84
Florida	477.2	4.7	176.3	3.9	0.91	0.76
Texas	435.6	4.6	299.9	6.6	0.64	0.91
Colorado	420.3	4.2	108.0	2.4	2.66	1.52
Total – top five states		63.0		31.2		
Illinois	394.2	3.9	225.0	5.0	0.85	1.09
Washington	329.1	3.3	72.7	1.6	1.56	0.76
Virginia	301.7	3.0	102.3	2.3	1.14	0.87
Connecticut	289.5	2.9	99.5	2.2	2.25	1.71
Pennsylvania	285.0	2.8	157.2	3.5	0.64	0.79
Total – top 10 states		78.9		45.8		
New York	232.2	2.3	363.8	8.0	0.36	1.25
Tennessee	203.9	2.0	171.1	3.8	0.98	1.85
Missouri	201.0	2.0	171.1	3.8	0.94	1.85
Ohio	189.6	1.9	184.5	4.1	0.45	0.97
New Jersey	169.6	1.7	152.6	3.4	0.55	1.11
Total – top 15 states		88.8		68.8		

Note: The source of data on private venture capital investment in this table is the National Venture Capital Association/VentureOne survey, whereas the data used in Tables 3 and 4 and Figures 6 and 7 are from the Price Waterhouse survey; the NVCA/VentureOne data are appropriate here because they do not include SBIC investments, whereas the Price Waterhouse data include SBIC investments

Sources: US Small Business Administration; National Venture Capital Association/VentureOne

Third, Florida and Kenney argue that 'simply making venture capital available will not magically generate the conditions under which . . . entrepreneurship can flourish' (1988b: 316–17). Venture capital is only one of a host of necessary conditions for entrepreneurial-led economic development (Florida and Smith, 1990). Moreover, as noted on several occasions in this chapter, financing is just one part of what venture capitalists do. According to Florida and Kenney (1988b), the success of new and early stage ventures is ultimately contingent on the support services that venture capitalists can provide either directly or through their networks. However, they are dubious about the amount of expertise that government agencies can provide, particularly when the investee companies begin to develop and require more management discipline.

Fourth, the ability of publicly supported venture capital funds in less favoured regions to achieve long-term financial viability is likely to be problematic. This criticism is based on Murray's (1994 and 1998) evaluation of the European Union

(EU)'s Seed Capital Fund Scheme which provided 24 funds across the EU with an interest-free loan towards their operating costs, and those funds located in Objective 1 and 2 Regions also received a financial contribution towards their investment funds. There are fixed costs involved in operating a venture capital fund which have to be paid out of the capital that the fund has raised from its investors. The smaller the fund, the greater the proportion of the capital raised that is accounted for by operating costs. Indeed, Murray (1994) estimated that because of their small size, the funds would exhaust their capital in 9–11 years, even if no investments were made, purely as a result of their operating costs. Three consequences arise: (i) the number of investments that each fund can make will be limited, thereby increasing the portfolio risk; (ii) there is a strong probability that the funds will run out of money before they can 'harvest' their existing investment portfolio, but without an investment track-record are unlikely to be able to raise additional finance from private investors; and (iii) the funds may be unable to provide follow-on finance to those firms in their portfolios that require a second round of investment.

Fifth, requiring the fund to invest in businesses located within a specified territorial unit (e.g. a state or local authority area) might constrain the fund's deal flow, causing investments to be made in non-competitive businesses with the consequence that it will perform badly. However, if there are no geographical restrictions on the fund's investments, capital may flow to other regions (Florida and Kenney, 1988a). Finally, there may be political pressure on the fund's investment decisions.

A superior approach may be to attempt to increase informal investment activity in such regions (Mason and Harrison, 1995). There is a clear consensus, first, that most business angels are unable to find sufficient investment opportunities (Mason and Harrison, 1994), and, second, that there are many 'virgin' business angels who could be encouraged to become active investors with appropriate support (Freear, Sohl and Wetzel, 1994). The most effective way of enabling entrepreneurs to access this supply of finance is by creating of business angel networks. These organisations enhance the communication flow between investors and entrepreneurs seeking finance and also provide training support to investors and help entrepreneurs to become 'investment ready' (Harrison and Mason, 1996). Moreover, firms which have raised finance from business angels and benefited from their involvement are likely to be more attractive to formal venture capital funds when they seek further rounds of finance.

Conclusions

> [O]ne of the major lacunae of industrial and territorial analysis . . . is the almost total absence of the financial dimension. . . . Regional economics is rarely worried by flows of finance outside questions of public finance, and especially local taxation. . . . Works on the local and territorial dimension of business and financial flows are rare. . . . (Courlet and Soulage, 1995: 302)

Research during the 1990s on the venture capital industry has resulted in a much more sophisticated understanding of the 'technical' aspects of the investment process. However, there has been much less progress in understanding the spatial aspects of the investment process since the pioneering studies of the late 1980s (e.g. Florida and Kenney, 1988a; Martin, 1989). This chapter has shown that there are spatial variations in venture capital investment activity which are shaped by explicitly geographical factors in the venture capital investment process. These include the venture capital firm's network of offices and informants, the limitation on the geographical focus of investment activity as a means of minimising risk and the need for spatial proximity to investee businesses for ease of monitoring and hands-on involvement. The effect is to constrain the spatial behaviour of venture capital firms (Thompson, 1989). In addition, there are various demand-side influences which also shape the geography of venture capital investments, including the nature of entrepreneurial activity in different locations and the knowledge of venture capital possessed by professional intermediaries which varies from place to place. However, given the existing state of knowledge, it has not been possible to go beyond this to examine the operation of the venture capital market in a spatial context. Fundamental questions concerning such issues as the spatial flow of funds, the characteristics of venture capital funds in different regions, spatial elements in the investment process (e.g. spatial preferences and perceptions of venture capitalists), spatial variations in the proportion of investment proposals accepted and rejected by venture capital funds, and spatial variations in investment returns have yet to be studied. Given such gaps in our understanding, further research into the geography of venture capital is undoubtedly required. The research agenda proposed by Thompson (1989) a decade ago remains a good starting point.

APPENDIX: DEFINITIONS OF US REGIONS

Alaska/Hawaii/Puerto Rico	Alaska, Hawaii and Puerto Rico
Colorado	The State of Colorado
DC/Metroplex	Washington DC, Virginia, West Virginia and Maryland
LA/Orange County	Los Angeles, Ventura, Orange and Riverside Counties, excluding San Diego
Mid-west	Illinois, Missouri, Indiana, Kentucky, Ohio, Michigan and Western Pennsylvania
New England	Maine, New Hampshire, Vermont, Massachusetts, Rhode Island and parts of Connecticut (excluding Fairfield County)
New York Metro	Metropolitan NY Area, Northern New Jersey and Fairfield County in Connecticut
North Central	Minnesota, Iowa, Wisconsin, North Dakota, South Dakota and Nebraska
North West	Washington, Oregon, Idaho, Montana and Wyoming
Philadelphia Metro	Eastern Pennsylvania, Southern New Jersey and Delaware
Sacremento/North California	North-eastern California
San Diego	San Diego Area
Silicon Valley	Northern California, Bay Area and Coastline
South Central	Kansas, Oklahoma, Akansas and Louisiana

South East	Alabama, Florida, Georgia, Mississippi, Tennessee, South Carolina and North Carolina
South West	Utah, Arizona, New Mexico and Nevada
Texas	The State of Texas
Upstate New York	Northern New York State, except the Metropolitan New York City Area

Source: Price Waterhouse (1997)

Chapter 9

Corporate Finance, Leveraged Restructuring and the Economic Landscape: The LBO Wave in US Food Retailing

Neil Wrigley

Introduction

The intense and remarkable wave of leveraged buy-outs (LBOs), leveraged recapitalisations and dramatically increased levels of corporate debt which characterised the mid to late 1980s, created an era of considerable importance in the recent history of industrial capitalism. For a short period, until macro-economic and regulatory conditions turned strongly against such high-leveraged transactions in 1989–90, the corporate restructuring process became firmly focused on the capital structure of the firm, with financial leverage being seen as a catalyst for organisational change, offering the possibility of improved firm value and operating efficiency. In particular, the work of Michael Jensen (1986, 1989 and 1991) was influential in supporting the view that leveraging firms to very high levels was an effective way to overcome 'corporate control failure', to open the management of large public companies to monitoring and discipline from capital markets, and to force them to take the hard decisions which would eliminate excess capacity and generate cash flow in the most productive manner.

Despite the considerable literature which has subsequently emerged in financial and industrial economics on the effects of high levels of financial leverage on firms' operating performance, and despite Clark's (1989) early plea for attention to be given by geographers to the enormous impacts of this form of corporate restructuring on the economic landscape, the spatial consequences of capital structure transformations of the firm in general, and of the LBO wave in particular, have remained remarkably under-researched topics in economic geography. The aim of this chapter is to redress that deficiency. In the context of one industry (food retailing) which became a prime target for high-leverage capital transformations, and one country (the United States) in which that industry assumed aggregate debt of more than US$20 billion and experienced a radical transformation of its financial and control structures, this chapter explores how debt and financial leverage translate into spatial outcomes.[1] In particular, it focuses on two key issues. First, how capital structure transformation affects the spatial organisation of the high-leverage firm, giving rise to a 'geography of divestiture', to local market concentration, and the avoidance of intra-market competition. Second, how capital structure transformation creates an *interaction* between the high-leverage firm and the decisions of rival firms, with important consequences for the nature of competition (exit, entry, pricing, etc.) in specific markets.

It will be shown that capital structure transformations signal new behaviour to the high-leverage firm's rivals – specifically a credible commitment by the firm to alter its investment behaviour – and that rival firms respond to such signals. As a result, capital structure transformations of the firm are viewed as having profound implications for the economic landscape, both directly through the spatial reorganisation of the activities of the high-leverage firm, and indirectly through the restructuring of markets by rival firms responding to the commitments implicit in those transformations. Shifts in the capital structures of firms recreate geographies in ways that are only now beginning to be understood.

More generally, the chapter represents an extended plea for issues of corporate finance to find a place within the emerging agenda of the geography of money and finance. It would be wrong to assume that such issues and their impact on the economic landscape are the priority of other areas of economic geography. For example, whilst economic geographers have long been concerned with issues of corporate restructuring and have made important contributions at both the conceptual level (e.g. Massey and Meegan, 1982) and via what Clark (1998c) terms 'the close dialogue' offered by case studies of industry restructuring, that literature (Clark and Wrigley, 1997b) has traditionally underemphasised types of restructuring which involve transformations of the capital structure and ownership configuration of firms. Likewise, the available literature on geographies of mergers and acquisitions has to a large extent ignored issues of capital structure reconfiguration. Nevertheless, so important are the spatial consequences of the transformation of the capital structures of firms that premature closure of the research boundaries of the 'geography of money and finance' must surely be avoided, and a sustained attempt made to claim space within it for critical interaction with the central theoretical debates of corporate finance and financial economics.

DEBT, LEVERAGE AND THE RETAIL FIRM

Before assessing the wave of high-leverage capital transformations of US food retailing in greater detail, it is important to first provide some essential background information on issues of corporate finance – in particular, on the question of the capital structure decision and the implications of high leverage for the retail firm.

THE CAPITAL STRUCTURE DECISION

One of the critical theoretical debates shaping contemporary corporate finance has been the capital structure controversy initiated by Modigliani and Miller (1958). Their capital structure 'irrelevancy' proposition essentially set out conditions under which the source of funds (debt or equity capital) used by a firm to finance its investment should have no impact on its profitability or market value. As that debate has unfolded over the last 40 years, however, a consensus has emerged which suggests that capital structure (the financing mix) *does* have a valuation impact (for details, see Balakrishnan and Fox, 1993; Hutchinson, 1995; Stern and Chew, 1992; also for a discussion in the context of the retail firm, see Hutchinson and Hunter, 1995; McCaffery, Hutchinson and Jackson, 1997). Essentially, the argument is that because of tax-shield benefits from employing debt capital (interest payments can be set against corporate tax liabilities), the firm has an incentive to increase the amount of debt in its capital structure but, as the debt/equity ratio rises, premiums are imposed by the suppliers of debt capital to compensate for agency costs[2] and potential financial distress costs. As these premiums eventually completely off-set the tax-shield benefits, but differ from industry to industry and firm to firm, it is the task of financial managers to uncover and target what the optimal financial mix for their firm should be.

Not all corporate finance theorists, however, accept this position. For example, Myer's (1984) 'pecking order' theory of capital structure places emphasis on a firm's preference for internal over external financing, and argues that a firm's observed capital structure is not the result of a deliberate policy to target a particular debt/equity ratio. Instead, as McCaffery, Hutchinson and Jackson (1997) note, it arises from a firm's 'past excursions into the capital markets when retained earnings, the most preferred source of financing, were insufficient to cover investment needs'.

Nevertheless, in the context of this chapter, what emerges strongly from the capital structure debates is the importance of debt capital, the debt/equity ratio, and financial leverage – defined in general terms as the ratio of total debt (or, sometimes, long-term debt) to firm value – on the operating performance of the firm. In particular, the relationship between increasing levels of debt capital and the probability of 'financial distress' (defined, most simply, as a situation in which cash flow is insufficient to cover current obligations) is viewed as being of central importance. Some believe that over-reliance on debt capital within the firm's financing mix increases the probability of financial distress and the likelihood of bankruptcy. This is essentially because, as debt increases, the fixed interest charges the firm must meet out of its operating income rise, increasing the firm's vulnerability to unanticipated market downturns. As a result, the firm may choose to leave some of its debt financing capacity unused to provide 'financial slack' which can be taken up during periods of financial difficulty. Others (e.g. Jensen, 1986) believe that increased leverage does not necessarily increase the probability of financial distress, rather it imposes discipline on managerial discretion which creates a more efficiently run firm that can carry a higher debt burden with equal or reduced probability of financial distress.

HIGH LEVERAGE AND THE RETAIL FIRM

At what point then, in terms of increased levels of debt capital, do we regard a firm as being 'high leveraged' and what are the implications? One of the minor problems in this regard is that there is no consensus on the measurement of 'leverage'. In the context of the retail firm, two commonly used measurements are:

(i) Total debt/total capitalisation = (total borrowings + obligations under capital leases)/(shareholders' funds + preferred stock + total debt).
(ii) Total debt/market value of firm = (total borrowings + obligations under capital leases)/(market value of equity + preferred stock + total debt).

Together with several minor variants, these produce somewhat different results with the debt-to-market-value measure, tending to yield more conservative debt-percentage figures. In addition, different industrial sectors have varied historically in their ability to sustain debt capital. Hence, what can be regarded as 'high' in leverage terms is to some extent sector-specific.

In an industry such as food retailing, however, which is intrinsically cash-generative and about which capital markets have traditionally held a 'positive view of the sector's ability . . . to service its debt commitments' (Hutchinson and Hunter, 1995: 76), firms with 'low' to 'moderate' levels of financial leverage have generally been regarded as having total debt-to-capitalisation levels in the range 25–50%, whilst those with 'high' levels have typically been considered to lie in a range above 75%.[3] In the case where shareholders' equity has relatively little book value, that figure will usually rise into the 90–100% range, and it can exceed 100% where that equity has negative book value.

If, as Jensen (1986) suggests, leverage imposes managerial discipline, acts as a catalyst for organisational change and creates a more efficiently run firm, then that firm will have incentives to manage its assets differently, to generate cash flow in a productive manner, to improve working-capital management, and to service its debt-interest payments in such a way that its competitive position is maintained or enhanced. In these circumstances, a self-reinforcing virtuous circle of operational and financial improvement can be created in which enhanced earnings permit, and are contributed to by, debt retirement and refinancing.

Conversely, the potential perils of too much leverage for the retail firm are illustrated in Figure 1. In this case, the high-leverage food retailer, faced with servicing much higher debt interest repayments than its more unleveraged competitors, is potentially disadvantaged in important ways. Its ability to maintain its capital expenditure programmes – to invest in new store development, store renovation, supply chain management, distribution/logistics and systems innovations – is reduced. Yet it may face intense competition in its markets from unleveraged rivals which can maintain or enhance their capital investment programmes, engage in competitive store openings, and/or invest in price/promotional activity. As a result, the firm may suffer a loss of market share to these competitors. However, in a high turnover but low operating profit margin industry such as food retailing, any decline in sales and loss of market share can have a very significant and rapid impact on operating efficiencies, in turn reducing the firm's cash flow. The result is likely to be a further reduction in the funds available to the firm for capital expenditure and competitive response, with lengthening store

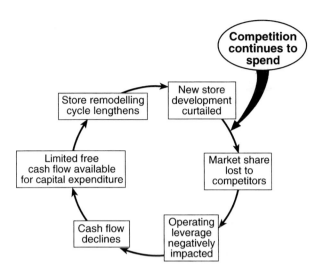

Figure 1 The Perils of High Leverage for the Retail Firm

renovation/remodelling cycles, deferred new store investment and a downward spiral in its competitive position.

Ultimately, if the high-leverage retailer becomes trapped in such a downward spiral, it faces the prospect of 'financial distress' – a situation in which cash flow becomes insufficient to cover current obligations such as debts to suppliers and employees, interest or principal payments under borrowing agreements, potential damages under litigation, and so on. Financial distress gives creditors a legal right to demand the reorganisation of the firm and a renegotiation of their contracts, but it is important to note that in these circumstances, claimholders of high- and low-leverage firms face somewhat different situations.

Wruck (1990) provides a neat visual summary (Figure 2) of these differences for the case of two firms – one with a high-leverage capital structure, the other with low/moderate leverage – in the same business, with identical initial firm values and liquidation levels. Although each firm is assumed to default when the net present value of its cash flow falls below the face value of its obligations, that trigger value is likely to be much higher for the high-leverage firm than the low-leverage firm. As a result, the firm value at risk of destruction (represented by Area A in Figure 2) if the high-leverage firm eventually liquidates (i.e. converts its assets to cash and distributes that cash to its claimants) rather than successfully reorganises, is greater than for the low-leverage firm, where the equivalent value at risk is represented by Area B. It follows, therefore, that a high-leverage firm's claim-holders are likely to have stronger incentives to demand rapid reorganisation of the firm to resolve financial distress than those of a low-leverage firm. In the context of the United States, and specifically Chapter 11 of the US Bankruptcy Code which involves a court-supervised process of reorganisation of the firm, this may mean an increased likelihood of private 'workout' outside the bankruptcy

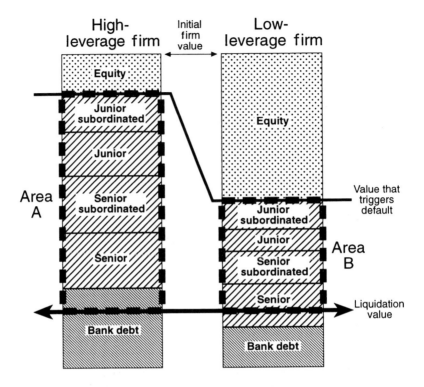

Figure 2 The Capital Structures of High- and Low-leverage Firms, and the Implications for Financial Distress. Source: adapted from Wruck (1990)

court or, at very least, increased incentive for rapid progress through the process using what are termed 'pre-packaged Chapter 11s'.

THE LEVERAGED RESTRUCTURING OF US FOOD RETAILING

Between 1985 and 1989 a wave of LBOs, leveraged recapitalisations,[4] and leveraged acquisitions swept through the US food retail industry involving both the leading multiregional firms and the smaller regional chains. Chevalier (1995a) has estimated that 19 of the 50 largest food retailers, accounting for almost 25% of US supermarket sales, undertook LBOs or leveraged recapitalisations during that period. Similarly, Cotterill (1993) has shown that half of the top 20 firms (see Table 1) were involved in high-leverage transactions during those years, ranging from the US$4–5 billion Safeway and Kroger LBO/recapitalisations (Denis, 1994), and the US$2.5 billion leveraged acquisition of Lucky by American Stores, to the US$1–2 billion range LBOs of Stop & Shop and Supermarkets General. The leveraged restructuring wave coincided with, and was stimulated by, the non-enforcement of anti-trust regulation in the United States under the Reagan

Table 1 The Top 20 US Food Retailers 1979 and 1989, and LBO/Merger/Ownership Changes during the Decade

	1979			1989	
Rank	Company	Market share (%)	Changes 1979–1989	Company	Market share (%)
1	Safeway	7.52	LBO-KKR, 1986	American	6.27
2	Kroger	4.95	Lev Recap – G Sachs, 1988*	Kroger	5.37
3	A&P	3.66	Acquired by Tengelmann, 1979	Safeway	4.08
4	American	3.36	Acquired by Skaggs, 1979	A&P	3.16
5	Lucky	3.19	Acquired by American, 1988	Winn-Dixie	2.61
6	Winn-Dixie	2.70		Albertson's	2.11
7	Grand Union	1.72	LBO-Mgmt, 1988; then acquired by Miller, Tabak Hinsch, 1989	Supermarkets General	1.79
8	Jewel	1.54	Acquired by American, 1984	Publix	1.53
9	Albertson's	1.47		Vons	1.48
10	Supermarkets General	1.30	LBO-Mgmt, 1987	Food Lion	1.34
11	Stop & Shop	1.03	LBO-KKR, 1988	Stop & Shop	1.32
12	Publix	0.99		Ahold**	1.03
13	Dillon	0.98	Acquired by Kroger, 1985	Giant Food	0.93
14	Vons	0.82	LBO-Mgmt, 1986	Grand Union	0.77
15	Food Fair	0.82	Bankrupt, exited 1986	H.E. Butt	0.74
16	First National	0.75	LBO, acquired by Ahold, 1985	Ralphs	0.73
17	Fisher Foods	0.73	Merged with Riser Foods, 1988; divested main division Dominick's	Fred Meyer	0.65
18	Giant Food	0.68		Bruno's	0.61
19	Waldbaum	0.60	Acquired by A&P, 1986	Dominick's	0.57
20	Fred Meyer	0.58		Hy-Vee	0.51

Notes: * In response to hostile takeover bid by Haft, Kroger with Goldman Sachs undertook a leveraged recapitalisation; recapitalisation resulted in debt levels similar to those typical of LBOs.
** Ahold includes First National, Bi-Lo, etc.

Source: Adapted from Coterill (1993)

Table 2 Change in Debt as % of Capitalisation in US Food Retailing during the LBO Wave

	1985	1986	1987	1988	1989
Donaldson, Lufkin and Jenrette sample*	46.5	56.7	52.6	70.5	71.0
JP Morgan sample**	N/A	N/A	48.8	68.3	67.9

* Sample of 34 major US food retailers, Donaldson, Lufkin and Jenrette Securities Corp., New York
** Sample of 15 major US food retailers, JP Morgan Securities, New York

administrations. Horizontal and market-extension mergers, of a type which had been restricted by the Federal Trade Commission (FTC) since the late 1950s under the provisions of Section 7 of the Clayton Act, became possible[5] (Wrigley, 1992), and 'deals could be done which in the past would have been challenged and probably stopped on anti-trust grounds' (Magowen, 1989).

During these high-leverage transactions, firms essentially borrowed against future cash flows to make payments to their shareholders – that is to say, projected future profits were capitalised and built into the financial base of the firm through increased debt. As a result, debt-to-capitalisation ratios across the industry rose abruptly as shown in the JP Morgan and Donaldson, Lufkin and Jenrette samples of firms in Table 2. The long-term debt component, in particular, increased sharply as bank loans and high-yield (junk) bonds and debentures became the financial instruments used to leverage the industry (Wrigley, 1998a).[6] Cotterill (1993) provides aggregate figures for the industry which indicate a rise in long-term debt, as a percentage of total liabilities and equity, from 24.9% in 1985–6 to 44.3% in 1988–9, and shows that an important consequence was a more than doubling in the level of debt-interest repayment as a percentage of annual sales in the industry during these years from 0.55% to 1.33%. Several of the high-leverage firms were, as a result, faced with servicing annual debt-interest repayment in the range of 2–3% of sales, compared to around 0.5% for the less-leveraged firms.

Three groups emerged as 'gainers' from the leveraged restructuring of the industry. Shareholders who sold their stock and relinquished control of the firms which became the targets of LBOs/leveraged acquisitions gained significantly via the substantial premiums which their stock commanded in the market. Cotterill (1993) suggests that these premiums were, on average, 85% above a benchmark price two months prior to the LBO/acquisition announcement. Similarly, shareholders in those firms which undertook leveraged recapitalisations gained equivalent amounts via the special dividends which were paid to shareholders (see Note 4). The corporate raiders and investment bankers involved in the leveraged restructuring also gained. Morgenson (1990) and Cotterill (1993) suggest that in the case of the Safeway LBO the corporate raiders who prompted the LBO earned US$140 million in three months, whilst Denis (1994) shows that the average percentage fees paid to the investment bankers and other professionals were between 5% and 6% of total transactions costs – in the case of the large Safeway and Kroger transactions, where the fee level was slightly lower at just 4%, this amounted to US$175–200 million. Finally, the top executives of the leveraged companies were often major gainers. During the high-leveraged transaction they were presented with significant opportunities for managerial 'cashing out' (liquidation of a large proportion of their pre-buy-out equity holdings), to the extent that in some cases they may have had incentives to participate in overpriced or poorly structured transactions (Kaplan and Stein, 1993). In addition, they were often placed in a position where they could exploit their privileged knowledge of their firm's operation to extract enhanced performance-related incentives to maximise post-leverage firm value. Following the Safeway LBO, for example, the new majority equity owners (Kohlberg, Kravis and Roberts) introduced executive compensation plans which tied bonus payments more firmly to performance measures and gave top executives far more incentive to maximise firm value. In addition, these executives were offered significant stock options, both at the time of the transaction and subsequently – options which by 1990 had shown a sixfold gain.

In contrast, many have argued that unionised labour emerged as one of the 'loser' groups during the leveraged restructuring of the industry.[7] Prior to the high-leverage transactions, many of the leading US food retailers were hampered in their attempts to control operating costs by significant unionised labour power. For example, of the 170 000 Safeway employees and 178 000 Kroger employees at the end of 1985, 90% and 59% respectively were unionised. Labour costs at both firms exceeded the industry average (Safeway paid hourly workers 33% more than the industry average, whilst labour costs at half of Kroger's stores exceeded those of its competitors), and both firms faced competitive pressure from smaller non-unionised chains (Wruck, 1992; Denis, 1994). Following the leveraged restructuring, however, the huge debt burdens built into the high-leverage companies forced them to confront labour and extract wage and work practice concessions. The key to this confrontation was the so-called 'credible threat', which significantly increased the bargaining power of the retailers. The high-leverage retailers could be seen to be irrevocably committed to servicing their debt burdens, and unions were faced with the knowledge that these firms had no option but to reduce operating costs significantly or to sell assets (parts of their store/distribution networks and/or manufacturing plant capacity). Threats by the retailers to liquidate parts of their operations or to sell to non-unionised firms, left the union with little option but to negotiate concessions.[8]

Compared to other industrial sectors in the United States, for which Kaplan and Stein (1993) reported an overall rate of 36% of high-leverage firms subsequently defaulting on their debt, the leveraged restructuring of the US food retail industry produced remarkably few failures, particularly amongst the major firms. The leveraged food retailers survived in part because of the industry's inherent capacity to generate consistent and substantial cash flows, in part (as will be discussed below) because of their ability to sell assets to pay down debt, in part because of enhanced bargaining power which the cash-hungry leveraged firms enjoyed *vis-à-vis* both unionised labour and food manufacturers (Marion, 1995), and in part because of their ability to pass some of the cost of servicing their debts to the consumer (Cotterill, 1990).

Typically, as Figure 3 shows for the case of Kroger, the high-leverage food retailers were progressively able to retire debt, refinance to lower coupon debt and

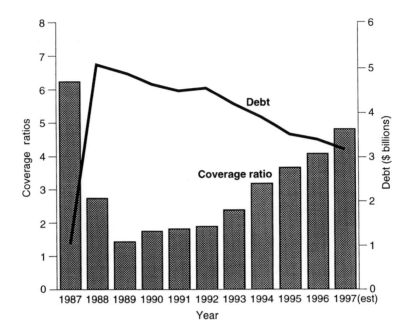

Figure 3 *Kroger: Debt and Cash-flow-coverage of Interest Expense Multiples, 1987–1997*

reduce short-term borrowing – in the process, lowering debt-to-capitalisation ratios and improving cash-flow-coverage of interest expense multiples (Wrigley, 1998b).[9] However, not all high-leverage food retailers escaped the downward spiral portrayed in Figure 1. Some firms suffered from overpriced or poorly structured transactions, assuming unreasonably high levels and/or more demanding forms of debt. Other firms became highly leveraged relatively late in the wave of such transactions, with little time to pay down debt before macroeconomic conditions (the onset of the recession in 1990) combined with regulatory restrictions placed on investments in high-yield bonds following the passage of the *Financial Institutions Reform, Recovery and Enforcement Act* in late 1989 (Jensen, 1991; Waite, 1991; Yago, 1991), and prosecution of the principal market-maker for junk bonds (Drexel Burnham Lambert), to produce an unexpected decline in the liquidity of asset markets (Denis and Denis, 1995; Wrigley, 1998a).

As a result, some of the high-leveraged smaller regional chains became trapped in the downward spiral outlined in Figure 1, became financially distressed and were forced into Chapter 11 and financial reorganisation (sometimes on more than one occasion). In certain cases, this was followed by liquidation (see the examples in Table 3). In addition, at least two of the LBOs of the larger food retailers of the mid 1980s shown in Table 1 (Grand Union, a firm principally operating in Vermont, metropolitan New York, and north-east New York state which effectively went through two LBOs in 1988 and 1989, together with Pathmark, the major part of the larger Supermarkets General LBO of 1987 and principally operating in New

Table 3 Examples of Bankruptcy/Financial Reorganisation amongst High-Leverage US Food Retailers

Firm	Chapter 11 and/or financial reorganisation	Result
Appletree	1991	Sale of assets/liquidation
Almacs/Victory	1993 and 1995	Sale of assets/liquidation
Harvest Foods	1994 and 1996	Sale of assets/liquidation
Megafoods	1994	Sale of assets/liquidation
Kash n' Karry	1995	Sold to strategic acquirer (Food Lion)
Homeland Stores	1996	Restructured debt, new union agreement
Grand Union	1995 and 1998	1996 equity infusion (US$100 million/control shift) 1998 second reorganisation of financial structure (default on interest payments) Debt to equity conversion?
Bruno's	1998	In Chapter 11 – 'Debtor in possession' cash infusion from bank.

York, New Jersey and Philadelphia) continue to struggle with major debt problems and reflect in their operations many of the perils outlined in Figure 1.

SPATIAL OUTCOMES OF THE LEVERAGED RESTRUCTURING

As part of his plea for attention to be given to the geographical consequences of the wave of LBOs and acquisitions in the late 1980s and of what, more widely, he termed the 'arbitrage economy', Clark (1989) noted that 'generally speaking, there appear to be three different ways in which an LBO affects the spatial organisation of production'. First, 'the sale of assets to finance the deal . . . a vital ingredient of any LBO' must inevitably imply a certain 'geography of divestiture' with 'vital implications for the concentration of competition in different geographical markets'. Second, the consequence is likely to be a 'spatial rationalisation of production and capacity', with LBO-related sellers rationalising 'to reduce the costs of servicing particular markets' and buyers rationalising 'to fit acquired facilities into existing networks of plants and outlets'. Third, LBOs will increase the pressure placed on surviving parts of the high-leverage firm 'to operate more efficiently relative to the debt burden assumed by management', and this can have significant consequences for what Karmel (1993) describes as 'non-shareholder constituencies . . . with competing claims on corporate assets and prospects', namely employees, suppliers, other creditors of the firm, customers, and communities in which the retail outlets, distribution facilities and (in certain cases) manufacturing plants of the firm are located. As noted above, however, in considering these three general and inter-related spatial impacts of leveraged restructuring, it is a useful device analytically to separate out the effects of capital structure transformations on the spatial organisation of *individual* high-leverage firms, from the *interactions* which such transformations create between high-leverage firms and the decisions of rival firms in particular geographical markets.

How Capital Structure Transformation Affects the Spatial Organisation of High-leverage Firms

As Clark (1989) suggested, the way high-leverage firms set about servicing their debt burdens is one of the keys to the geographical manifestations of the 'arbitrage economy'. Several methods of generating the cash necessary to service debt payments in excess of current earnings before interest, taxes, depreciation and amortisation (EBITDA) are possible. As Denis (1994) indicates, these include, but are not limited to, sale of assets, reductions in capital expenditures and increases in operating performance. Another possible source of cash, reductions in pensions contributions or 'pension revision' (the termination and stripping off of the excess assets of pension funds) is shown by Clark (1989 and 1993b) to have been important in the early phase of the LBO wave in the United States, peaking in 1985, but to have become far less important in the late 1980s as regulation tightened.

Asset Sales and Market Consolidation

Paying down debt via asset sales was of considerable importance to the high-leverage US food retailers, although the scale and rapidity of the sales varied considerably. Safeway, for example, rapidly disposed of US$2.4 billion of assets (1000 stores) in the two years following its LBO in 1986. Kroger, in contrast, sold only US$351 million of assets over a similar period following its leveraged recapitalisation in 1988, whereas American Stores divested US$1.4 bill of assets (574 stores) over a six-year period following its leveraged acquisition of Lucky in 1988, in the process reducing its debt-to-capitalisation ratio (Figure 4) from 77% to 49%, and increasing its cash-flow coverage of interest expense multiple from 2.67 to 7.07 times. As a result of these asset sales, however, a distinct 'geography of divestiture' and significant market consolidation emerged. The sales tended to be concentrated within the more marginal/peripheral parts of the high-leverage retailers' businesses, as 'core focus' strategies (Clark and Wrigley, 1997a) were widely adopted. The result was an increase in market consolidation as the divesting firms concentrated on regional/local markets which they could effectively exploit as a result of market dominance or co-leadership.

Some indications of the scale of the market consolidation are provided by Franklin and Cotterill (1993) using CR_4 statistics (the ratio of the top four firms' sales to total sales in each market) and Hirschman–Herfindahl Index statistics (the primary measure of concentration used by the US Department of Justice in anti-trust cases) from a special tabulation of the 1987 US Census of Retail Trade. Powerful concentration effects during the 1980s are revealed, with a significant upward shift in four-firm supermarket concentration in the Metropolitan Statistical Areas retaining the same definition as in the 1977 Census.

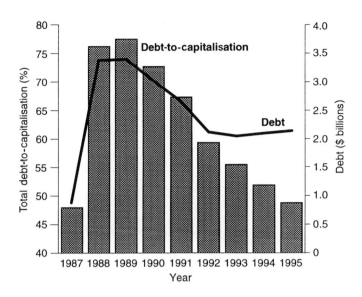

Figure 4 American Stores: Debt and Debt-to-capitalisation (%), 1987–1995

Not only did the 'geography of divestment' imply local market concentration, it was also accompanied by a 'geography of avoidance' characterised by a surprisingly low degree of intra-market competition between the major firms. Using detailed information on retail market shares in the 1980s developed by Selling-Area Markets Inc. (the major rival to AC Nielson during the decade), data donated to Purdue University on the liquidation of the company in 1991, Connor (1997) provides evidence of the degree of market overlap between the top 10 food retailers in 1990, at the end of the leveraged restructuring wave. Figure 5 and Table 4 summarise these market overlaps across the 54 US 'grocery marketing areas' (GMAs) defined by Selling-Area Markets Inc., which together accounted for 75.4% of the total US grocery sales in 1990. Figure 5 reveals that it was the South East region which had the highest levels of intra-market competition between the major food retailers at the end of the leveraged restructuring wave, whereas it was the North East and Mid-West that had the lowest levels.

Taken overall, the amount of intra-market competition is shown to be remarkably low. In the case of American Stores, for example, operating at the time in 13 of the GMAs, its intra-market contacts with Safeway in 1990 occurred only in the San Francisco GMA (via its Lucky chain), the Baltimore/Washington GMA (via its Acme chain) and the Seattle GMA (via its Buttrey chain). That is to say, they occurred in just 23% of the GMAs in which American Stores operated – a figure which was further reduced at the end of 1990 following the disposal of its Buttrey Food & Drug chain. In practice, only the San Francisco GMA represented a significant area of competition between the firms. Moreover, as Figure 6 illustrates, despite the fact that the top three food retailers in 1990 (American Stores, Kroger and Safeway) operated in 39 (72%) of the 54 GMAs, intra-market contact of any

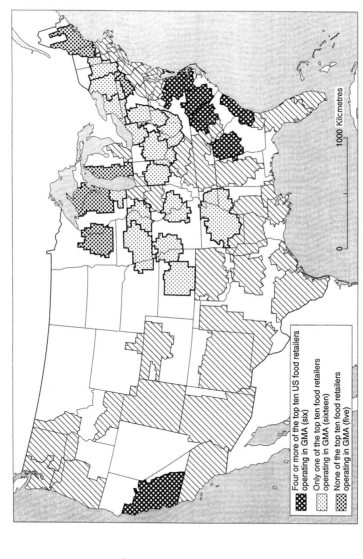

Figure 5 Grocery Marketing Areas (GMAs) with Highest and Lowest Levels of Intra-market Competition (Market Overlap) between the Top 10 US Food Retailers in 1990. Source: Connor (1997) with Mapping and Secondary Analysis by the Author

Table 4 Intra-Market Competition: The Degree of Market Overlap (%) between the 10 Leading US Food Retailers, 1990

Firm	GMAs firm operates in	American Stores	Kroger	Safeway	A&P	Winn-Dixie	Albertson's	Ahold	Publix	Vons	Food Lion
American Stores	(13)	–	15	23	31	15	38	15	0	8	0
Kroger	(25)	8	–	12	20	32	20	8	0	0	12
Safeway	(8)	38	38	–	25	13	75	0	0	0	13
A&P	(14)	29	36	14	–	43	0	29	0	0	21
Winn-Dixie	(15)	13	53	7	40	–	7	27	13	0	47
Albertson's	(11)	45	45	55	0	9	–	0	0	9	0
Ahold	(10)	20	20	0	40	40	0	–	0	0	30
Publix	(2)	0	0	0	0	100	0	0	–	0	50
Vons	(1)	100	0	0	0	0	100	0	0	–	0
Food Lion	(7)	0	43	0	43	100	0	43	14	0	–

Notes: * The denominator is the total number of GMAs in which the firm in the left-hand column operated; there are 54 GMAs. For example, American Stores operated in 13 GMAs in 1990, and had intra-market contacts with Safeway in three of these, that is in 23% of the GMAs in which it operated

Source: Adapted from Connor (1997)

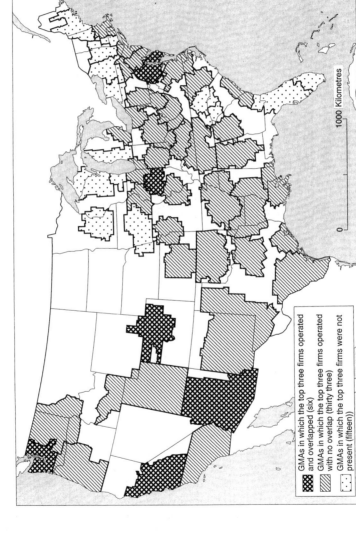

Figure 6 Grocery Marketing Areas (GMAs) Operated in by the Top Three US Food Retailers in 1990 and GMAs in which there was Intra-market Competition (Market Overlap) Between Two or More of those Firms. Source: Connor (1997) with Mapping and Secondary Analysis by the Author

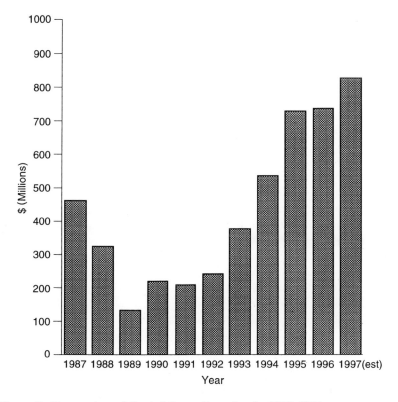

Figure 7   Kroger: Annual Capital Expenditure Levels, 1987–1997

form between these firms occurred in only six of those GMAs, and there were *no* GMAs in which all three firms significantly competed. As Connor (1997) notes, avoidance of such intra-market competition is a strategic option, open to spatially diversified firms, which can be termed 'forbearance'.

Reductions in Capital Expenditure

Figure 7, for the case of Kroger, illustrates the pattern of capital expenditure typical of many of the high-leverage food retailers. The significant reduction associated with Kroger's 1988 recapitalisation is apparent. Indeed, Denis (1994) has estimated that Kroger reduced its ratio of capital expenditure to total assets by 46% in the four years following the recapitalisation. Although there is evidence that Kroger may have substituted somewhat larger reductions in capital expenditure in place of asset sales, even those firms that divested considerable amounts of assets simultaneously reduced capital expenditure in the years which followed their high-leveraged transactions. Safeway, for example, reduced its ratio of capital expenditure to total assets significantly in the two years following its LBO. As a result, despite appearing to follow a strategy of 'removing leveraged-induced cash-flow constraints as quickly as possible through asset sales in order to increase capital expenditures' (Denis,

Table 5 The Annual Capital Expenditure Levels ($ million) of Two Leading US and UK Food Retailers, 1990–1997

Year*	1990	1991	1992	1993	1994	1995	1996	1997(Est.)
Kroger	220	208	241	376	534	726	734	825
American Stores	374	379	477	653	565	801	900	1000
J. Sainsbury	1209	1187	1220	1192	763	1176	1254	1163
Tesco	1476	1321	1012	1207	1195	1032	1175	1287

Notes: UK retailers capital expenditures converted at constant US$1.55 to £1
* Year ending 31 December (Kroger), and January (American Stores), February (Tesco) and March (J. Sainsbury) of following year

1994) it was not until 1990, four years after the LBO, that its capital expenditure to total assets ratio exceeded the immediate pre-LBO levels.

Figure 7 also illustrates the recovery in capital expenditure which has taken place since about 1993–4 amongst many of the high-leverage firms. Successful retirement of debt and refinancing to lower cost forms of debt has allowed these firms to increase investment in their store base and systems technology, and this, in turn, has been encouraged by increasing return on investment rates. Nevertheless, as Table 5 demonstrates, for two of these firms (Kroger and American Stores) in which the recovery in capital expenditure has been most pronounced, in international perspective US food retailing in the immediate post-leveraged restructuring phase was characterised by remarkably low levels of capital expenditure as the task of meeting debt-service requirements became all-consuming. Indeed, in comparison to the leading UK food retailers, Tesco and Sainsbury (see Wrigley, 1998c), these differences are dramatic. In the early 1990s, Kroger and American Stores operated between three and four times *more* stores than Tesco and Sainsbury, yet the average annual capital expenditure levels of the US retailers were in the order of just 20% of their UK equivalents. An important spatial consequence of attempting to spread such limited capital expenditure over large store networks was that certain markets in the United States became deprived of investment and suffered deterioration in their store base. Philadelphia provides an example (Wrigley, 1997). American Stores under-invested in its Acme chain, the market leader in Philadelphia, until the mid 1990s; and Pathmark, the second-ranked chain in Philadelphia, could only spend at levels which reflected its increasing post-LBO financial difficulties. The question remains, however, in what types of market could such under-investment by the leveraged food retailers occur without incurring the response from rivals and the sort of downward spiral in competitive position shown in Figure 1?

How Capital Structure Transformation Creates an Interaction Between High-Leverage Firms and the Decisions of Rivals

Committed to servicing their huge debt burdens by divesting assets and cutting expenditure programmes, the 'liquidity constrained' position of the high-leverage firms offered major opportunities to less-leveraged rivals. In particular, opportunities were created for aggressive market expansion, market entry, and predation via pricing, by these less-leveraged firms, with important consequences for the nature of competition in specific markets. Identifying the nature and extent of these interaction/market-rivalry effects of capital structure transformations has recently become an important focus of research in economics.[10] At a general level, it appears that changes in capital structure signal new behaviour to a high-leverage firm's rivals (a credible commitment by the firm to alter its investment behaviour), and that debt encourages leveraged firms to behave less aggressively in markets, to which rival firms respond by behaving more aggressively (Kovenock and Phillips, 1997). However, the extent of predation and, in particular, the nature of competition and pricing effects in markets (Chevalier, 1995a and 1995b) are difficult to determine with precision, one problem being that 'competitive price responses in multiproduct markets with non-identical firms serving non-identical consumers are difficult to predict' (Binkley and Connor, 1996).

Market Entry and Market Expansion by Rival Incumbent Firms

The major work in this area is that of Judith Chevalier. Beginning with existing theories of capital structure and market competition which posit that transformations of the capital structure of firms change the 'toughness' of market competition, Chevalier (1995a) examines the effect of debt on local market competition via an extensive econometric study of samples of the high-leverage US food retailers and their major rivals during the LBO wave. Her results suggest the following.

1. Announcement of an LBO or leveraged recapitalisation by a food retailer increases the expected future profits of that firm's rivals – a result consistent with the view that when a firm radically increases its leverage there is likely to be a decrease in the toughness of competition in the markets in which it operates.
2. Rival non-LBO (lower-leverage) incumbent firms in the market find expansion attractive.
3. Other large food retailers – the potential 'queue' of entrants in the sense of Cotterill and Haller (1992) – find entry into local markets dominated by firms that undertake LBOs attractive. However, because market entry takes time to organise, and can be expected to lag changes in market conditions more than expansion by non-LBO incumbent firms, this effect shows in Chevalier's empirical results only in the case of entry into local markets dominated by firms that undertook LBOs relatively early in the leveraged restructuring wave.

Pricing, Predation and Competition

In the second part of her study of debt and market competition during the LBO wave, Chevalier (1995b) extends her focus to the rather more contentious area of price competition and predation in local markets amongst leveraged and unleveraged chains. In this context, a widely held view is that high-leveraged firms are likely to pass some of the costs of servicing their debts to consumers in the form of increased prices, and that 'unleveraged competitors tend to follow the lead of leveraged chains with large market shares to increase their own profits' (Cotterill, 1993). However, as Figure 1 suggests and as Chevalier (1995b) notes, 'rival chains may not increase their prices following the LBO if the LBO presents an opportunity for predation'. Therefore, although prices may rise in some markets following LBOs, they may fall in other markets where predation is likely. The question is which are those markets and what are the likely competitive results?

Chevalier's view is that predation is only likely to be significant in markets in which relatively unleveraged rival firms exist that are financially unconstrained enough to engage in such activity, and in which at least one rival exists with a strong incentive to prey and enough stores in enough locations in the market to be able to do so successfully. Her extensive econometric analysis of data from over 70 US cities bears out this view. Her results suggest the following.

1   A high-leverage firm has incentives to raise prices following an LBO/leveraged recapitalisation, but those price increases will tend to be accommodated if rival firms in the same market are also highly leveraged.
2   Predation is more likely in markets in which a single relatively unleveraged rival exists with a large market share.
3   Price decreases, as a result of predation following an LBO, are associated with a reduction in the profitability of the local market for the high-leverage firm and an increase in the likelihood of market exit by that firm.

In accepting the overall thrust of these results, it must be borne in mind, however, that market area based price competition studies of the food retail industry pose some particularly difficult analytical problems. As Binkley and Connor (1996) have established using the same statistics from the Inter-City Cost of Living Index of the American Chamber of Commerce Researchers Association employed by Chevalier (1995b), two very different groups of commodities underlie any single index of metropolitan supermarket prices: a set of packaged, branded grocery products, and a non-branded set of fresh produce items. Such multiproduct markets allow discriminatory pricing, permit complex competitive responses, and produce unexpected rivalry effects – particularly in heterogeneous markets in which non-identical competitors serve specialised consumer segments. Binkley and Connor's (1996) work suggests that via the application of innovative 'non-cooperative oligopoly models' (Connor, 1996) to increasingly micro-based data sets, more complex pricing and predation effects amongst leveraged and unleveraged firms may ultimately be uncovered, reflecting the multifaceted competitive responses likely within multiproduct geographically segmented markets.

CONCLUSIONS

Although increasing attention has been paid to the geographies of money and finance by economic geographers in the 1990s (e.g. Corbridge, Martin and Thrift, 1994; Leyshon and Thrift, 1997; Clark, 1993b and 1997c; Tickell, 1994 and forthcoming; Laulajainen, 1998), encompassing themes ranging from the 'geopolitics of money' and the 'geographies of financial structure and financial exclusion' to the 'cultural geographies of money', much of this work has focused on what might be termed the 'geography of financial institutions, systems and markets'. Issues of corporate finance and how they shape the space economy have, for the most part, been ignored. Yet, as the financial leveraging of US corporations in the late 1980s demonstrated, and continues to demonstrate on some readings of the contemporary US economic boom, those issues have profound impacts, not least on the economic landscape. If, as Leyshon (1995b) has suggested, we have reached merely the 'end of the beginning' in the historical evolution of writing on the geography of money and finance, then premature closure of what is regarded as the agenda of that writing must be avoided and space claimed for critical interaction with some of the central theoretical debates of corporate finance and financial economics.

In this chapter, the capital structure decisions of firms and the spatial consequences of transformation of those capital structures have provided the focus. The use of debt financing to radically increase corporate leverage in one industry which became a prime target for high-leveraged capital transformations has been explored to illustrate how financial leverage translates into spatial outcomes. Capital structure transformations of the firm have been shown to have vital implications for the economic landscape, both directly, through the spatial reorganisation of the activities of the high-leverage firm, and indirectly, through the restructuring of markets by rival firms responding to the commitments implicit in those transformations. Geographies of divestiture, market consolidation and avoidance; issues of spatial predation, market entry, expansion and exit; and the complexities of competitive price response by rival firms in multiproduct geographically segmented markets are just some of the outcomes. The hope must be that the developing sub-discipline of the 'geography of money and finance' remains, as Porteous (1995) concluded, sufficiently 'wide open' to incorporate issues of corporate finance which clearly have profoundly important geographical implications.

ACKNOWLEDGEMENTS

This chapter forms part of wider research on the post-LBO reconfiguration of US food retailing, which is supported by the Leverhulme Trust under a 1997 Individual Research Award. The author is grateful to Gary Vineberg of Merrill Lynch, New York, Ed Comeau of Donaldson, Lufkin and Jenrette, New York, and Ron Cotterill of the University of Connecticut for on-going discussions on the nature of the US food retail industry. Thanks also go to John Connor of Purdue University for material on research in progress, and to Michelle Morganosky, University of Illinois, Robert Hutchinson, University of Ulster, Ron Martin, Colin Mason and Gordon Clark for constructive comments on the manuscript.

CHAPTER 10
LOCAL MONEY: GEOGRAPHIES OF AUTONOMY AND RESISTANCE?

Roger Lee

Monetary relations have penetrated into every nook and cranny of the world and into almost every aspect of social, even private life. (Harvey, 1982: 373)

Economic reasoning underpins a form of knowledge which has dominated the sociological analysis of modern society. But the supposition that individuals maximise cannot withstand closer inspection even in the one area of human activity to which it should most convincingly apply, namely the transaction of money. (Dodd, 1994: vii)

. . . and there is to this day a village in Scotland where it is not uncommon, I am told, for a workman to carry nails instead of money to the baker's shop or the ale house. (Smith, 1776: 18)

INTRODUCTION: UNIVERSAL MONEY?

Money is the most geographical of economic phenomena. On the one hand, it gains its credibility, or lack of it, from the places which produce it and regulate its circulation. On the other, it is a means through which value may be sustained, challenged and exchanged across space and time: its very existence is predicated on the space/time constitution of social reproduction. As a medium of circulation, money generalises value across countless commodities and so becomes necessary if generalised exchange is to be enabled as the norm. Thus, Adam Smith (1776) points out that in comparison to the direct exchange or bartering of values, money is convenient in its readiness for use – it avoids difficulties of division which can often make the exchange of values themselves impossible, and it gets around the problems of weighing and assaying metals which, precious or less than precious, may be used as a form of money. Money has the power to bridge space and time and, for Smith, has 'become, in all civilised nations (*sic.*), the universal instrument of commerce' (1776: 21).

This apparent universalism of money reflects its effectiveness in facilitating trade and barter. Historically, the increasing complexity of social divisions of labour within, and the geographical extent of, economic development have been associated with the emergence of various forms of money acting as a convenient means of exchange. Early monies often took the form of a physical commodity – something highly valued by the people who agreed to use it (e.g. iron nails in Scotland, dried cod in Newfoundland, salt in ancient Rome and corn in Massachusetts in the seventeenth century). Coins made of precious metals (especially gold and silver) came to replace these commodity money forms and, by the 1700s, paper money had been created. Each nation-state had its own official currency or officially sanctioned currencies, not least to try to maintain discipline and control over taxation and economic transactions (Ogborn, 1998). However, as late as 1913 (when the US Federal Reserve Act was passed) numerous local currencies – based on everything from timber to land – still existed in the United States. Yet such non-standard local monies tended to be marginalised in the industrialised economies as the norms conveyed by singular national currencies became ever more significant. Several decades later, as we enter the age of 'credit card capitalism', electronic money and the cashless society, it seems that the process of monetary universalisation has entered a new historic phase, that of 'stateless' global money (see Martin, 1994).

As Greco (1994) argues, 'the proper kind of money used in the right circumstances is a liberating tool that can allow the fuller expression of human creativity'. However, as he goes on to emphasise, money has not lived up to its potential as a liberator. This is due in part to strict state controls over its creation and to the political manipulation of its distribution. – not least through the powerful economic role of central banks charged with preserving the value of money and with designing and implementing policies to achieve that goal. Conventional money tends to be misallocated at its source, it goes disproportionately to those who already have lots of it, and it systematically redistributes wealth from the poor to the rich. The manifestations of the malfunctioning of conventional money are all too familiar: inflation, bankruptcies, unemployment, speculative bubbles, stock market crashes, increasing indebtedness and poverty.

There is an even more important influence at work here. In one sense, money is socially neutral – it is simply a highly effective mechanism to exchange and store value. Yet, in performing these tasks, money also carries with it the social norms and relations which provide the context in which economic activity may take place (Lee, forthcoming). As a result, the norms of evaluation sustained by these social relations are brought to bear on the conduct of economic activity. Within capitalist society, for example, the dominant evaluative norm is that of profitability. Activities failing to sustain levels of profitability acceptable to the owners and managers of financial capital are disciplined through evaluations of them made through increasingly globalised financial markets. On the basis of advice offered by highly sophisticated financial analysts, participants in these markets, such as pension fund managers, are capable of switching capital away from certain activities and places and towards others. One consequence of this lack of commitment to place is that the economic activities, the people and capital employed in them and the localities in which they take place can become disincorporated from, or re-incorporated into, the new circuits of socio-economic reproduction deemed acceptable by such evaluations.

Under these circumstances, it is, perhaps, hardly surprising that the very process of monetary universalisation, and the malfunctioning that tends to accompany conventional money, have stimulated the origins of what, given the space/time convergence facilitated by money, seems paradoxical, namely the re-emergence and growth of *local money*. It is with this re-emergence – a process occurring not for the first time (Davies, 1994) – that this chapter is concerned.

The next section discusses the apparent paradox of local money. This is followed by a discussion of the influence of the increasing geographical scale of conventional monetised economic transactions on the construction of the geographies of social and economic life. Extending this theme, the notion of 'monetary networks' which act as systems of information about the conditions under which monetary exchange is made possible, and the consequent distanciation of the 'community of money', are then discussed. Against this background, the chapter then moves on to consider the mutually formative influences of changes in the geographical scale made possible by monetary transactions, and the functioning of monetary networks on the relationships between local money and local economic and social geographies. The conclusion returns to the question of the relationship between local and non-local forms of money. Although local money and local economic and social geographies may be interpreted both as reassertions of autonomy and as resistance, they may also offer tantalising possibilities for the state in its seemingly growing determination to divest itself of welfare and social provision and its promotion of self-help by those excluded from the economic mainstream.

THE QUESTION OF LOCAL MONEY

Local money is money which is differentiated in some way, but particularly by relative geographical location, from conventional official currencies. Historically, local monies have often been physically distinguishable from the official, for example the nails Adam Smith referred to, cigarettes and tobacco in circumstances of deprivation within prisons or during times of material shortage exerted by the contingencies of war, or significant local commodities. In the contemporary era of universal currencies, however, one of the most rapidly growing forms of alternative local money is the local exchange and trading system (LETS).

The first LETS was created by the community of the Comox Valley in British Columbia in 1983 (see Meeker-Lowry, 1996). The concept involves a system of exchange of services and skills amongst the members of a community. The form of money is essentially the valuation of the various services and skills that these members provide for one another. The unit of exchange (e.g. the 'Green Dollar' in the Comox Valley case, or 'Tales' in Canterbury, England; for more examples, see Croall, 1997) need not have any physical existence as such and does not need to change hands. It may simply be a register of transactions, credits and debits of the services offered and rendered. The basic point is that the unit of exchange remains where it is generated and used, and provides a continually available source of local liquidity. Thus, the ultimate resource of a local community – the productivity, skills and creativity of its members – is not limited by lack of money. These resources themselves are the 'backing' behind the local money. Numerous local variations on this theme now exist across the United States, Canada, the UK and Australia.

There are numerous other 'local' monies (see Offe and Henze, 1992; Tibbett, 1997). Most of these adopt either a different form to that of the official currency or involve trading in commodities directly without the intermediary of money.

Perhaps the most general occurrence of 'local' money is that of 'earmarking' (Zelizer, 1994). Parcels of conventional money may be set aside for particular use, thereby allowing individuals and groups to segment their economic activities and so operate within different systems of value whilst continuing to use official currencies. This may simply allow different criteria of expenditure to be applied within a household budget but, more generally, it may also permit and encourage norms of economic activity unconformable with evaluations of the wider monetary networks and the financial systems associated with them.

Money (of any form) is rather more than a geographically constituted instrument for the storage and exchange of value. It is a *social construct* requiring a trust and belief in its efficacy amongst the community using it as a means of exchange, as a unit of account or a store of value, as well as an acceptance of the social values carried with it into economic activity. Local monies arise from the desire or necessity of a group or community to decouple from such values and their associated practices of socio-economic reproduction. In order to make a living, people must necessarily engage in the material processes of consumption, production and exchange. Yet, any such engagement is simultaneously a discursive as well as a material process (Lee, 1989 and forthcoming). It involves not merely the sustenance of consumption and production over space and time, but also the negotiation and practice of acceptable or imposed social relations of reproduction (Strange, 1988) which define the nature of value and so provide a framework within which the material practices of economic activity may be evaluated. As a result, in conducting geographically extensive forms of economic activity, people are necessarily drawn into social relations shaped by influences beyond the local. The ability of money and monetary criteria 'to penetrate every nook and cranny' of these economic geographies, as Harvey puts it, has the effect of transmitting, contesting and reinforcing particular norms and evaluations of them.

Thus, the move towards local monies and local community currencies is motivated not only by the desire to bridge the quantitative gap between what individuals earn and what they need to survive financially and socially, although this may, of course, be an important factor. It is also a community-building tool, a process of resistance against unacceptable values, a means of revitalising depressed and impoverished local economies and communities which have become marginalised from the monetary and economic mainstream, and a way of constructing alternative economic geographies founded on different social relations and the conceptions of value that they bring (Offe and Henze, 1992). Community currency encourages wealth to stay within a locality rather than flowing out to where conventional monetary returns are deemed higher. It provides a means by which local people can escape from the brute economic rationality of the market, and elevate need and use values above the exchange values of the conventional capitalist system.

The triple character of money – as a practical means of socio-economic transaction, as a community of belief and as a socially embedded process of evaluating economic activity and relations – involves the construction of a network of monetary relations articulated through the establishment not only of a spatial and temporal frame (conventionally, the territorial unit of the nation-state) within which money may be used, but also of a politically acceptable means of legitimating and validating that money, together with mechanisms or processes for the sharing of information about these conditions throughout the network. More specifically, such a network must be predicated on credible forms of economic regulation, whether formal or informal, through which the value and legitimacy of money may be maintained within socially acceptable limits – no matter how politically manipulated such limits, or their acceptability, may be.

Local or community money represents a dislocation of such relationships or, rather, a re-articulation of them, and therefore raises issues concerning the constitution of local economies and communities that go well beyond the functions of money in exchange. From the perspective of the logic of exchange implicit in the spread of easily convertible conventional forms of money over geographical space, local monies or community currencies appear to represent an anomalous resistance to the ever-widening spatial and temporal horizons offered by networks of conventional monetary relations. Yet, as already noted, there is a long historical geography of such alternative currencies (Rotstein and Duncan, 1991; Offe and Henze, 1992).

Local currencies work, by definition, as components of local socio-economies. However, because they serve the essentially practical purpose of enabling exchange between producers and consumers within locally grounded circuits of economic and social reproduction (their function as a local store – and especially as an accumulative store – of value is more problematic), they may also act as a basis of autonomy from, and even resistance to, dominant economic, social and political norms, and as a framework across which alternative sets of economic and social relations might be constructed (Offe and Henze, 1992). The creation, organisation and functioning of local currencies challenge the acceptance of 'official' currencies and so may have profound political effects, spilling over into – or even stimulated by – opposition to established, nationwide values and practices. Of course, given the power of dominant conventional monetary relations, the purely economic significance of alternative currencies is always likely to be relatively limited.[1] However, the immediacy and accessibility of local socio-economies, and the consequent widespread possibilities of participating in them, offer not only the possibility of autonomy and resistance, but also a practicable and material means through which individuals can participate directly in the socio-economic regulation and welfare of their own local community. In this sense, local monies offer a potential form of social policy with which to address questions of social, economic and spatial exclusion.

MONEY AND THE CONSTRUCTION OF ECONOMIC GEOGRAPHIES

The historical geography of money is one of a continual and increasing distancing of money from tangible representations. Leyshon and Thrift (1997) trace this process through five major monetary forms or practices:

pre-modern money, used primarily for specific purposes and mirrored today in the different interpretations of money as gifts, profits, wages, bribes, etc.;

commodity money, circulating mainly as coins and related in value to the precious metals from which they are made, usually in closely regulated mints;

money of account, linked to the need to escape the geographical and temporal confines of commodity money. Such geographies were the origin of the discount markets, whereby merchants would offer immediate but discounted access to commodity money. Within the extended problems of space-time distanciation created through the growth of international trade, bills of exchange along with the rise of merchant banks, initially enabled the issuance of IOUs for goods in transit, and subsequently became securitised as banks were prepared to convert them into their discounted monetary equivalent against the value of the commodities being traded;

state credit money, in which the state becomes the guarantor of public debts founded on its ability to issue money and to pursue monetary policy, as institutionalised though central banks and international regulators;

virtual money, or book entry money, facilitated by the development of electronic information technologies, and which represents the further development of fictitious capital.

The effect of the development of these money forms and practices has been to integrate local places and economies into ever wider circuits of monetary relations. These complex geographies, within which money is produced, transmitted and used, shape its meaning in terms both of the local cultural understandings and practices of the economic in the conduct of social life, and of the regulatory stances towards money. However, money is a product of geography not only because it is socially constructed in place, but also because it is about flows, circulation and transformation (Swyngedouw, 1996). Such transformations give meaning to money. Money offers a solution to the problem of how to represent and measure value in a standard fashion and, thereby, provides a language for social communication and a means of exchange across space and time.

Money and Local/Non-local Relations

The increasing universalism of money acts to bring places and times into closer connection (see Cronon, 1991; Leyshon, 1995a). As money facilitates exchange and so expands the potential spatial extent of economic relations, local monies increasingly come into contact with each other and so require mechanisms of exchange or convertibility, or even merger or takeover of one by the other. Such convergence between local monies further extends the geographical spaces over which values may be traded. As a result, the comparisons implicit in such geographically extensive exchanges become based on information from more distant places. The local circumstances surrounding the production, price or quality of values are, thereby, evaluated increasingly by criteria – including those associated with the uneven geographies of the timing of economic cycles – derived from non-local conditions. The extent to which values are formed by non-local circumstance reflects the geographical fungibility of money and the nature and spatial distribution of economic and regulatory power. Thus, the development of universal monies has simultaneously involved the erosion and obliteration of locally based valuation systems, and the imposition of externally based measures of value and exchange.

Conventional money both enables and constrains the construction of new socio-economic geographies, not least through its ability to transmit active information almost instantaneously across space and time. Monetary information is active in the sense that it both carries and evaluates knowledge of, for example, local economic circumstances and prospects. It thereby offers standardised templates of evaluation and value which constrain the possibilities of local alternatives. Comparative evaluations may well favour certain local economies, but, equally, may place others under more intensive competitive pressure. Not only that, but local producers and consumers may, with the spread of powerful forms of money (especially global monies), become tied to the practices and norms associated with their production and regulation, and so lose a further element of autonomy in the conduct of their own economic activity.

An excellent example of this is the debate on the likely effects of economic and monetary union (EMU) in Europe – a debate which is replete with examples of the national will to sustain autonomy. National currencies are linked to notions of national sovereignty and autonomy, even nation-hood itself. Joining the European Monetary System, and the single currency (the Euro), involves ceding monetary authority and control to a supranational authority (European Central Bank). Some see this as effectively giving up national sovereignty and identity. Others, in contrast, emphasise the potential benefits that membership of a wider, common economic and financial space will bring.

The optimists argue that the existing geographical differentiation associated with multiple currencies in Europe imparts problems of economic instability arising out of the massive volumes of foreign exchange dealings taking place across European currency markets and the transaction costs and financial unpredictabilities of multiple currencies in a region characterised by heavy cross-country flows of knowledge, money, capital and commodities. The pessimists, on the other hand, argue that inside EMU, national economic policy-makers will not only lose a lever of control over their own economic spaces but, as they will no longer be able to manipulate interest rates or exchange rates, they will also be forced to accept a monetary policy founded on European-wide conditions, with all the locally differentiated disjunctions (stemming from non-coincident business cycles, for example, or the relative stickiness of labour markets across Europe) that such a supranational scale of monetary regulation implies. In essence, the argument is that the imposition of a uniform and universal monetary space on highly differentiated national and regional socio-economic geographies not only removes a significant element of national and local autonomy, but is likely to collapse precisely because of this geographical differentiation.

Thus, the new economic geography of EMU may well be liberating for some, reducing barriers to trade and greatly expanding the size of capital markets. However, for others, it may be threatening, reducing their control over the circumstances in which they struggle to make a living or constraining their economic and political autonomy. In other words, what is at issue here is the contradiction that exists between the local social constitution and control of money in and through place, and its mobility – with all the interactions that this mobility may engender – almost independent of, or at least indifferent to, place.

As capitalist economic geographies become increasingly global and globalised, the geographical extent of monetary relations and the speed with which they may be entered into, become ever more important. Technical transformations in data processing and telecommunications and the liberalisation of monetary regimes in the face of globalisation and information technology permit almost instantaneous responses within money markets as it takes no more than a few seconds to conclude a deal between partners. Yet, precisely because of this collapse of space and time (Ohmae, 1990; Warf, 1995; Cairncross, 1997), trust in money requires a reliable system of monetary regulation. Ever since the emergence of state-based monopolies of official currencies in the nineteenth century, regulation has taken place primarily within *national* economic spaces through the control and management of *national* currencies. Such regulation is being displaced today both to the international sphere (the emergence of EMU is a prime example) and to (global) financial markets themselves. The latter gain their significance through their domination of the money supply in the form of credit which endows them not only with great regulatory power over money and its movement but, even more significantly, with evaluative power over circuits of economic and social activity.

Thus, with the adoption of money as a universal equivalent, local relations of trust in exchange, which are limited in extent by face-to-face contact, are displaced by abstract notions of value upheld by the reputation of banks and the wider financial system, and supported in turn by the state. The advantage of such a displacement is, as already noted, that exchanges can be entered into across great expanses of time and space. Local dependencies (for example on conditions of supply and demand and on local merchants) are broken down. However, the corollary of this is that the money system imposes a series of universal values derived from a multitude of exchanges emanating from distant places. Local economic geographies and communities are, therefore, pushed towards wider systems of norms, rationality and direction, and, as a result, become vulnerable to the effects of financial and economic crises that may originate in distant places.

In today's global economy and global monetary network, localities can thus become caught up in, and be severely affected by, financial devaluations and revaluations that occur on the other side of the world and which are wholly beyond local control or comprehension. In this sense, global money exerts a powerful exogenous influence on local economic relations and prosperity. Under such circumstances, official monies have few, if any, local obligations. This is not to say that official monies are immune to the exigencies of local places. Clearly, this is not the case. Money is sensitive to (perceived) local differences in economic return, and will quickly abandon one place for another. Of course, financial and currency crises still often have a distinctly local origin. However, it is precisely the *non-local* nature of modern money that enables it to move, virtually without effort, from place to place, paying scant regard to the local consequences. Financial crises originating in particular places may, therefore, be transmitted to other places irrespective of differences in local circumstances.

MONEY AND SOCIAL REPRODUCTION

In responding to the geographies in which it is formed, money acts as a powerful form of evaluation across space and time. In generalising exchange and reducing it to a common standard of measurement, money also performs a potentially universal measure of value. This universality is limited only by the geographical extent of belief in the ability of money symbols to represent real values. Such a 'community of money' (Harvey, 1985: 4) facilitates the incursion of exogenous forms of money into local and non-monetised social relations, thus displacing them and enabling and enforcing dependence on an unknown network of other social actors. As the material link between producers and consumers, money provides a social means of validating and valuing the expenditure of labour time and so provides a net which links, regulates and enforces participation in exchange and the calibration of production. Local circuits of social reproduction become linked into 'an identical system of market valuation' so that the 'reproduction of social life' may be procured 'through an objectively grounded system of social bonding' (Harvey, 1989b: 102). Thus, the advent of a money economy apparently dissolves 'traditional' society based on personal and concrete exchanges to 'one in which we depend on impersonal and objective relations with others' (Harvey, 1989b: 100). As exchange develops, money appears as a power external to, and independent of, producers. Money conceals the complex social reality of production and, in so doing, greatly increases the flexibility of the geographical organisation of production.

The permeation of money into most aspects of social relations, and its ability for geographical insinuation, implies the spread of a common system of valuation which forces conformity with social norms of worth and so makes autonomy from, or resistance to, such social norms – the propounding of (local) alternatives – extremely difficult. The introduction of money disrupts traditional patterns of exchange, enables the circulation of a wider range of products and asserts a new system of valuation. A recent dramatic example of such a revaluation was demonstrated by German reunification. The convoys of lorries transporting millions of Deutschmarks to banks in East Germany on the eve of re-unification in 1990 set in motion an adjustment process within (and well beyond) the eastern part of the country which has been profound, long-lasting, violent and, as yet, incomplete.[2] These difficulties reflect the fact that the inefficiencies of the East Germany economy were unable to compete with the West Germany economy at the rate of exchange agreed and universally applied in the reunification process.

Embraced by monetary relations which are widely accepted elsewhere (that is that the form of money being used is freely and fully convertible), circuits of social reproduction tend to conform with the norms and values transmitted by those monetary relations. The spread of money creates financial spaces conformable with wider systems of monetary valuation. However, although the spread of money in this way may well override local systems of valuation, it does not necessarily obliterate them. Indeed, ironically, the degree to which the variety of contemporary money symbols (notes, coins, plastic and electronic cash) are now accepted as self-contained forms of value actually opens up potential space for the creation of yet other, alternative systems of production and exchange based on local currencies outside the formal economy (Croall, 1997). More generally, the persistence and expansion of what Offer (1997) calls the 'economy of regard' – economic practices which govern a substantial part of social life, but which take place outside market relations and beyond the power of formal financial criteria – enables local monies and alternative meanings of money to be nurtured and sustained. Money is not merely a passive means of communication. Its possession endows social agents with the power to shape and influence geographies of economic accumulation and social reproduction. However, the apparent power of money to change geographies derives not merely from particular qualities of money itself or the forms of exchange that are enabled by it, but from the socially embedded *monetary networks* critical to its existence. As much as the facilitative power of money itself, it is the nature of these networks which conditions the possibilities of, and constraints on, local money.

Monetary Networks

Consider the differences between barter and monetary exchange. In barter, an individual or group sustains its place in a social division of labour and, hence, in a circuit of social reproduction by repeatedly seeking out a set of trading partners with whom it might barter the commodities that it has produced for all the other commodities it needs to sustain socio-economic welfare. A great deal of time and effort will be needed not only to locate such trading partners, but also physically to conclude the barter deals. What is more, information about, and assessments on the trustworthiness of, potential exchange partners have to be renewed each time bartering takes place.

Money, however, appears to dispense with this problem. 'It saves on the time and effort needed to search for potential co-transactors and to compromise or extend the relationship when the requirements of each transactor do not match' (Dodd, 1994: xxii). The mobility, divisibility and lack of particularity of money with respect to different forms of value also makes the practical business of exchange so much easier. You may receive money for your values and exchange it for other values in quite different proportions elsewhere and at a later date. Yet, what is happening here is not merely that exchange can take place in a far more flexible fashion over a larger geographical scale and over a much extended time-frame. It is also that these characteristics are enabled by a monetary network which makes monetary transactions possible and which creates and disseminates information allowing the perpetual extension and reproduction of exchange.

[M]oney does not carry or transmit the information required in barter, but replaces it with information of its own: that it can be reused in the future, that it will be accepted by other members of a society or social group, and that it truly represents its face-value and will continue to do so over time. (Dodd, 1994: xxiii)

However, this information is valid only within the geographical area over which the network extends. Thus, what is distinctive about monetary exchange is the network which comprises its conditions of existence.

The point about money in exchange, therefore, is that it displaces the need for information about the transactions themselves (information which is so essential to bartering) with information about the spatial and temporal extent of the network, its system of standardised accounting, the regulatory framework within which it operates, the legal framework of trading, and the behaviour and expectations of others in the network. If autonomous forms of local money are to be effective, all of these processes must be replicated within locally constructed monetary networks. Whereas in barter, geographical and temporal extension may reduce the degree of trust individuals may place in the system, in the case of monetary networks, expansion is normally an indication of its reliability, effectiveness and trustworthiness. Yet it is precisely this expansion beyond what is perceived as the 'local' that can threaten more localised monetary networks which do not have the resources or power of the more extensive networks. Furthermore, with the displacement of information from the commodities being produced and traded to the wider, non-local network of monetary relations facilitating exchange, the determinants of trade shift from the character and use values of commodities towards a geographically extensive rationality based on assessments of exchange value made through non-local (indeed, increasingly global) markets driven by effective (monetised) demand rather than by social need.

Trust in the money system which opens up this far less certain world – a world no longer capable of a degree of control by dealing directly with known individuals – has to be secured in other ways. One possible route is through the more-or-less well-articulated constitutional requirement of states to secure and maintain such trust. Another is through the geographically ever-more extensive systems of monetary regulation and control (Leyshon and Thrift, 1997). However, such responses may serve merely to substitute one framework of distanciated meaning for another. Another alternative is to attempt pro-actively to re-establish trust and control via the creation of alternative community currencies.

LOCAL MONEY AND ECONOMIC GEOGRAPHIES

Although the rise and circulation of non-local money may corrode local distinction and local determination, it should not be imagined that the logic of change is always in the direction of uniformity. It is possible for different monetary metrics – national (or global) and local – to co-exist and this possibility derives, in part, from the different scales at which economic production and social reproduction take place.

LOCAL MONEY AND MULTIPLE SCALES OF SOCIAL REPRODUCTION

On the basis of his survey of the rise of European capitalism, Braudel points to the co-existence of several 'layers' of economic exchange, to the fact that 'there were not one but several economies' (economic geographies), namely:

> ... the so-called market economy, ... the mechanisms of production and exchange ... and (of course) markets.
>
> ... lying underneath the market economy ... is that elementary basic activity which went on everywhere and the volume of which is truly fantastic. This rich zone, like a layer covering the earth, I have called ... *material life* or *material civilisation*. ... [A] proper term will one day be found to describe this infra-economy, the informal other half of economic activity, the world of self-sufficiency and barter of goods and services within a very small radius (emphasis added).
>
> [L]ooking up instead of down from the vast plane of the market economy, one finds active social hierarchies constructed on top of it: they could manipulate exchange to their advantage and disturb the established order. In their desire to do so ... they created anomalies, 'zones of turbulence' and conducted their affairs in a very individual way ... [t]his is where [capitalism] takes up residence and prospers. (Braudel, 1985: 23–4)

Whatever the veracity of this particular characterisation, the idea is a powerful one (and given greater significance by Braudel's recognition of the increasing significance of 'material life' in contemporary economic geographies). It suggests that circuits of economic and social activity cannot be reduced to singular constructions, and that circuits founded on 'local exchange' may be made to interact in macroeconomically purposeful ways.

In early seventeenth-century Spain, for example, the international geographies of circulation of large-denomination gold and silver coins forced the monarchy to behave competitively and not to seek short-term revenue to finance its wars. Local coinage was, however, a local monopoly – a 'petty geography' – and the monarchy raised seigniorage rates and issued large quantities in order to generate large revenues. The real stock of petty money grew, but was depreciated into some form of nominal equilibrium (Motomura, 1994). More recently, Perlin (1993) employs the phrase, 'the invisible city' to act as a metaphor for:

> the invisible realm of overlapping and intersecting processes which make up international and local exchange relationships, systems of surplus wealth distribution and appropriation, commercial and political connections. (Haldon, 1994: 309–10)

In such ways, local currencies facilitate local geographies of expanded economic reproduction and the surplus extraction that goes along with it. At the same time, local money may co-exist with higher-order currencies linked to non-local circuits of exchange. Local moneys reflect local practices of production and transference of value but may, in certain circumstances, be congruent with higher-order currencies. This is the case, for example, within the business practices of South Asia and China, where local circuits of social reproduction are able to connect with and co-exist alongside non-local exchanges between Asia and Europe (Haldon, 1994: 316–17). Thus, given the multiple scales at which economic and social reproduction takes place, the 'dissolution of difference' (Leyshon, 1996) that is supposedly engendered through the emergence of global financial markets and spaces may actually serve to open up opportunities for local alternatives.

Alternative processes of consumption, production and, above all perhaps, valuation, can combine in complex ways and at different geographical scales. There may be many reasons for this complexity – not least the active processes of inclusion/exclusion and empowerment/disempowerment that are inherent in capitalist accumulation. However, it may also be an expression of deliberate attempts to preserve or construct local circuits of exchange in parallel with, or even in place of, higher-order (national or global) circuits. Thus, local currencies and community monies may take on the nature of local political projects.

Local Money and Alternative Economic Geographies

The very idea of resistance may be a powerful force in negating the singularity and universalism of money. The contested gaze of the world economy and the resistant practices that may ensue (Hines, 1997) give rise to the emergence of alternative sets of social relations and, with them, alternative communities of local money. Economic geographies are culturally, socially and politically embedded, and so are continually constructed and reconstructed (Lee and Wills, 1997). Arguing that 'globalisation is an historically specific project of global economic (financial) management', McMichael (1996) traces the links between globalisation, its dislocative effects on state structures and local economies, and the consequential re-emergence of local self-determination:

> The erosion of state capacities to manage economic growth and welfare disorganises class coalitions formed around developmentalism, including the dismantling of public patronage systems. . . . As states decentralise, the opportunity for local political renewal presents itself, often quite compellingly. As global integration intensifies, the currents of multiculturalism swirl faster. Under these conditions, . . . the politics of identity tends to substitute for the civic (universalist) politics of nation building. Also regions and communities see self-determination as more than a political goal. It extends to the idea of cultural renewal, which includes recovering local knowledges. (McMichael, 1996: 42)

One aspect of this re-assertion of self-determination is the creation of community monies. Thus, Tibbett suggests that:

> [W]hile financial globalisation has gathered pace, there has also been a major increase in barter and alternative currency schemes of various types, from commercial barter schemes to community-based Local Exchange Trading Systems (LETS) and alternative paper currencies. (Tibbett, 1997: 127)

Although alternative currencies may arise out of local economic crisis and malfunction rather than resistance, there is, as Tibbett points out, 'a long tradition of dissent from the orthodox viewpoint about the nature and purposes of money' (1997: 129). Much of this dissent derives from the potential for contradiction between money as a store of value and as a means of exchange. Positive interest rates not only shift money from direct, reproductive investment towards indirect accumulation through speculation, but also increase costs and so reduce the level of economic activity and trade; whilst 'hot' money, lent short against high levels of interest, can undermine whole economies if long-term debts are called in (as occurred in the South-East Asian financial crisis during the late 1990s). Thus, the removal of restrictions on monetary flows, or the introduction of monetary integration which increases the geographical scale over which singular monetary regimes are practised, not only reduces local sovereignty and autonomy, but also stimulates the constant outflow and leakage of money from localities.

It is hardly surprising, then, that a growing proportion of people in countries such as the United States, Canada and Britain are participating in community-based schemes such as local cooperatives and LETS (Bayliss, 1998; Meeker-Lowry, 1996) and use non-official (alternative) monies; or that, for example, in Britain one-quarter of the adult population has no relationship with any formal financial institution. The emergence of alternative currencies tends initially at least to be stimulated where and when particular regions and groups experience exclusion from mainstream circuits of socio-economic reproduction. Thus, Tibbett (1997) argues that the current revival in alternative currencies and bartering has been stimulated in large part by a perceived shortage of the means of exchange in certain localities and amongst certain groups. However, the growth of such local monies – which allow the partial insulation of local communities and social groups excluded from formal circuits of finance by the effects of increasingly profit-driven and globally orientated forms of financial evaluation (Lee, 1998a) – is not merely a form of protection. They also provide a means of generating local employment (Mulgan, 1994). Where possible, it makes sense to substitute local production and labour for imports when the human resources and skills are available locally. Moreover, in today's world of free trade and hypermobile capital, when companies can close down local operations and switch jobs to countries where labour is cheaper, taxes lower or environmental regulations less restrictive, there is a compelling case for internalising a larger proportion of local production, enterprise and employment within individual communities.

In fact, the geography of LETS (Lee, 1996; Williams, 1996) reveals not only the influence of social exclusion, but also the desire to be excluded and to establish alternative forms of economic and social reproduction. Local currencies are also a means of political resistance, a vehicle for (re)constructing local communities in opposition to the dictates of global capitalism. They offer an opportunity to move away from valuing people's time and labour only in conventional (global) monetary terms, and to build community relationships within a system based on mutual respect and reciprocity (Meeker-Lowry, 1996). For such projects to succeed, it is necessary that communities understand the role that today's international product and financial markets play in shaping and controlling local economies, and how alternative locally based systems of production and exchange can be constructed and operated amidst these wider economic and financial spaces. For local communities intentionally to remove themselves from these larger markets in favour of creating stronger local ones is a political statement with implications and effects that extend beyond those localities. At the same time, such 'local currency activism' may serve to strengthen local social and community relationships and cohesion.

The practicality and accessibility of alternative local currencies in enabling exchange between producing and consuming individuals, along with the low cost of the mechanisms through which such exchanges may be tracked and recorded, increase the possibilities of at least the partial autonomy of such local projects from the logics and malfunctions of the general financial system.[3] The aim of community currencies is not to produce everything a locality or region needs; clearly, that is not possible. Likewise, LETS often involve the 'revaluing' of official currencies within the local community. For example, within what may be called 'mutual markets' (Lee, 1998b), official currency is used but in a 'local' or 'marked' fashion motivated by a desire to operate certain activities and to mobilise a range of local skills outside the conventions of market norms. Yet further, local currencies provide a means of revaluing the normal socio-economic valuations and exchanges associated with official money. Thus, in the Tomkins County, Ithaca scheme in the United States, each Ithaca Hour is equivalent to US$10, the approximate average hourly wage in that locality. These Hours are used locally for purchasing food, services (e.g. carpentry, plumbing and car repair), even childcare and to pay rent. However, although the general idea is to value each person's work equally, some groups (mainly professionals, such as lawyers and dentists) are allowed to charge a higher rate of Hours for their particular services (e.g. double or triple Hours) to reflect the highly specialised nature of their skills and the very high values they command in the normal monetary economy. Nevertheless, expressed in conventional dollar terms, those rates are still well below what they would be in the normal marketplace. In effect, through the use of Hours, the official dollar is revalued to fit in with the local system of exchange and values.

Beyond the Market: LETS as a Vehicle for Change

Motivations for participating in local currencies are diverse (Croall, 1997). Certainly, participants in two LETS in Kent, in the UK, during the mid 1990s had remarkably realistic notions of the significance of their involvement in material terms; no participant was earning more than 10% of their reproductive income and none saw the possibility of exceeding 25% of such income through LETS.[4] Nevertheless, one motive for commitment (for both individuals and institutions) to LETS is the possibility that they present a means for re-incorporating individuals into inclusive circuits of social and economic activity, and hence are tools for social change outside normal market mechanisms (Lee, 1996). They may be used as a 'soft' extension of those circuits (Williams, 1996) or as a way of expressing explicit counter-cultural alternatives (North, 1998). Yet others participate in LETS as a meaningful ecological gesture, as a way of reducing the pressure on natural and environmental resources. Certain LETS in Britain, for example, are part of a wider movement towards permaculture and sustainability. They operate not only to sustain and extend the labour of those engaged in production, but to act as a network of knowledge and distribution for the commodities so produced.

Cahn and Rowe (1992) argue that the world of the family and close friends, where people help and care for each other without thoughts of remuneration is rapidly disappearing. Whether or not such a world ever really existed, what is certain is that many of the local social services and benefits (including childcare, transport and care for the elderly and the disabled) have been monetised. People have perforce become purchasers of community and care rather than participants in providing them. What is more, the availability and distribution of the resources necessary for participation – time, living space or transport, for example – favour those already best able to pay for such care. The result is that the poor are unable to purchase or gain access to such services, many of which are of the basic welfare variety. The local 'Time Dollars' schemes that have sprung up across the United States (from Miami to Boston to San Francisco) represent a deliberate attempt to re-establish non-monetary community-based care and welfare systems. Many of these schemes, as well as a number of LETS schemes in the UK pursuing similar objectives, have been launched with the help of state and local authorities. Time Dollars have even been incorporated into conventional medical care systems that provide services that dollars alone cannot buy. Such local schemes are aimed at demonetising welfare so as to bring it within the reach of all members of the local community, regardless of their financial circumstances and resources.

However, social exclusion is not simply a question of ability to pay. Linkage into networks of communication and knowledge – of jobs, for example – is largely denied to those excluded from formal labour markets, whilst knowledge of, and ability to market, second-hand goods are increasingly commodified. At the same time, the spaces for re-integration into the wider economic geographies after separation or exclusion from them, for whatever reason, are limited when the search for shareholder value under stringent financial evaluations of economic activity dominates employment decisions, when voluntary work is formalised, and when 'care in the community' is used politically to justify reductions in public expenditure. Under such circumstances, LETS and other forms of local currencies may be part of a larger strategy to rebuild and assume a measure of self-determination in local and regional economies, of finding ways in which wealth generated in an area can be kept within it, and of providing an informally and locally determined space within which personal self-confidence may be re-established. The savings and earnings of rural places and urban neighbourhoods tend to flow out into a few international metropolitan financial centres. A local currency is, in part, a way that communities can retain some proportion of their wealth (as contained in skills and services), gain some measure of self-reliance and secure some independence from the national monetary system. As community currencies can be spent and traded only in the locality in which they are issued, goods and services that are dependent on resources from outside the locality still have to be paid for in conventional monetary terms. This can help people to assess the resources available locally against those imported from other places (Meeker-Lowry, 1996).

Of course, such local monetary schemes are far from unproblematic. For example, at what spatial scale are they best constructed? Is there some ideal geographical scale? Some proponents favour large regionwide systems; others advocate highly localised (neighbourhood-level) schemes. Equally important, is there not a possibility that LETS and similar local currencies will become just like 'official' currencies, just another commodity to be earned and spent? Will there not be increased pressure to differentiate the values of different people's time and skills, thereby threatening to undermine the equity principles that underpin such schemes? Should local businesses be encouraged to participate in local currency systems? What are the difficulties in valuing the material goods those businesses provide and will the participation of businesses tend to lead to the resurfacing of conventional producer and consumer behaviour and values? What are the consequences for local business remaining outside such schemes and what is the impact on local formal employment of the emergence of a strong business-orientated local currency? More generally and, perhaps, most crucially, what processes of 'othering' shape inclusion and exclusion in local community-based currencies, and how are internal relations of power and control to be negotiated within them?

Conclusions: Local Resistance or Local Sustenance?

This brings us back to the beginning: with local money operating as an expression of local social relations and the meanings that they endow on social life. Harvey concludes his historical geography of money by suggesting that:

> we can best interpret the different forms money takes – the money commodity, coins, convertible and inconvertible paper currencies, various credit moneys, etc – as an outcome of the drive to perfect money as a frictionless, costless and instantaneously adjustable 'lubricant' of exchange while preserving the 'quality' of the money as a measure of value. (Harvey, 1982: 251)

In enabling exchange over space and time, money contributes towards a non-local division of labour and so helps shape the uneven geography of development, which in turn underpins the geography of international monetary relations: it expresses its own inherent geography. However, this geography is more complex than such an interpretation allows. Almost because of 'the drive to perfect money', the social use of local money inserts new and possibly disruptive and subversive geographies into prevailing circuits of social and economic life. Yet, these local monies themselves represent highly diverse, and possibly temporary, alternative ways of organising local economies and communities.

Perhaps this is the key issue – that the significance of local money lies primarily in the fact that it is *local* in space and time, rather than in the particular economic and social functions that it performs, or whether or not it might challenge an increasingly insistent and universalising global money. The functions that community currencies perform are, in truth, no different from those of normal, official money. What is different is the context – the social relations – in which those functions are performed. The challenge of globalisation is not so much that it has unleashed an ineluctable force of socio-economic convergence, homogenisation and uniformity which erodes local difference and self-determination. Instead, it is that the increasingly global standards of economic evaluation exerted through financial markets are producing a fragmentation and hybridisation of the socio-economic landscape (e.g. UN, 1998). Socio-economic and spatial inequalities have widened substantially. At the same time, fuelled by their neo-liberal commitment to global competition and monetary austerity, nation-states are actively seeking to curtail and recast their public service and welfare provisions to local communities (World Bank, 1997). Such conditions have served to stimulate the re-assertion of local social, economic and civic self-determination. LETS and similar local currency schemes have a potentially important role to play in this process of the re-assertion of the local.

However, ironically, the growth of LETS and similar alternative community monies, and the local social support systems they involve, may themselves become attractive to national governments wishing not only to reduce expenditure on social security, but also to rewrite the relations between welfare and the state, to 'flexibilise' work and to promote notions of 'social entrepreneurship' (Bayliss, 1998). As an increasing proportion of jobs become temporary and part-time and as the mobility of capital to seek out labour becomes increasingly global, not only is the gap between work-rich and work-poor households likely to grow, but the socialisation offered by work is likely to fragment. Under such circumstances, social reproduction may increasingly take place outside the private and public sectors. The accessibility of local circuits of social and economic activity founded on locally organised alternative currencies then takes on a much greater significance in sustaining, rather than resisting, the dominant order, for example in helping to reduce the costs of labour power and, at the same time, offering forms of welfare and related social services at no cost to the national state.

The thrust of contemporary governmental rhetoric is increasingly (and ever more universally) aimed at the local decentralisation of social regulation and the need for local communities to assume greater responsibility for their own local economic prosperity, employment and welfare. Notwithstanding the loss of tax revenue and social security contributions implied by non-monetary LETS-type schemes, the potential social welfare gains for the state could be substantial. By being incorporated as a means of helping to sustain local social order in the face of an increasingly fragmented world of employment and incomes, LETS schemes may be drawn in to aid the state in its own welfare-reducing endeavours. Not only do local monies present a challenge to dominant forms of social relations, they also offer a form of socio-economic organisation all too complementary to prevailing discourses on the restructuring of welfare states and the fragmentation of work.

Part V

Money and the Retreat of the State

CHAPTER 11

THE HYPERMOBILITY OF CAPITAL AND THE COLLAPSE OF THE KEYNESIAN STATE

Barney Warf

INTRODUCTION

During the late twentieth century, the global economy has witnessed a series of startling transformations (Dicken, 1992): enormous technological changes unleashed by the microelectronics revolution, trade agreements such as World Trade Organisation (WTO), the North American Free Trade Agreement (NAFTA) and the European Union (EU), mounting international competitiveness, the rise of the East Asian newly industrialised countries (NICs), and an explosion of financial services linked through telecommunications.[1] These events have markedly altered the constraints and priorities of national and local governments. The welfare state, long the bedrock of Fordist national economic policy, has been under assault worldwide, and the social contract between labour and capital has been firmly rewoven. Led by the United States and the UK, national governments have repealed numerous social programmes around the globe, including even the bastions of state welfarism in Scandinavia.[2] Throughout the EU, the shift to the Euro has induced attempts to reduce social spending, even in Italy and France, where social programmes have long been regarded as sacrosanct. This political transformation, often buttressed by right-wing governments, has accentuated the onslaught against the poor and working classes by corporate élites, often using the neoclassical argument that the global economy is necessarily Pareto-optimal.

Simultaneously, a fundamental reworking of the relations between capital and space has taken place, a change that underpins, accompanies and reinforces the contraction of the Keynesian welfare state. A new geographic flexibility, an ability of firms to move across local and national borders with ever greater ease, can be readily contrasted with the traditional lack of mobility (as expressed in many textbooks in terms such as 'locational inertia', or a calculus in which transportation costs figure prominently). Freed from many of the technological and political barriers that hampered it, capital has become not merely mobile, but *hypermobile*. As Castells (1996 and 1997) emphasises, the 'network' society of contemporary capitalism, dominated by a 'space of flows' rather than a 'space of places', has led to new political formations, forms of identity and spatial associations. Swyngedouw (1989) makes a similar argument for the positionality of regions in a volatile, post-modern 'hyperspace'. In climbing out of the crisis of Fordism, capital has replaced the Keynesian national 'spatial fix' (Harvey, 1989b) with a highly fluid, globalised counterpart. (Although the spatial fix argument reeks of crude functionalism, it does capture the deeply geographical level at which social and political processes are constructed and played out.) At the local level, the global economy has led to heightened competition among places for capital, a process generally manifested in popular calls for a 'good business climate', deregulation, privatisation, tax concessions, subsidies, relaxations of environmental controls and reductions in social expenditures. In short, the emergence of a globalised, post-Fordist production regime has accompanied, and at the same time been underpinned by, the collapse of the national Keynesian state and its replacement with a series of localities intent on auctioning themselves off regardless of the costs.

It must be stressed that this transition is far from irreversible and is hardly propelled by some teleological inevitability. Indeed, much of what appears to be the annihilation of Keynesian welfarism is in fact a change to different forms of state intervention (Martin and Sunley, 1997). Military spending in the United States, for example, resembled Keynesianism in its effects on national aggregate demand (Markusen, 1986; Markusen and Yudken, 1992). Precisely because the shift to post-Keynesianism is not pre-determined, this process is politically charged and highly contested in different ways and by different groups, including the politically critical middle classes.

The origins of this transformation lie in the 1970s and 1980s (Thrift, 1986; Lash and Urry, 1987), with the collapse of the Bretton-Woods agreement in 1971; the end of the gold standard and subsequent shift to floating exchange rates in 1973; the oil crises of 1974 and 1979 and the ensuing worldwide recessions; the explosive growth of Developing World debt, largely driven by recycled petro dollars; the deterioration in the competitive position of industrial nations, as reflected in their growing trade and/or budget deficits; the concomitant rise of newly industrialising countries; the emergence of 'flexible' specialisation, the microelectronics revolution and computerised-production technologies; the global wave of deregulation, privatisation and the lifting of state controls in many industries; new trade regimes such as the NAFTA, the EU and WTO, which free capital to move across national boundaries; and the integration of world financial markets through telecommunications systems. In the 1990s, one might add to this nexus of structural forces and changes the collapse of the Soviet bloc and the steady integration of formerly 'socialist' nations, including China, into the world economy.

This chapter focuses on the relations between the accentuated mobility of capital and the contraction of Keynesian systems of state regulation. First, it argues that post-Fordism gave financial capital the capacity to move effortlessly across the globe – the latest chapter in the 'annihilation of space by time' that has accompanied the historical emergence of capitalism. Second, it argues that this new-found flexibility was largely responsible for the reduction of Keynesian systems of regulation and the rise to hegemony of conservative factions preaching the wonders of the free market. Third, it traces the 'spatial fix' that capital offered in place of the national regulations of Keynesianism, a series of localities in heated competition with one another for capital with little regard to the costs or consequences. The conclusion attempts to bring these themes together in light of their geographic relevance.

Post-Fordism and the Global Hypermobility of Capital

A central part of the late-twentieth-century global economy has been an explosive growth of information and financial services, which are often celebrated as a principal harbinger of the so-called 'Fifth Kondratieff' (Hall and Preston, 1988). The rapid escalation in the supply and demand of information services has been propelled by a convergence of factors, including dramatic cost reductions in information-processing technologies induced by the microelectronics revolution, national and worldwide deregulation of many service industries, such as the Uruguay Round of the General Agreement on Tariffs and Trade (GATT) negotiations, and the persistent vertical disintegration that constitutes a fundamental part of the emergence of post-Fordist production regimes around the world (Gill, 1989; Hepworth, 1990).

The increasing reliance of business and financial services, as well as numerous multinational manufacturing firms, on telecommunications to relay massive volumes of information through international networks has made telecommunications a fundamental part of regional and national attempts to generate a comparative advantage (Gillispie and Williams, 1988). The reliance on such technologies corresponds with the increasingly information-intensive nature of commodity production in general (necessitating ever larger volumes of technical data and related inputs on financing, design and engineering and marketing). Such technologies also enable the spatial separation of production activities in different nations through globalised subcontracting networks. The decreases in the price of communications technologies, the high price- and income-elasticities of demand, and the rapid development rate of these services have all promoted their proliferation. In addition, the high level of uncertainty that accompanies the international markets of the late twentieth century (to which the analysis of large volumes of data is a strategic response) is a key driving force (Hepworth, 1986 and 1990).

Telecommunications facilitated new volumes of inter-regional trade not only in information services, but also in capital (Thrift and Leyshon, 1994). The origins of contemporary global capital markets lay largely in the flourishing Euro market of the 1960s, propelled to new heights by the persistent US trade deficits and petro dollars of the 1970s. Originally, the Euro market comprised trade in assets denominated in US dollars but not located in the United States; today, it has spread far beyond Europe and includes all trade in financial assets outside of the country of issue (Martin, 1994). Unfettered by national restrictions, the Euro market has been upheld by neoclassical economists as the model of market efficiency. One of the Euro market's prime advantages was its lack of national regulations. Indeed, US banks invested in the Euro market in part to escape the restrictions of the 1933 Glass–Steagall Act, which prohibited commercial banks from buying and selling stocks. Further, the Euro market lacked any reserve ratio requirements until 1987, when the world's central bankers met at the Bank for International Settlements in Basle, Switzerland, and agreed on global reserve standards (the so-called 'Basle Accord'; Kapstein, 1994).

In the context of post-Fordism, Harvey argues that:

> Flexible accumulation evidently looks more to finance capital as its coordinating power than did Fordism. This means that the potentiality for the formation of independent and autonomous monetary and financial crises is much greater than before, even though the financial system is better able to spread risks over a broader front and shift funds rapidly from failing to profitable enterprises, regions and sectors. Much of the flux, instability and gyrating can be directly attributed to this enhanced capacity to switch capital flows around in ways that seem almost oblivious of the constraints of time and space that normally pin down material activities of production and consumption. (Harvey, 1989b: 164)

Today, banks and securities firms have been at the forefront of the construction of an extensive network of electronic funds transfer systems that have come to form the nerve centre of the international financial economy, allowing banks to move capital around on a moment's notice, arbitraging interest rate differentials, taking advantage of favourable exchange rates and avoiding political unrest (Langdale, 1989; Warf, 1995). Reuters, with 200 000 interconnected terminals worldwide through systems such as Instinet and Globex, alone accounts for 40% of the world's financial trades each day (Kurtzman, 1993: 47). Other systems include SEAQ in London, Soffex, the Swiss Options and Financial Futures Exchange, the Computer Assisted Order Routing and Execution System at the Tokyo stock exchange, and TSE and CATS in Toronto. Such networks give banks the ability to move money around the globe at stupendous rates (the average trade takes less than 25 seconds); subject to the process of digitisation, information and capital become two sides of the same coin. Supercomputers used for that purpose operate at teraflop speeds, or one trillion computations per second.

In the securities markets, global telecommunications systems have facilitated the emergence of 24-hour trading, linking stock markets through computerised trading programs. Accordingly, trade on the New York stock exchange has risen from 10 million shares per day in the 1960s to more than 1 billion per day in the 1990s. The volatility of trading, particularly in stocks, has also increased as hair-trigger computer trading programs allow fortunes to be made (and lost) by staying microseconds ahead of (or behind) other markets. Heightened volatility, or the ability to switch vast quantities of funds over enormous distances, is fundamental to these capital markets; speculation is no fun when there are no wild swings in prices. In short, the ascendancy of 'electronic' money has shifted the function of finance from investing to transacting, institutionalising volatility in the process.

Liberated from gold, travelling at the speed of light, as nothing but assemblages of zeros and ones, global money – Harvey's (1989b) 'fictitious capital' – performs a syncopated electronic dance around the world's neural networks in astonishing volumes. The world's currency markets, for example, trade roughly US$800 billion every day, dwarfing the US$25 billion that changes hands daily to cover global trade in goods and services. Every two weeks, the sum of funds that passes through New York's fibre optic lines surpasses the annual product of the entire world

(Kurtzman, 1993: 17). Salomon Brothers, which routinely buys 35% of US government bonds, runs the equivalent of the nation's total bank holdings through its computers every year. CS First Boston, the world's leading bond trader, trades more money each year than the entire GNP of the United States. The Chicago Mercantile Exchange trades roughly US$50 billion daily, while the New York bond market trades something in the order of US$150 billion daily (Kurtzman, 1993: 77).

National borders mean little in this context: it is far easier to move US$1 billion from New York to Tokyo than a truckload of grapes from California to Arizona. As Kurtzman notes, money 'may start the morning as dollars, travel through the exchange markets of London to emerge as German marks by midday, and wind up in the afternoon in Chicago as an option contract on an index of 500 stocks' (1993: 92).

As large sums of funds flowed with mounting ease across national borders, national monetary policies became increasingly ineffective. In a Fordist world system, national monetary control over exchange, interest and inflation rates is essential. In the post-Fordist system, however, those same national regulations appear as a drag on competitiveness. In any event, national control has been markedly reduced by the global markets of hypermobile capital. 'As private capital began increasingly to circuit globally on a deregulated basis, Keynesian nation-states progressively lost control of one of the most important macroeconomic levers – the setting of interest rates' (Peck and Tickell, 1994: 291). In the United States, for example, the Federal Reserve, alarmed about the prospect of inflation, raised the reserve ratio of banks seven times in 1994 and 1995, only to find that its control over the national money supply had diminished to the point of virtual irrelevance. Worse still, national monetary policies may work at cross-purposes; for example when Germany's Bundesbank sought to dampen inflationary pressures by lifting interest rates in the early 1990s, it angered French policy-makers, who found their own rates rising in tandem (the French, however, feared rising unemployment, not inflation). This is not to say that nation-states enjoy no leverage whatsoever over such factors, but the steady convergence of short-term interest (but not profit) rates worldwide undoubtedly reflects the ease with which capital transcends national borders.

The mobility of capital on a global scale since the Second World War has been accentuated by the generalised reduction in protectionism, including the GATT (the primary mechanism used worldwide to minimise national threats of protectionism), and the WTO, as well as a series of regional trading blocs such as the EU and NAFTA (Gibb and Michalak, 1994). Fordism and multilateralism can be viewed as mutually reinforcing systems; the GATT, after all, played a major role in the post-war boom by facilitating international trade. In this light, multinational corporations could be seen as the result, not the cause, of multilateral market liberalisation. Similarly, the emergence of post-Fordist production regimes has been accompanied by regional trading blocs. The ability of global capital to undermine existing trade regimes led to widespread disenchantment with the GATT, to which regional trade blocs are a strategic response. Thus, trade policies reflect the complex interplay of global capital and nation-states: regional trading blocs are attempts at state (re)regulation at the international level as well as attempts by multinational firms to preserve flexibility on a continental scale (Gibb and Michalak, 1994; Michalak and Gibb, 1997).[3]

The transformation to global, financially propelled post-Fordism has accentuated the status of 'world cities', including London, New York and Tokyo, as well as, to a lesser extent, Paris, Toronto, Los Angeles, Osaka, Hong Kong and Singapore (Sassen, 1991). Each of these cities seems to be as attuned to the rhythms of the global economy as to the nation-state in which it is located. Although other cities certainly can lay claim to being national cities in a global economy, these world urban centres have played a disproportionate role in the production and transformation of international economic relations in the late twentieth century. It is important to note that urban specialisation based on finance is not a new phenomenon – Amsterdam was the 'Wall Street' of the seventeenth century (Rodriguez and Feagin, 1986) – but the magnitude and rapidity of change in such markets is without precedent. London, for example, has become largely detached from the rest of Britain (Budd and Whimster, 1992; Hutton, 1995). Similarly, New York rebounded from the crisis of the mid 1970s with a massive influx of petro dollars and a prolonged bull market on Wall Street (Mollenkopf and Castells, 1991; Shefter, 1993). Tokyo likewise is one of the world's largest centres of capital accumulation (Fujita, 1991). In each city, a large agglomeration of banks and related firms generates well-paid jobs; in each, high incomes for a wealthy stratum of traders and professionals have sent real estate prices soaring. In the recession-plagued 1990s, these cities have exhibited severe vulnerability to the oscillations of the global economy, including corporate downsizing, widespread lay-offs, falling real estate prices and increasing poverty.

Telecommunications also threaten the agglomerative advantages of these dense urban regions, particularly the reductions in cost and uncertainty accomplished through face-to-face communications. The North American Securities Dealers Automated Quotation (NASDAQ) system, for example, has emerged as the world's fourth largest stock market; however, unlike the New York, London or Tokyo exchanges, NASDAQ has no trading floor, and instead, connects millions of traders worldwide through telephone and fibre optic lines – in essence a 'placeless' market (see Chapter 6). Similarly, Paris, Belgium, Spain, Vancouver and Toronto have abolished their trading floors in favour of screen-based trading. Such events point to the relatively short – and decreasing – time horizons within which places may enjoy comparative advantages; for example, Markusen and Gwiasda (1994) argue that New York's pre-eminence reflects only its historical experience, not a current comparative advantage.[4]

The new, hypermobile capital markets also include emerging centres of off-shore banking, a reflection of the shift from traditional banking services (loans and deposits) to lucrative non-traditional functions, including debt repackaging, foreign exchange transactions and cash management. The growth of off-shore banking, usually in response to favourable tax laws, has stimulated banking in such previously unlikely places as Panama, Bahrain and the Netherlands Antilles. Given the extreme mobility of finance capital and its increasing separation from the geography of employment, off-shore banking can be expected to do relatively little for the nations in which it occurs. For example, Roberts (1994) illustrates the case of the Cayman Islands, which made its restrictions on capital flows exceptionally lax. However, the economic benefits to the Cayman Islands are marginal: now the world's fifth largest banking centre in terms of gross assets, more than 600 foreign banks employ only about 1000 people (fewer than two each).[5]

Another reflection of hypermobile capital is the globalisation of clerical functions through the internationalisation of back offices, which perform low-wage, unskilled tasks such as data entry and claims processing. Off-shore offices are established not to serve foreign markets, but to generate cost savings for firms by tapping labour pools where wages are much lower than in the home country. Notably, many firms with off-shore back offices are in industries facing strong competitive pressures to enhance productivity, including insurance, publishing, and airlines. Off-shore back-office operations remained insignificant until the 1980s, when advances in telecommunications such as transoceanic fibre optics lines made possible greater locational flexibility just when the demand for clerical and information-processing services grew rapidly. The mechanics of off-shore offices are relatively simple. Such functions may be either subsidiaries of multi-national firms or operate under contract with the home-based businesses. Inputs, usually documents or magnetic tapes, are sent by air to off-shore processing facilities (for example, via Federal Express). After processing, generally a few days at most, the results are returned via air couriers, dedicated telephone line, satellite or fibre optics. Thus, the capital investments in such operations are minimal.

Off-shore back offices have sprouted up rapidly in various parts of the world. Several life insurance companies have erected back-office facilities in Ireland, with the active encouragement of the Irish Government (Warf, 1995). Despite the fact that back offices have been in Ireland only a few years, Irish development officials are already worried, with good reason, about potential competition from Greece and Portugal. Likewise, the Caribbean has become a particularly important locus for US back offices, partly due to the Caribbean Basin Initiative instituted by the Reagan Administration, and the guaranteed access to the US market that it provides. Most back offices in the Caribbean have chosen Anglophonic nations, particularly Jamaica and Barbados. American Airlines has paved the way in the Caribbean through its subsidiary Caribbean Data Services (CDS), which began when a data-processing centre moved from Tulsa to Barbados in 1981. In 1987, CDS opened a second office near Santo Domingo, Dominican Republic, where wages are half the level of those in Barbados (Warf, 1995). Hence, the same flexibility that allowed back offices to move out of the United States can be used against the nations to which they relocate.

THE COLLAPSE OF KEYNESIANISM

The relations between Fordism–Keynesianism and the global economy were always problematic. Essentially, the global economy under Fordism consisted of a series of linked national economies, not a seamless world devoid of barriers to the flow of capital. 'One of the fundamental tensions of the Fordist regime was the uneasy interface between *national* forms of regulation and the *globalising* dynamic of accumulation' (Peck and Tickell, 1994: 289, emphasis in original). As capital circulated on an increasingly global scale, and with drastically shorter time horizons, the Fordist–Keynesian production/regulation regime collapsed, with profound political effects. De-industrialisation, recession, a slowing of productivity growth and a deteriorating trade balance set the stage for the rise of increasingly powerful, politically conservative coalitions. Put differently, the emergence of the post-Keynesian state reflects the inability of the welfare state to adjust to the realities of late-twentieth-century capitalism (Martin and Sunley, 1997). Annual real productivity growth in the United States, for example, fell from 2.4% between 1945 and 1973 to 1.4% between 1974 and 1994. In this context, social programmes initiated in the 1960s, such as the War on Poverty, Great Society and Social Security, when expectations of continued growth made their fiscal burden seem small, became intolerably expensive, with costs driven further by the surge of military spending in the 1980s.

Unable to tame the lions of inflation, unemployment and interest rates in the 1970s, Keynesian interventionism became increasingly discredited, and, with it, numerous social-democratic political parties and programmes. Ideologically, the conservative factions and parties that took advantage of these circumstances preached an interlocking web of notions centred on the faith in the free market and a distrust in state intervention, including supply-side economics and the 'trickle down' of the benefits of economic growth from rich to poor. This shift was particularly evident in the United States and the UK. In the UK, monetarism took precedence during the Thatcher Administration in the 1980s, whereas in the United States, monetarism jostled with Keynesian fiscal policy during the 1980s under the leadership of Paul Volker at the Federal Reserve Bank and the ideological justification offered by supply-side economics. In the name of restoring national competitiveness, in national political discourse considerations of equity quickly gave way to those favouring efficiency.[6] By the 1990s, throughout the West, orthodox welfarist prescriptions had been thoroughly discredited. The need to substitute an 'enterprise culture' for the debilitating dependence of welfarism was linked to a determination to deflate the swollen, fiscally profligate public bureaucracy (Jessop, Bonnett, Bromley and Ling, 1988). Closely associated with this was a continuing infatuation with tax cuts, including Proposition 13 in California and Proposition 21/2 in Massachusetts in the late 1970s, both of which reflected middle-class frustrations over escalating property taxes, and the Reagan Administration and Thatcher Government's corporate and personal income tax reductions of the 1980s.

A central facet of the emerging post-Keynesian state was deregulation. In the United States this included the Securities Acts Amendments of 1975, the Airline Deregulation Act of 1978, the Motor Carriers Act of 1980, the Depository Institutions Deregulation and Monetary Control Act of 1980, and the break-up of AT&T in 1984. Banks enjoyed a relaxation of inter-state banking regulations (Holly, 1987), the removal of restrictions governing pension and mutual fund portfolios, the abolition of fixed commissions on stock market transactions, and the approval of foreign memberships on stock markets. In the UK, the 'Big Bang' of 1986 signalled the start of a wave of deregulation of financial markets and institutions. The consequences have been dramatic. The removal of state controls opened the way for new competitors, which often entailed declining profits in several industries (including aviation, shipping and trucking), and new sources of investment funds (such as pension and mutual funds). The stock markets, fuelled by an influx of foreign capital, saw dramatic gains in New York and London.

Conservative *laissez-faire* dogma also led to widespread efforts to sell government assets and privatise public services. The ideological origins of privatisation can be traced to Milton Friedman, the most celebrated neoclassical advocate of the wonders of market competition. By selling off inefficiently operated publicly owned and operated assets and services (e.g. by contracting out), privatisation allowed governments to provide services without producing them. The UK is regarded as the innovator of privatisation, largely due to the much higher levels of nationalisation in the UK than occurred in the United States. Efficient economic management of industry was to be entrusted to the business acumen of the entrepreneur, not to public bureaucrats. In the UK by the mid 1990s, half of the publicly owned shares of industry had been transferred to the private sector (see Chapter 13; Gaffikin and Morrissey, 1990). British Telecommunications, for example, was sold *en masse* to private investors. Similarly, in Japan, the break up from 1985 onwards of Nippon Telegraph and Telephone opened up its long-held monopoly to competition (McIntyre, 1994). Privatisation in the United States was less extensive than in the UK, although the 1987 sale of the US Federal Government's share of Conrail was an important step in this direction.

Thus, Harvey (1989a) and Leitner (1990) tie the rise and fall of the Keynesian state to the conterminous disappearance of Fordist production; the Keynesian state was largely legitimated by the benefits of Fordist production, particularly the provision of collective goods that depended on a continuously rising productivity of labour. As productivity growth declined, political agendas for restructuring the economy by cutting state funds for collective consumption gained support under the credo of 'restoring competitiveness', the transfer of social surplus from the sphere of production to reproduction became politically popular, and right-wing criticism of those most dependent on welfare (e.g. the poor, women, minorities, immigrants and the elderly) increased. Examples of this include the Republicans' Contract with America, current 'welfare reform' and workfare, and attempts to reduce Medicare and Medicaid benefits.

In short, the transition from national to global markets made the viability of nationally based state welfarism questionable. In place of nationally based Fordist–Keynesian regimes, capitalism has become increasingly structured around a series of subnational localities relatively unfettered by state bureaucracies, in which each place is free to auction itself to the highest bidder. Hidden within this transformation, however, is a fundamental reworking of the relations between capital and place, one that sets hypermobile capital against increasingly vulnerable localities (Block, 1996: Chapter 23).

LOCALITIES IN THE AGE OF HYPERMOBILITY

Amid the generalised resurgence of interest in localities by geographers and others, Cox and Mair (1988) articulated a powerful thesis of local dependence, in which locality criss-crosses class relations, tying workers and firms to places through intricate webs of relations and the local reproduction of social relations. According to this view, firms find themselves effectively trapped by their reliance on a particular client base, their own sunk capital, the immobility of investments in the built environment and the non-substitutability of local exchange linkages. Labour, too, is tied to places through home-ownership and dense networks of personal ties. Mutually reliant on the same place, various forms of capital form local coalitions, often with the active support of politicians and labour, to promote local 'boosterism' in their competition with other places.

However, the degree of this spatial 'fixity' is unequal as between capital and labour. Local dependence is not a universal feature of capital; indeed, 'footlooseness' is a prominent theme among many readings of economic landscapes, especially those concerned with the most influential sectors. In the context of hypermobile capital, local dependence becomes problematic to the point of losing its meaning altogether. Many financial firms, such as mortgage banks and insurance firms, do rely heavily on place-specific investments and clients. However, for large money-centre banks, securities firms, commodity brokers, foreign exchange speculators and other institutions whose business extends globally, not locally, dependence may not be so advantageous. The very nature of finance capital, with its tenuous ties to the landscape, redefines local dependence, revealing it as one contingent way in which capital relates to places, rather than as a universal necessity. Giddens notes that:

> The vast extension of time-space mediations made structurally possible by the prevalence of money capital, by the commodification of labour and by the transformability of one into the other, undercuts the segregated and autonomous character of the local community of producers. (Giddens, 1981: 121)

Similarly, Logan and Molotch make the point that:

> Capital becomes difficult to trap because it dissolves, moves, redefines its internal relations, transforms itself into something else. Unlike the experience for people, 'homelessness' serves capital well. Homeless money is liquidity, and liquidity is an advantage, not a tragedy. (Logan and Molotch, 1987: 252)

Places – and people – are always at a definite disadvantage when confronting financial capital: local dependence is an asymmetric relation that can work against them as well as for them in inter-community competition. Indeed, in an era in which markets can be linked in microseconds, the notion of local dependence as a universal structural feature of capital–labour relations is outmoded.

Desperate for jobs, many localities vie for one another with ever greater concessions to attract firms, including foreign ones, in an auction resembling a zero-sum game. The effects of such competition are hardly beneficial to those with the least purchasing power and political clout, often leading to pressures from employers and local governments to lower local wages and inhibit union organisation. Thus, every economic and political initiative is interpreted in the light of its ability to 'generate jobs'. The luxury of local resistance is increasingly unsustainable. Sympathy for the poor, the homeless and the disadvantaged melts away under discourses that emphasise 'self-reliance' and 'individual initiative'. Unions, beleaguered by decades of de-industrialisation, offer little alternative. Left to auction themselves to the highest corporate bidder, localities find themselves in a 'race to the bottom' in which entrepreneurial governments promote growth – but do not regulate its aftermath – via tax breaks, subsidies, training programmes, low-interest loans, infrastructure grants and zoning exemptions. As Peck and Tickell note:

> Workforce training, the erosion of social protection, the construction of science and business parks, the vigorous marketing of place and the ritual incantation of the virtues of international competitiveness and public–private partnership seem now to have become almost universal features of so-called 'local' strategies. In this sense, the local really has gone global. (Peck and Tickell, 1994: 281)

In public policy terms, the transition to a post-Fordist, hypermobile regime of accumulation has induced a switch from 'managerial' to 'entrepreneurial' planning (Fainstein, 1991). Governments, national and local, have become less concerned with issues of social redistribution, compensation for negative externalities, provision of public services, and so forth, and have become obsessed with questions of economic competitiveness, attracting investment capital and the production of a favourable 'business climate'.[7] Logan and Molotch (1987) have emphasised the growth of connections between local governments and private capital promulgated under the rhetoric of 'growth coalitions', alliances obscured by the pretence that privately led and market-driven growth is inherently optimal for all parties concerned. In the same vein, Fainstein (1991) has shown that as planners have come to involve themselves more directly in economic development, market rationality and local competitiveness have replaced comprehensiveness and equity as the primary criteria by which planning projects are judged. Thus, long-term capital budgeting, master planning and a concern for the environment have gradually given way to short-term concerns of job generation, looser regulations and tax relief. Thus, planning is concerned more with promoting development and less with regulating its effects.

Local successes, however, are bound to be transitory. Indeed, if nation-states lack the power to set economic policy in the face of hypermobile capital, localities are sure to fare much worse.[8] The competition among localities is one in which a few benefit but most find themselves virtually powerless. Such a devaluation of the power of places – the precise opposite of the local dependence thesis – leads ultimately to the negation of even the strongest places. Thus, the more capital cooperates, the more places compete. As Peck and Tickell suggest, 'the more vigorously localities compete with one another, the more pronounced their subordination to supra-local forces becomes' (1994: 304).

Conclusions

As numerous authors have noted, late-twentieth-century capitalism has entered a new phase, one that unintentionally emerged from capital's search for a solution to the pressing problems of the 1970s. Globalised, 'flexible' (Knudsen, 1996) and 'disorganised' (Lash and Urry, 1987), this regime of production is characterised by a renewed aggressiveness of corporations and right-wing political factions against the working classes and poor of the industrialised (and industrialising) world. Part of the sea-change was capital's acquisition of a spatial freedom qualitatively greater than that characteristic of manufacturing. The fluidity of finance capital on a global basis revolves, on the one hand, around widespread deregulation; it is precisely the lack of regulation at the global level that allows capital such flexibility (Michie and Greive-Smith, 1995). On the other hand, this mobility reflects the introduction of telecommunication networks and the digitisation of information, which give many firms markedly greater freedom over their locational choices, as is evident in recent studies of back offices and off-shore banking. In short, in the late twentieth century, finance capital is not simply mobile, it is *hypermobile*. The hypermobile world is one in which large sums of money never appear as commodities, but are perpetually ensnared in a 'paper chase', whereby capital chases capital in a continual surge of speculative investment that never materialises in physical, tangible goods. In dramatically reducing the circulation time of capital, telecommunications have linked far-flung places together through networks in which trillions of dollars move instantaneously across the globe, creating a geography without transport costs.

The transition to global post-Fordism ushered in a fundamental reworking of the relations between capital and space. In the deregulated, volatile and hypermobile environment of the late twentieth century, capital's new 'spatial fix' has sharply accentuated the power of capital to set communities against one another. Telecommunications allow banks and insurance firms, for example, to play off region against region and centre against centre. The reduction in technological barriers has been augmented by trade agreements that reduce the political restrictions to capital (but not labour) mobility. Of course, threats to leave localities are not new in the history of capitalism (Pfister and Suter, 1987), but never before have firms been able to carry them out with such effectiveness. Nevertheless, it would be simplistic to claim that hypermobility renders space unimportant, *contra* the claims of O'Brien (1992) and Ohmae (1990). Indeed, the fact that both Fordism/Keynesianism and post-Fordism/post-Keynesianism (if these may be linked so bluntly) have varied in nature across and within national borders testifies to the significance of geography in their construction and unfolding. Further, the durability of world cities as centres of capital accumulation reflects the significance of place, if not space.

Hypermobility has not eliminated the differences among places nor equalised their competitiveness, but it has pressured national and local governments to accommodate global capital in ever more earnest ways. Deregulation has levelled the regulatory 'playing field', but other factors, including pools of expertise, specialised information and institutional thickness have risen in significance (Thrift and Leyshon, 1994).

The emergence of hypermobile capital poses serious questions for the traditional (i.e. Keynesian) functions of the nation-state. State regulation and control over capital flows remains essential for commodity production. Given the enormous increases in volatility and liquidity that banks and securities firms have acquired, and the huge quantities of funds that flow freely across national borders in the world's stock, bond, currency and other markets (which dwarf world trade in agricultural and manufactured goods), the regulation of national money supplies is highly problematic, leading to diminished control over interest, inflation and exchange rates. Contrary to much received opinion, post-Keynesianism presupposes not the disappearance of the nation-state, but its re-articulation together with a redefinition of its functions. Indeed, rather than a simplistic contradiction that views global capital in opposition to the nation-state, the emergence of a global, post-Fordist economic system and a system of post-Keynesian states should be seen as mutually presupposing.

At the local level, this trend has disturbing implications for the internal politics of individual places. In the context of slow productivity growth and 'expensive' public services, right-wing groups and politicians preaching the gospel of the free market have acquired considerable influence, a discourse that explicitly privileges efficiency over equity considerations. It is little wonder, then, that Conservatives have enjoyed such popularity throughout Europe and North America, and even in strongholds of social welfarism such as Scandinavia.[9] Post-Keynesian policy prescriptions and the entrepreneurial state, centred on the master narrative of global neoliberalism, have done little but sharply accelerate the competition among communities for jobs and public and private investment, leading to a downward spiral, as a result of which few places reap the benefits of growth for long, and almost inevitably do so at the expense of someone, and some place, else.

It is critical to view these global and local transitions at different spatial scales as parallel refections of one underlying process. At the international level, this switch may be observed in the replacement of import substitution by export promotion, including export processing zones, which were pioneered in East Asia but have now been adopted around the world (Jenkins, 1991; Mitchell, 1995; Brohman, 1996). Locally, the triumph of capital over place is manifested in the entrepreneurial state and associated concessions to firms to reduce the costs of production. In this environment, the global and local topography of regulation becomes critical. Internationally, the clear winners so far have been global cities, whose vast tentacles of control extend to every corner of the world, and which have survived the latest round of global time-space compression, and the centres of off-shore banking, which have competed by offering low tax islands in the seas of global competition. Given the notorious instability of capitalism, however, and its perpetual tendency to melt 'all that is solid into air' (Berman, 1982), the only certainty of the next few decades will be that both of these sorts of places will find their own tenuous hold slipping away as capital moves on to greener pastures.

Chapter 12

The Retreat of the State and the Rise of Pension Fund Capitalism

Gordon L. Clark

Introduction

Hobsbawm (1994) refers to the twentieth century as a 'short century', dating its onset with the First World War and its conclusion with the 'end of the Soviet era'. He suggests that the early years of this century were dominated by events that owed their origins to nineteenth-century European imperialism, while the last decade of the twentieth century seems to be best understood as the beginning of a new era whose features and logic will only be fully revealed next century. Between 1914 and 1989, wars, revolutions, economic catastrophes and economic growth engendered extraordinary experiments in social and economic organisation, many of which are now thoroughly discredited by virtue of the lives lost forcing through their implementation.

What, if anything, has this to do with the retreat of the state and the private provision of urban infrastructure? The answer to this question is hard to summarise simply and precisely. Recognising that the chapter is devoted to explicating the argument, the author believes that many responsibilities and functions collected together under the banner of the 'welfare state' are either being returned to the private sector or are being systematically discounted in terms of their real value as governments underinvest or fail to maintain, budget-to-budget, the original value of those functions and entitlements. To illustrate, a recent report of the UK National Institute of Economic and Social Research (Pain and Young, 1996) showed that previous Conservative Governments failed to maintain the inherited stock of infrastructure, suggesting a policy of deliberate dis-investment in the basic capital goods of society. In a similar vein, it is also apparent that the value of state social security has declined in many countries. In the UK, this is largely the result of a switch in bench-marking the value of state-provided pensions from wage growth to changes in the cost of living.

It has been suggested that the change of a Conservative to Labour Government in the UK, and a possible shift from a Republican-controlled to a Democratic-controlled Congress in the United States, could reverse these trends. This is unlikely, notwithstanding the 'stakeholder society' espoused by Hutton (1995) and others. For some, it is difficult to believe that the policies and institutions of the last 50 years or so could wither and die. The welfare state was conceived in the darkest moments of the Great Depression and was forged as an institution in the aftermath of the Second World War. At its peak, its many functions and responsibilities literally institutionalised the social relations of Western societies. For many, the state was both the proper provider of public infrastructure and, given the vagaries of the market, the *only* institution capable of providing a comprehensive system of public goods. However, events of the last two decades across the Western world have shown that neither the normative nor the positive claims justifying the state provision of infrastructure are now compelling. Recent moves to radically reform (UK) and even dismantle (US) welfare programmes suggest, in fact, that the post-war consensus that legitimised state intervention in market capitalism is in tatters. The state is in retreat on many fronts, from providing the most obvious forms of infrastructure such as bridges and roads, to the less obvious forms of urban development, including employment and housing.

Set against these trends is another, equally compelling trend. Since the early 1980s, Anglo-American private pension assets have 'attained stupendous size and importance' (Langbein, 1997: 168), eclipsing all other forms of private savings and transforming the nature and structure of global financial markets.[1] This phenomenon is not found, however, in most continental European countries; in those countries, government-funded social security systems remain the central pillar of retirement planning, albeit threatened by bankruptcy (Clark, 1998a). In the Anglo-American economies, the growth of private pension assets reflects, in part, the rapid post-1950 expansion of employment and increased participation in employer or multi-party-sponsored private pension plans. The fact that private plans must be fully funded, that pension benefits have come to be an important component in many employees' wages and that the baby boom generation has moved into its peak earning years means that the net flow of assets to pension funds has become a tidal wave.

What are the prospects for urban infrastructure and development funding given, on the one hand, the retreat of the state and, on the other hand, the rise of 'pension fund capitalism'? This chapter documents and accounts for the retreat of the state, the phenomenal growth in pension assets and the prospects for pension fund investment in urban development. In essence, it suggests that the retreat of the state has been accompanied by a variety of displacement strategies shifting responsibility for many inherited nation-state functions to market agents and other tiers (higher and lower) of government authority. In the case of urban infrastructure and development, in part, the state has displaced responsibility for its financing and provision to the financial sector (its institutions and decision-making systems). In this respect, two issues are emphasised in the later sections of this chapter: first, pension fund trustee investment decision-making, and, second, increased social and geographical polarisation and the likelihood of greater accountability applied to public and private pension funds. Note that the chapter makes wide-ranging connections between countries, abstracting from particular institutions and economic trends. A fuller treatment would uncover the nuances and differences the author ignores by virtue of the logic of his argument.

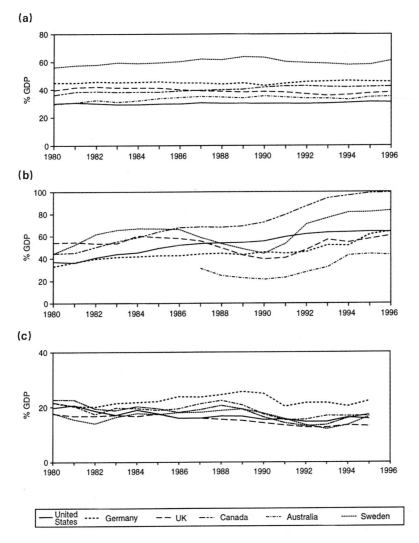

Figure 1 Fiscal Constraints and National Savings for the UK, Australia, the United States, Canada, Germany and Sweden, (a) General Government Current Receipts; (b) General Government Gross Financial Liabilities; (c) Gross National Savings, 1980–1996. Source: Organisation for Economic Cooperation and Development (1997b)

STATE INCOME AND EXPENDITURES

To set the scene, Figure 1(a), (b) and (c) summarises patterns of state receipts and liabilities for the four principal Anglo-American countries (the UK and Australia, the United States and Canada), Germany and Sweden from 1980 to 1996. It has been argued elsewhere that Anglo-American countries are closely related in terms of their financial markets and the inherited common law tradition which is the basis of their regulatory institutions (Clark, 1998a; La Porta, Lopez-de-Silanes, Shleifer and Vishny, 1997). The other two European countries are included for the purposes of contrast. Over the last two decades, it appears that the Anglo-American states have had very similar flows of receipts as a proportion of GDP. In this respect, these countries are different from Germany and significantly different from Sweden. However, compared to the United States it is apparent that the state's share of receipts has slightly increased in Australia, decreased in the UK and significantly increased in Canada. A second observation is that state indebtedness, having stabilised and even declined during the 1980s, surged in the early 1990s as a consequence of the global recession.

Perhaps not surprisingly, three of the four Anglo-American states' shares of national income are either constant or declining. The exception is Canada, which increasingly appears more like Sweden in terms of its expenditure profile than the other Anglo-American countries. Indeed, its recent poor economic performance is more akin to the continental European countries than its Anglo-American cousins.[2] In all cases, as the flow of state revenue closely tracks national economic performance, economic downturns tend to add significantly to states' indebtedness as state expenditure is maintained or even increased. Comparing state receipts to expenditures, the latter is more volatile than the former, especially over the early to mid 1980s. More recently, however, and not withstanding the recession of the early 1990s, expenditures have been less volatile as states have gradually cut into the rate of growth of their revenue and, especially, their expenditures.

The macro patterns of revenue and expenditure are revealing. Even more revealing is the decreasing discretion of nation-states with respect to their allocation of expenditures to competing budgetary categories. Figure 2 summarises budget outlays for the UK, Australia and the United States, comparing 1980 to 1994. Where possible, categories have been merged and re-labelled to facilitate comparison. For the UK, three trends are apparent. First, the categories of social security, health and education, although important in 1980, dominate 1994 expenditure allocations. Entitlement expenditures are also of increasing importance for Australia and the United States. Second, defence and 'other functions' have taken the brunt of budget re-allocations in the UK, just as defence has been squeezed in Australia and the United States. If there was a 'peace dividend' from the end of the Cold War, it has been swallowed up by burgeoning entitlement expenditures. Nevertheless, US defence expenditures are still important when compared to the other two countries. Third, in the UK housing, transport and public safety consume practically all that is left of the public purse. This is also the case in Australia.

Much has been written about the rapid growth in entitlement expenditures, and how their growth in the context of limits on revenue and total expenditure have conspired to radically reduce governments' spending discretion. There is no need here to belabour the point. However, it is important to acknowledge an implication which follows from the apparent squeeze on the available public resources in the Anglo-American world. The capacity of governments to switch significant resources from current committed expenditures to new needs is small, and often only symbolic. Inevitably, government departments and the public services they

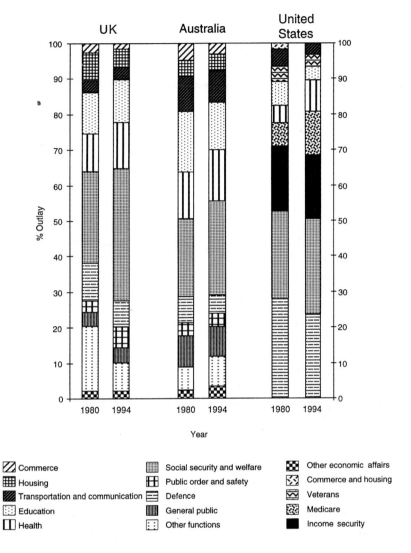

Figure 2 Where it Goes: Changes in Budget Outlays for the UK, Australia and the United States, 1980 and 1994. Source: Organisation for Economic Cooperation and Development (1997b) and US Budget

represent must campaign year-to-year to hold their past budget allocations let alone to maintain their real value. Their capacity to garner significant new resources to fund investment in areas of need such as transport, employment, housing and urban infrastructure is caught in a cruel vice affecting all government expenditures. In the UK, at least, this has resulted in a long-term decline in infrastructure investment, and has greatly contributed to the wholesale privatisation of many government enterprises and services. Furthermore, there is no evidence that the underlying financial constraints will change in the near future, whatever the political orientation of the party in government. The recent election of the New Labour Government confirms the significance of these constraints as well as the increasing conservatism of electoral politics.

There is a complex and highly inter-connected relationship between economic growth, state receipts and expenditures. This relationship is also an essential constraint on state capacity. The logic is as follows, recognising that its actual manifestation varies between countries as macroeconomic circumstances vary.

1. On the revenue side, the close relationship between the flow of state revenue and national growth and international economic competitiveness means that any new resources must be derived from either higher than average rates of economic growth or higher levels of deficit expenditure.
2. All the evidence suggests that higher than average rates of economic growth are followed by episodes of recession, so that expenditures based on additional receipts due to higher than average rates of economic growth are often unsustainable.
3. Deficit financing requires higher interest rates to attract private capital, thereby limiting the rate of economic growth and subsequently cutting into the flow of receipts. As a consequence, declining receipts tends to add to state indebtedness given the impossibility of avoiding committed entitlement expenditures.
4. In the long term, as current revenue and expenditure objectives have dominated treasury planning, unmet investment needs have accumulated to the point at which whole sectors (including the utilities) and crucial functions (e.g. transport and urban infrastructure provision) have been so impoverished that their poor performance threatens the rate of economic growth.

One implication which follows from this analysis is that high interest rates made necessary by government debt financing also affect currency exchange rates with international trading partners. Although short-term currency speculation is likely, over the longer term high relative exchange rates tend to affect the competitiveness of traded goods and services, setting in motion the possibility of a slow-down in the rate of domestic economic growth. Thus, in this context, massive urban infrastructure investment projects, such as rebuilding and re-equipping the London Underground, can be seen as too expensive for any government to finance over the long term. The alternatives are dis-investment, user-charges, privatisation or some form of private investment and public partnership. The Keynesian state we inherited after the Second World War is increasingly the 'pauper state' (Froud et al., 1997).[3]

THE PROBLEM OF THE COLD-HEARTED VOTER

A simple-minded argument would be that the state is in retreat because of declining resources in relation to contemporary needs. However, this seems to imply that state resources have actually declined, whereas in fact, state receipts have expanded step-by-step with economic growth. In some instances, states have gone beyond available resources, expanding expenditures to either sustain inherited functions and/or gain short-term political advantage. One consequence has been increased indebtedness. Another consequence has been global financial markets' speculation regarding government budget policies, interest and currency exchange rates. As suggested above, there are long-term economic limits to state indebtedness. In this respect, the hard political lesson of the 1970s and 1980s has been that '[e]xcept for the briefest of periods, in conditions of dire crisis or brief and dramatic opportunity, the fulcrum of political realism for any modern society is always an understanding of the conditions for its economic flourishing' (Dunn, 1990: 7).

The retreat of the state is a product of three interlocking forces. Notwithstanding the growth of state income, politically inspired attempts at expanding state expenditures beyond that justified by the rate of growth often set in motion macroeconomic forces which ultimately threaten the stability of national economic performance. Significant changes in budget priorities, including new budget expenditures on urban infrastructure, must be found more often than not in the re-allocation of current and expected resources. This uncomfortable fact has been exacerbated by the rapid increase (compared to the rate of economic growth) in resources allocated to entitlement programmes. Many states have been unable to limit the rate of increase in social security and health expenditures which are driven, in part, by the ageing of their populations. Furthermore, the priority assigned in many countries to recurrent programme-related expenditures as opposed to long-term investments has led to significant short-falls in infrastructure provision and even dis-investment in the basic infrastructure of Anglo-American economies. The retreat of the state is a combination of reduced fiscal discretion and narrowed functional scope.[4]

Reduced fiscal discretion and functional scope are also profoundly related to the current practice of democratic politics. For many voters, accustomed to a paternal and relatively resource-rich state, reduced fiscal capacity has translated into declining standards of public services, increased loads on the available services and even the abandonment of public service provision in favour of market provision and individual responsibility. At every turn, when service deterioration has been seen to be driven by dis-investment, it seems that the only available option has been privatisation. In the UK, this is evident in the energy sector, the water sector, railways, airports and many other related infrastructure-dependent public services (see Chapter 13). In some cases where privatisation has proved impossible, secondary markets have been created to allow those with more income to by-pass the existing state-based rationing system and to purchase services directly. Although clearly a policy that leads to gross inequalities in terms of the timing and quality of overall service provision, for previous Conservative Governments it was a step towards greater competition within the state sector and the possibility of declining or, at least, constant per unit service costs. The experience of the UK National Health Service over the last 10 years exemplifies this argument.

Twenty years ago, analysts debated the causes of the fiscal crisis of the state (O'Connor, 1973). Various left-of-centre and right-of-centre explanations of the over-burdened state were proffered. Some analysts on the left focused on the necessary, increasing functions of the state in relation to the crisis-ridden nature of capitalist economies. This argument was justified, more often than not, by reference to the Great Depression and the immediate post-war era of economic reconstruction. Those on the right tended to focus on rent-seeking behaviour, arguing that there was a close and even exploitative (in relation to the taxes paid by voters) relationship between state apparatuses and their 'clients' in the private sector.[5] The increasing scope of state functions, coupled with the close relationship between the state and private sectors by virtue of the increasing significance of state expenditures encouraged theorists to suppose there was a near equivalence or balance between the state and markets. When linked with the widespread use of long-term planning and management in the state and private sectors, Shonfield (1965) heralded the advent of a new era of 'corporatism'.

If corporatism represented the Anglo-American world of the 1960s, Reagan and Thatcher attacked and destroyed its legitimacy. In Clark and Dear (1984: Chapter 3) a different logic is suggested, working from consensus-oriented functions, through economic reproduction functions and social integration functions, and finally executive functions. The supposition is that each type of function has its own clients and institutional forms by virtue of the resources and powers associated with those functions – an argument not dissimilar from those of analysts who have focused on rent-seeking behaviour. Although not specifying a hierarchy of functions, Clark and Dear emphasise those related to consensus building, arguing that the nature and status of democratic politics plays a crucial symbolic role in legitimising the unequal distribution of resources. It is assumed that functions related to economic reproduction are essential for the maintenance of state power, and that social integration functions come after satisfying consent and economic reproduction. Reassessing this logic in the light of recent history, it seems that the state remains pre-occupied with consensus and the conditions of economic reproduction. The most vulnerable functions, however, have been those related to social integration.

Dispute over the legitimacy of the state presupposed an alternative, even if idealised with respect to the communist world of the time. It is now apparent that the end of the Cold War has both legitimised the capitalist state in relation to any alternative and has discounted the claims of many poorer voters for state resources. Consensus can be sustained without a wide range of social integration functions. Although many state functions were inherited by current governments from the post-war welfare state, rationalising and discounting the value of those functions dominates democratic politics. The historical legacy of the Great Depression and the immediate post-war era has been forgotten by the ascendancy of a new generation of middle-class voters with little knowledge of those circumstances. In fact, many of those living in the major centres of the Anglo-American countries believe that the state as an economic agent has become a long-term burden on the international competitiveness of the private sector. Even though privatisation has often been prompted by unmet capital needs, governments of all political persuasions suppose that privatisation is necessary to improve the performance of the private sector itself. Thus, states' economic functions have been redefined, in part, in relation to the imperatives of the emerging global economy (*contra* Boyer and Drache, 1996).[6]

There are, of course, some who argue for increased nation-state taxes to sustain the inherited range of state functions, or at least those functions that are able to garner the support of a broad spectrum of voters. Yet, despite the force of their argument (Hutton, 1995, in the UK; Kuttner, 1997, in the United States), they have had little success in countering the movement towards the narrowing of states' functional scope. At the most general level, arguments for general tax increases in order to support the inherited scope of state functions have been discounted by counter-arguments regarding the probable negative effects of tax increases on long-term economic growth. At the same time, polling data suggests that many voters do not trust governments to use new revenue in ways that would enhance the standards of existing public services. Given chronic shortfalls in operating revenue in state sectors such as health and education, it is apparent that any new revenue derived from any general tax increase would simply be absorbed into current expenditure. Furthermore, in an era of low inflation and low rates of individual income growth, many middle-class voters consider taxes to be a burden on their real income. There is a political presumption against general tax increases, recognised as such across the political spectrum (Glynn, 1996).[7]

It is also obvious that middle-class voters tend not to support state expenditures except in areas that directly affect their well-being. In part, middle-class voters increasingly perceive themselves to be 'consumers' of public goods and expect a level of service consistent with the quality of service available in the market for related and even competing goods. In part, too, middle-class voters are increasingly hostile to cross-subsidies or transfers of state revenue to other groups of public service consumers. In this respect, the functional coherence of the welfare state is under heavy attack from the fragmentation of the electorate into rival groups of 'public goods' consumers all of which seek to maximise their share of existing resources while discounting the claims of those perceived to be net costs to the state. Accentuating this trend has been the emergence of the identity politics of multiculturalism and post-modernism as opposed to the politics of class (the origins of the welfare state).[8]

In this new world of the cold-hearted and isolated voter, the functional scope of the state is dependent on spontaneous issue-based coalitions rather than collective solidarity sustained by long-term class alliances. Political fragmentation threatens the coherence of state functions, replacing the substance of social cohesion with the symbolism of consensual agreement. In this new political world following the Cold War, the nation-state has actively sought ways to displace inherited responsibilities to the market or to higher or lower tiers of government organisation. If state legitimacy is threatened by these displacement strategies, such threats are, more often than not, fractured by deep divisions in civil society about the proper conception (let alone design) of state responsibilities and institutions. State legitimacy is secure.[9]

THE RISE OF PENSION FUND CAPITALISM

Writing in 1965, Shonfield argued that 'modern capitalism' was dominated by large economic organisations, some public and others private. He described a world in which collaboration was common between the public and private sectors, and dispute resolution was managed between nominally equal partners (including unions). In comparison with the economic world prior to and immediately after the Second World War, modern capitalism was, as a consequence, believed to be more stable and predictable. Economic volatility had been tamed by state macro-economic policy-making, and the increased significance of planning and management in all large enterprises. By his account, the modern state could claim equal standing with industry by virtue of its control of banking and finance, in some cases or 'the existence of a wide sector of publicly controlled enterprise' (1965: 66) in other cases. He argued that the state was a major economic agent because 'government's expenditure has been enormously enlarged', therefore directly affecting 'a large segment of each nation's economic activities' (1965: 66).

Thirty years later, and notwithstanding political rhetoric to the contrary, the modern state is a much smaller fraction of Anglo-American economies. In part, this is because of the economic and political limits imposed on state revenue and expenditures. In some cases, such as the UK, the smaller role of the state reflects the significance and scope of privatisation since the mid 1980s. At the same time, the economic structures of all of the Anglo-American economies have been more or less altered by the relative decline of employment in national industrial and manufacturing sectors set against the rise of firms in international information-intensive industries. For some, this is the new world of post-Fordism; a common, global movement towards flexible production and accumulation.[10] For others, including Webber and Rigby (1996), the transformation of Anglo-American economies over the last 30 years reflects the inevitable capitalist dynamics of capital accumulation and uneven development. Here, however, the author emphasises a distinctive aspect of the transformation of Anglo-American economies over the last 20 years which is often only accorded the briefest of mention, namely the rise of 'pension fund capitalism' and the world of finance with which it is intimately associated (Clark, 1997b and 1998a).

To illustrate, Table 1 summarises UK personal financial assets in nominal terms for 1980, 1990 and 1996.[11] Absolute values are less important than the relative changes between categories over the entire period. In this respect, three investment categories stand out in terms of their relative growth rates: pension/life assurance assets (ten-fold increase), unit trust assets (two-hundred-fold increase) and UK securities (ten-fold increase). By contrast, bank deposits (seven-fold increase), building society deposits (four-fold increase), and national savings (five-fold increase) registered significantly lower rates of increase. These patterns are also apparent in the United States for household assets (Table 2). Up to the end of 1994, US retail deposits increased less than three-fold, bonds of various kinds five-fold, corporate securities about three fold, pension/life assurance assets more than six-fold and mutual fund assets by an amazing three-hundred-fold. Mutual funds are an important tax-preferred option for individual retirement savings, in

Table 1 Personal Sector Gross Financial Assest for the UK, 1980, 1990 and 1996 (31 December) in Current £ Billion

Financial asset	1980	1990	1996
Bank deposits	37.4	157.3	219.6
Building society deposits*	57.6	160.0	210.0
Government securities and debt	13.8	10.4	12.3
Miscellaneous instruments	4.4	15.3	24.0
National savings	12.1	35.6	61.3
Notes and coins	8.3	13.6	18.9
Overseas investments**	4.5	12.0	15.9
Pension/life assurance	106.6	528.0	1080.3
Trade credits (DOM)	14.8	42.0	51.8
Unit trusts	3.0	18.2	60.0
UK securities	38.4	171.1	336.4
Accruals adjustments	5.0	24.4	36.0

Source: *United Kingdom National Accounts – The Blue Book* (1997). London: HMSO, Table 12.2 (p. 138)

* Includes deposits with other financial institutions
** Includes direct investments and securities

Table 2 US Households' Financial Assets 1980, 1990 and 1994 (31 December) in Current $ Billion

Financial asset	1980	1990	1994
Deposits	1490	3239	3531
Credit market instruments (incl. bonds)	423	1494	1961
US Government securities and bonds	491	1418	2358
Municipal securities	104	574	432
Corporate and foreign bonds	31	195	390
Mortgages	84	147	217
Open-market paper	38	63	47
Mutual fund shares	46	467	1491
Corporate and non-corporate equities	2806	4382	7514
Life insurance/pension fund reserves	1176	3788	7003
Personal trusts	265	522	834
Security credit	16	62	158
Miscellaneous assets	74	224	376

Source: *Statistical Abstract of the United States* (1997). US Department of Commerce, Washington, DC, Table 777. Derived from Board of Governors Reserve System, *Flow of Funds Accounts*, March 1997 diskettes

some cases supplementing (and even replacing) existing employer-sponsored pension schemes (the '401(k) option').[12] Similar trends can be identified for Canada and Australia; the former is more like the United States and the latter is more like the UK in terms of the distribution of assets between savings categories.

In the Anglo-American world, bank deposits and related financial institutions are of declining significance compared to market transactions and related financial institutions. This trend has been widely noted and is considered a major difference between the Anglo-American financial systems and the banking-based financial systems of continental Europe (Allen and Gale, 1994; Boot and Thakor, 1997). The available evidence on European pension fund assets as a proportion of GDP suggests that most continental European countries have small to negligible assets. French, German and Italian pension fund assets are each only about 5% of GDP compared to 93% for the UK. Note that pension fund assets are 89% for the Netherlands, 87% for Switzerland and 57% for the United States.[13] In this respect, three observations can be made about the distinctiveness of Anglo-American pension fund capitalism. First, the three largest continental European economies have very modest pension assets. Second, governments' restrictions on investment options used by existing continental European pension funds have constrained the rate of growth of these funds over the last 20 years (this is especially important in Switzerland). In combination, the size and scope of many European equity markets are much smaller than might be predicted by the size of their economies. Not surprisingly, the nature and rate of financial innovation is also significantly less than the Anglo-American economies (La Porta, Lopez-de-Silanes, Shleifer and Vishny, 1997).

Two other observations should be made about national variations in retirement financing and likely long-term liabilities. Data from the Organisation for Economic Cooperation and Development (OECD) (1997b) on national savings show that the advanced economies (except Germany) experienced marked declines in savings rates from 1980 to 1996 (Figure 1). For example, Australia's savings rate declined from 21.6% in 1980 to 16.8% in 1995. Over the same period, UK and US savings rates declined from approximately 17% to 14% and 20% to 16%, respectively. Even in the case of Germany, the 1995 savings rate was slightly less than the 1980 savings rate (22%) despite significant increases during the 1980s (peaking at 32% in 1986). The OECD also estimates that by the year 2010 the elderly dependency ratio will be 18.6% for Australia (compared to 16.0% in 1990), 20.4% for the United States and Canada (19.1% and 16.7% in 1990, respectively), 25.8% for the UK (24.0% in 1990), 29.1% for Sweden (27.6% in 1990) and 30.3% for Germany (21.7% in 1990) (Leibfritz, Roseveare, Fore and Wurzel, 1995). By the year 2020, the dependency ratios of Germany and Sweden will be about 35.0%, compared to an expected European OECD average of 30.8%.

The early 1980s recession and subsequent slow growth rates, and then the early 1990s recession which continues in one form or another in continental Europe, if not in the Anglo-American countries, have, in combination, limited the capacity of many nation-states to accumulate public savings. In this respect, the recent growth of pension assets in the Anglo-American economies may have profound positive consequences for their future economic performance relative to much of continental Europe. Not surprisingly, these trends have prompted significant debate inside and outside the academic world about the long-term funding of state social security programmes and the status and significance of pension schemes now and in the future.[14] In response, many countries are contemplating the introduction of mandatory pension systems based on employer and employee contributions to sponsored pension funds. Such systems have been introduced in Australia and Chile, and are the object of discussion in Western and Eastern Europe, including Germany (see Diamond, 1994, on recent developments).[15]

URBAN INFRASTRUCTURE AND ECONOMIC DEVELOPMENT

If the state is in retreat with respect to urban infrastructure and development, and if financial institutions and markets are now the dominant sources of capital, how are those resources to be mobilised? What is implied by the private financing of public infrastructure? These are significant and far-reaching questions that require extensive investigation.[16] These questions are, in part, the object of academic scholarship as well as being of practical concern to pension fund trustees and their investment advisors. The study here is of the dominant logic underpinning the pension fund investment process and a brief overview of recent moves towards pension fund investment in urban infrastructure and development. The first part of the story is common to many Anglo-American countries. The second part draws upon recent US experience.

To set the issue in context, consider Figure 3, where annualised rates of return reported for commonly traded market investment products from 1947 to 1996 are set against their measured risk or volatility as indicated by standard deviations. The data is for the United States, but is indicative of other Anglo-American countries' experience, including the UK, Australia and Canada up until 1998. Embedded in Figure 3 are a series of theoretical propositions that currently underpin pension fund investment strategies. Based on the pioneering work of Markowitz (1952) and Sharpe (1964) on modern portfolio theory, analysts and practitioners expect that rates of return are commensurate with their risks. The higher the risk, the higher the expected return. As indicated by Figure 3, this relationship can be 'mapped' over a set of asset classes: from treasury bills (practically risk-free investments) to small stock portfolios (high-risk investments). Note also that the larger the stock portfolio, the more likely it is to reflect the performance of the whole market. In this sense, diversification is a valued strategy because it spreads the risk of poor performance of selected stocks over the performance of all stocks.

Many theorists assume the market is efficient in the sense that investments are properly priced in accordance with all the available information. There are two implications which follow from this assumption. One is that it is impossible to systematically beat the market. Although some investment managers may perform better than average by chance or by virtue of 'local' information, this kind of performance is very difficult (even impossible) to sustain in the long term. The other implication is just as drastic. Active investment strategies are, more often than not, less rewarding than passive investment strategies. Indeed, it is apparent that once the costs of investment management are factored into the analysis of net

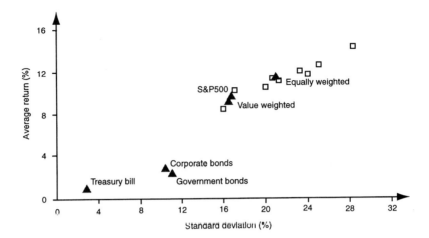

Figure 3 Mean Versus Standard Deviation of Real Returns, 1947–1996. Source: Cochrane (1997: 5). Note: Triangles are Equally Weighted and Value Weighted; S&P 500; Three-month Treasury Bill; 10-year Government Bond; and Corporate Bond Returns. Unmarked Squares are NYSE-size Portfolios

returns, the majority of managers do not perform as well as the standard benchmarks such as the S&P 500 or the FT-SE 100. This does not mean that all investors are equally rational or that all investors are equally informed. It is also apparent that some investors, especially those with significant institutional resources, may take advantage of the poorer judgement, information and trading competence of other investors (Shleifer, 1998). The point is simply that active investment must be justified by reference to benchmarks of performance, relative expertise and the availability of information (Grossman, 1995).

Modern portfolio theory and the efficient markets hypothesis are essential reference points for pension fund trustees' investment decision-making. Although it would be surprising to find that trustees always acted in accordance with these theoretical propositions, the language of decision-making is thoroughly impregnated with related terms and concepts (Clark, 1998b). At the same time, the flow of contributions and the current and expected flow of paid benefits are also important variables. Following US legislation in the 1970s and revisions to UK legislation following the Maxwell scandal, all Anglo-American countries have rigorous reporting and funding standards for most forms of pension plans. In part, these standards require the current funding of expected liabilities and the diversification of investment strategies. With respect to Figure 3, this means that trustees first allocate assets amongst asset classes, such as equities, bonds and cash equivalents, and then choose amongst competing investment managers in accordance with their past performance and expertise. Although it seems that stock portfolios are more desirable than other types of investments, bonds, for example, do offer trustees greater certainty with respect to the rate of return. Even cash equivalents may be desired by trustees concerned to meet short-term liabilities. These are especially important considerations for plan sponsors that offer defined benefit plans.

Now let us return to the issue of urban infrastructure and economic development. In order to give the analysis specificity, let us assume that two kinds of investments are contemplated, one in infrastructure (e.g. a link-road or hospital emergency unit) and one in development (e.g. a start-up firm with close links to a particular community). Whereas in the past the state may have contemplated both kinds of 'investments' as part of its urban and regional strategy, fiscal and functional constraints make both impossible. Let us also assume that the first is offered as a bond promising a 20-year rate of return net of inflation, whereas the second is offered as a share of ownership (equity) with no contractually agreed rate of return but a proportionate share in any profit with the option to sell out after five years to other partners. For the sake of argument, let us also assume that pension fund trustees and their investment advisors seek opportunities to diversify their portfolios with no special conditions imposed by the state.[17]

One issue facing trustees, therefore, is how best to assess the value of the investment options. With respect to the infrastructure option, the obvious benchmark is the risk and return profiles of other competing bonds. Yet how should we assess the particular risk of this kind of investment? This issue is explored in depth in Clark (1997a and 1998b) and Clark and Evans (1998). Lacking a well-established market in urban infrastructure bonds and lacking experience in the performance of such investments, trustees may draw an analogy with existing cases and assign a relevant risk somewhat higher than traded bonds or retreat from the option altogether. If the risk is higher, the contractually agreed rate of return would have to reflect that risk. With respect to the development option, the obvious reference point would be a small portfolio of venture capital investments (the upper right-hand corner of Figure 3). However, again, the fact that the investment would be a private placement with no option for short-term trade means that the pension fund would be locked in to the project. Therefore, a higher risk would be attached to the investment, implying a higher expected rate of return.

Notice, though, that the annualised rate of return on a small portfolio of traded equities hides considerable temporal specificity. Apparent high rates of return identified in Figure 3 for US equities (of all kinds) are the product of the accelerating run-up in all Anglo-American markets over the past seven, five and three years leading up to mid 1997. As rates of return are almost impossible to predict over the short term, and because down-turns are highly likely after bull markets, to make an investment decision with respect to the immediate past performance of equities (on the upside) or bonds (on the downside) runs the risk of rejecting investment options that at another time would seem entirely compatible with the broad range of expected risks and returns. It is also clear that the accelerating markets of the 1990s attracted assets from other investment classes, notably bonds, as well as more specialised investment options, because of a perceived lack of short-term risk and a perceived opportunity cost of not being 'in the market'. The implication is stark: notwithstanding the 1998 'crisis', infrastructure bonds may find few takers because of a more general presumption in favour of equities, and the development equity may find few takers because of the significance assigned to immediately past high rates of return on traded securities. Nevertheless, herd behaviour is very common in financial markets and in investment decision-making in general (Zeckhauser, Patel and Hendricks, 1991). The Asian 'crisis' may alter the balance in favour of bonds.

Other trustees may pursue contrary investment strategies relying on expertise and special knowledge to discriminate between 'good' and 'bad' urban investments. Expertise is expensive. Therefore, the investment management industry is highly differentiated with, at one end of the spectrum, large international and multi-functional investment firms controlling vast amounts of traded assets and, at the other end of the spectrum, small highly specialised firms offering product-specific and project-specific expertise based on their access to non-traded information. At one end of the spectrum, dis-intermediation is the order of the day, whereas, at the other end, especially designed intermediaries are crucial for mobilising assets and transferring them to investments that fall outside the accepted conventions of the industry. For large firms specialising in market-related investment products, efficiency is a product of the design and management of markets and the capacity of firms to identify and take advantage of short-term asymmetries. For small firms specialising in custom-designed intermediation, management efficiency is a product of scale economies with respect to transaction costs and reporting mechanisms that allow investors to confirm the veracity of their partners. Not surprisingly, both kinds of firms have to be close to the markets for finance and expertise. The industry is remarkably spatially and functionally centralised (Clark, 1997b; Martin and Minns, 1995).

One aspect of financial intermediation in relation to urban infrastructure and economic development concerns the role played by public sector pension funds. The available evidence from Greenwich Associates (1996) suggests that only the largest private defined benefit plans in the United States have a significant interest in so-called 'alternative investment products', such as infrastructure and venture capital. As is the case in so many other issues, such as proxy voting and corporate governance, private pension plans tend to rely on their investment managers rather than taking an interest in the active assignment of pension fund power. In the United States, at least, perhaps one reason for this particular difference is the reliance of public funds on local economic performance for the continuing flow of participants and the flow of pension fund contributions. It also seems that defined contribution plans are preoccupied with the short-term returns. Therefore, we should accept that some types of pension funds are more likely to invest in urban infrastructure and development than others, and that the increasing preference of private employers for defined contribution plans and the like may mean that urban investment is largely a public sector pension plan phenomenon.

However, there has been considerable criticism of public pension funds investment in the urban realm. In part, it has been observed that many pension funds have neither the internal expertise nor the resources to hire the external expertise necessary to discriminate between competing development options. Although it seems that the demand for alternative investments is increasing rapidly amongst many of the largest pension funds, the supply of viable projects remains limited because of the high level of transaction costs involved.[18] Other critics contend that the often close relationships between pension fund trustees and the providers of non-traded urban investment options leaves trustees open to charges of recklessness and even corruption when deals go wrong. As a result of the public significance of many large urban infrastructure and development projects, and of the inevitable long-term commitments implied by such deals, the motives and actions of pension funds are often open to public scrutiny. The prospect of scrutiny has made trustees even more risk-averse, reinforcing the perceived high-risk nature of such investments.

One result of scrutiny has been a shift away from long-term investment relationships to project-by-project assessments ruled by the law of contract. Another result of scrutiny has been the development of third-party urban investment trusts, where pension funds remain distant from the actual development and management of particular investments. In this model, pension funds can be thought of as partners in investment clubs (Clark, 1997a). In the author's opinion, these criticisms of pension funds and their investment in urban development projects are entirely warranted. However, it would be misleading to suppose that these criticisms are only applicable to investment in urban infrastructure and development projects. Although no doubt often attached to particular projects and pension funds, these criticisms are indicative of the extensive nature of agency relationships in the pension fund investment industry and the decentralised nature of government regulation of the relationships between trustees and plan beneficiaries. It is very rare for plan participants to have a voice in trustee investment decision-making. Anglo-American governments have relied on the trustee institution to regulate the interests of beneficiaries. The principal-agent problem (Pratt and Zeckhauser, 1985), is deeply embedded in the whole system of Anglo-American pension fund management.

In general, urban infrastructure and development are a very small fraction of even the largest funds. Nevertheless, those funds with either a close attachment to an industry or a place have become more interested in community investment. As noted above, some large US public sector funds do participate in a range of development options, including venture capital. There is a significant number of specialised investment consultants that design and broker these investments for public sector funds. In the United States, at least, federal pensions legislation does not apply to states' government-sponsored pension funds allowing for, some might contend, a wider range of investment options (as well as the possibility of underfunding!). Industry-based jointly trusteed pension funds that are concentrated in certain segments, sectors or regions are also increasingly interested in what are termed 'economically targeted investments' (ETIs). Although regulated by federal law and related expectations regarding the ultimate value to beneficiaries, these investments have flourished, even if their relative significance remains minor and disputed (on the experience of Wisconsin, see Levine, 1997).

IMPLICATIONS AND CONCLUSIONS

It is clear that Thatcher and Reagan and their notional ideological allies around the Western world have made a lasting difference to the structure and size of

Anglo-American governments relative to their national GDP. However, the argument goes beyond ideological warriors – the period following the Cold War will be a 'cold-hearted era' in which the functional scope of government spending is squeezed by a combination of political limits imposed by the middle classes on state income, reduced discretion in the spending of the available state income, and economic limits imposed by financial markets on state deficit spending. In other words, forces already at work in the 1980s have combined in the last decade of the twentieth century to de-legitimise the welfare state. Implicated in this story is the baby boom generation whose fealty to the axioms underlying the welfare state have been under-cut by the collapse of the enemy without and by threats to the stability and predictability of the national economy within.

The rise of 'pension fund capitalism' as a set of Anglo-American market-based financial institutions is also an important reference point in understanding the full extent of the retreat of the state. Although previous sections emphasised limits on the state's functional capacity, the rise of pension fund capitalism has undermined the nation-state's status as a significant economic agent. Compared to the 1960s, when Shonfield could reasonably argue that the state and markets were balanced in terms of their respective sizes and financial resources, new market institutions now dominate the state in terms of their command of global finance. Thus, any understanding of the likely topography of urban infrastructure and development over the coming decades must begin with the financial industry in general and with the pension fund investment decision-making process in particular. This is a real and profound change in what we have believed to be the proper status of the state. As a consequence, we must better understand how and why trustees make investment decisions as well as the incentives and impediments to financial innovation in the urban and non-urban realms.

With the advent of pension fund capitalism, urban structure will be increasingly an investment good managed with respect to the interests of pension funds and their beneficiaries. If Figure 3 were to remain as the essential reference point for pension fund investment strategy, it is likely that the urban fabric of Anglo-American societies would be systematically discounted by underinvestment over the coming generation with selective private investment replacing comprehensive investment by the state. There need be no connection between the goals of funds' investment strategies and the economic and social coherence of urban society. Indeed, as has been suggested above, the very idea of social integration as a necessary element in the legitimacy of the state has been lost with the collapse of an (albeit primitive) alternative. This does not mean that the state is irrelevant as a regulatory institution. Financial institutions are neither benign in effect nor systematically self-organising in ways consistent with one another's interests or the interests of communities. This much has been recognised by the union movement in the United States, and reflected in the controversial but largely failed push to encourage economically targeted investments. The apparent centralisation of funds and the investment management industry in the major urban centres of Western economies may have far-reaching implications for urban and regional inequality.

If urban investment is to be a private responsibility, and if individual decision-makers are to be responsible for urban structure in the twenty-first century, it is possible that new political movements may want to regulate the scope of pension fund investment. The manner in which such movements may appear over the coming years is not obvious at this time, notwithstanding the AFL-CIO's (American Federation of Labor – Congress of Industrial Organisations) ETI strategy. In addition, it is not clear how such movements would affect the polity to such an extent that the state would be mobilised to regulate pension funds in these ways. Most nation-states have lagged behind market-based financial innovations, preferring to idealise market and trustee decision-making rather than develop a centralised regulatory response. However, it is clear that pension entitlements are not evenly spread; the middle classes are the principal beneficiaries. As a consequence, over the coming decades, there are likely to be gross inequalities in retirement welfare between the middle classes and others, together with increasing evidence of gross inequalities in terms of urban infrastructure. Pension funds may reasonably argue that their responsibility is simple and exclusive – to plan their participants' welfare. In this respect, any argument to the effect that pension funds generally owe society more than that which the state is now perceived to owe society may be quite problematic. Even if particular types of pension funds may believe the interests of beneficiaries are related to sustaining urban development, there are many who dispute the legitimacy of any investments that take into account the wider context of members' interests.

At the same time, as the investment industry becomes more concentrated in terms of the management of pension assets and more divorced from particular jurisdictions as the flow of resources pushes them out into the global economy, plan beneficiaries who together do have particular jurisdictional loyalties may come to see their interests and the interests of the industry opposed at every turn. As pension funds offer more investment choices to their members and presume that members should assume greater responsibility for their welfare, the politics of community loyalty may be internalised into the finance industry through coalitions of beneficiaries and non-beneficiaries. The responsibilities of trustees may be perceived as wider and more comprehensive than previously imagined in law (Langbein, 1995). The trustee institution may be integrated into the regulatory structure of the nation-state and the supra-nation-state even if political debate about its status is muted by the displacement strategies of the state. Here, perhaps, are the rudiments of an intellectual strategy: '[w]e must then rediscover in the small variations on which legal thought has traditionally fastened the beginnings of larger alternatives we can no longer find where we used to look for them' (Unger, 1996: 2).

Acknowledgements

This chapter was sponsored by the Australian Housing and Urban Research Institute, and by the Australian Research Council (Grant C5-9700312). It reflects conversations with Mike Berry about the prospects for public finance and the advice of John Evans with respect to pension fund investment management. Interviews with Sir Alastair Morton and Deborah Roseveare (OECD, Paris) also helped to put the issues in context. Research assistance was provided by Melanie Feakins. Jan Magee processed the manuscript. Ron Martin made very useful and constructive comments on the chapter, as did participants in an Oxford Economic Geography Workshop presentation. None of the above should be held responsible for any errors or omissions.

CHAPTER 13

SELLING OFF THE STATE: PRIVATISATION, THE EQUITY MARKET AND THE GEOGRAPHIES OF SHAREHOLDER CAPITALISM

Ron Martin

INTRODUCTION

A key feature of twentieth-century capitalism, especially after 1945, was the increasing penetration of the private market economy by the state. One aspect of this penetration was the construction of extensive regulatory structures by which the state was able to influence the operation of economic markets and the economic behaviour of firms and households. A second was the state's direct manipulation of key fiscal and monetary aggregates, and the use of elaborate tax and benefit systems, in order to stabilise the economy and secure social harmony. Third, most capitalist states also embarked on major programmes of 'nationalisation', the transfer into state-ownership of various basic services and industries. Although the extent of state-ownership and the specific services and industries 'nationalised' varied from country to country, in almost every case 'public sector' activities became a integral component of the 'mixed' capitalist economy. Likewise, although the exact political rationale underpinning state-ownership also differed between countries, some common aims and objectives prevailed. It was widely believed, for example, that only the state could ensure the universal provision of modern mass (collective) public services, utilities and social infrastructures (e.g. water, gas, electricity, education, health, housing, transport, postal services and telecommunications), public facilities which, if left to the private market, were almost certain to vary significantly in availability and cost between different social groups and different geographical areas. Other activities (ranging from coal and steel, through motor vehicles and aerospace, to banking) were often regarded as too 'strategic' to be left to the private sector and were likewise taken into state-ownership. From a regulation-theoretic viewpoint, the large-scale state-ownership of services and industries formed a logical component of the Fordist regime of economic production and accumulation that characterised post-war capitalism: it provided an important mechanism by which the economy was regulated and the costs of key services and utilities essential for economic growth were socialised.[1]

Since the beginning of the 1980s, however, the post-war model of state capitalism, with its large public sector, has been rapidly unravelling. With the resurgence of economic liberalism in the late 1970s and 1980s, the Keynesian-welfare (Fordist) mode of state intervention and regulation has been widely replaced by 'free-market' oriented policies and strategies. The 'privatisation' (or de-nationalisation) of the state has been a central element of the new market orthodoxy. The word 'privatisation' seems to have first been used by Drucker (1969), but is now used to denote diverse forms or 'modalities' of disengagement and withdrawal of the state from the economy – from 'deregulation' in the broadest sense, through the 'marketisation' and 'commercialisation' of public sector services, to (at its most radical) the selling-off of state activities. It is this latter aspect of privatisation, the 'selling-off of the state', that this chapter will address and, thus, the word 'privatisation' is used in this sense here. Whereas in the post-war period states were busy nationalising private assets in a wide range of industries, services and utilities, over the last two decades they have been equally busy selling off those nationalised assets.

A considerable economic literature has been devoted to assessing the effects of selling off public sector activities. Some see privatisation as a means of raising the economic performance, profitability and efficiency of public sector activities and monopolies (Vickers and Yarrow, 1988; Chapman, 1990; Bös, 1991; Martin and Parker, 1997). Other economic work has concentrated more on the distributional consequences of privatisation for taxpayers, consumers and employees (Milne, 1992; Ramanadham, 1995). Still others have been concerned with issues of corporate governance in privatised activities (Caves, 1990; Foster, 1992). By contrast, there has been very little attention directed to the impact of privatisation on the financial landscape.

Under state-ownership, public sector activities are financed in part or wholly by general taxation, and are largely isolated from the external equity market, except for raising debt finance. This means that such activities are shielded from the scrutiny (and speculation) of market analysts and investors. At the same time, they are protected from bankruptcy and potential takeover. Ostensibly, every member of the (adult) population 'owns' a small fraction of public sector activities, regardless of his or her location, and the charge for using these activities likewise does not vary from area to area. Being tax-funded, public sector activities typically involve financial transfers from the richer regions of a country to the poorer ones.[2]

The divestiture of a state activity not only brings a one-off windfall income to the state, the sale of shares immediately integrates privatised public services, utilities and enterprises into the external capital market, thereby creating new maps of share-ownership, investment, disinvestment and dividends. Huge sums of money and billions of shares are involved in privatisation sales, so that the spatio-institutional structure of the capital market through which the flotation of public sector activities is made may influence both the scale and the geographical pattern of the public's take-up of privatisation shares. In its turn, the geographical distribution of privatisation shareholding will have implications for the geographies of property rights, wealth, income, consumption and saving. This issue is of interest not only in its own right, but also because one of the political-ideological motives behind privatisation has been that it provides an opportunity to spread individual shareownership across society, to create a 'popular capitalism'. The main focus of this chapter is to examine how far and in what ways privatisation has achieved this aim. The discussion centres on the UK experience, not only because it is the country that has set the pace for selling off state assets and has been a model that countries the world over have sought to emulate, but also because the creation of shareholder or 'popular' capitalism has indeed been one of the key motivations – and claimed successes – of the UK's privatisation programme. To set the discussion in context, however, the chapter begins with a brief overview of the historical evolution of capital-ownership structures.

THE EVOLUTION OF SHARE-OWNERSHIP UNDER CAPITALISM

In his theory of industrial capitalism, Marx predicted that modern societies would become increasingly divided between a small class of 'capitalists' who own all of the productive assets, and a large class or 'proletariat' of workers who own little or nothing (other than their labour power). In the early nineteenth century, this characterisation had some validity. In the UK and elsewhere, industry and commerce consisted mainly of small firms and enterprises run by individual or family owners. Investment in new plant, machinery and premises was funded either by these individual or family owners recycling profits back into the business, or by borrowing from banks, which at that time were mainly local or regional.[3] The ownership and management of capital were thus fused in the individual proprietor or the family, and the creation and investment of new capital were highly localised geographically.

Over the last century and a half, several developments have fundamentally altered the form of capital-ownership highlighted by Marx – developments that have separated ownership from management and led to the dispersal and 'delocalisation' of capital-ownership. The precise evolution of share-ownership has, however, differed between different countries. In the UK, the first industrial nation, the introduction of limited liability with the Companies Act of 1865 dramatically reduced the downside risk of equity investment, and this encouraged 'outside' shareholders to invest in business, thereby lifting an important financial constraint on the growth of industry and achievement of scale economies. The dispersal of ownership proceeded rapidly, as was also the case in the United States. In these countries, the era of owner-manager and family capitalism gave way to a form of industrial organisation based on the joint-stock company. By 1929, for example, only 22 of the 200 largest US non-financial companies were single- or family-owned (Berle and Means, 1932). By the mid twentieth century, in the UK and the United States individual and household shareholders owned two-thirds or more of the equity market. Geographically, this development of individual shareholding entailed a spatial dispersal of capital ownership, in the sense that a significant proportion of a given locality's capital stock was typically owned by individual investors well outside that area. At the same time, the trading of shares

Table 1 The Share-ownership Structures of Some Major Countries (% of Total Listed Stocks)

		Households	Financial institutions	Non-financial companies	Government institutions	Foreign owners
France	1977	41	24	20	3	12
	1992	34	23	21	2	20
Germany	1970	28	11	41	11	8
	1993	17	29	39	3	12
United Kingdom	1969	50	36	5	3	7
	1993	19	62	2	1	16
Japan	1970	40	35	23	0	3
	1993	20	42	28	1	8
United States	1981	51	28	15	0	6
	1993	48	37	9	0	6

Sources: Berglöf (1997) and International Capital Markets Group (1995)

became increasingly concentrated in a few urban centres with stock markets, and mainly within the national capitals of London and New York.

In other countries, however, the development of share-ownership was different. In Japan, for example, individual shareholding also expanded, but never to the extent of dominating the two major forms of ownership, namely by companies and banks. By the 1960s, company share-ownership had evolved into a 'relational' kind, whereby a large proportion of the stock of one enterprise was held by other enterprises, sometimes involving reciprocal ('inter-mingled') ownerships. A similar 'relational' form of cross-company shareholding had also become important in Germany; for example, in 1960 private individuals owned just over one-quarter of the capital stock, whereas enterprises owned more than half (Schneider-Lenné, 1992).

One feature common to most advanced capitalist countries in the post-war period, and especially since the early 1960s, has been the rise of institutional shareholders (see Table 1 and Figure 1). Again, this took different forms in different countries. In the UK it was associated with the marked post-war expansion of pension funds, life insurance and home mortgages. Specialised financial institutions were formed to handle these various funds. In one sense, the rise of institutionalised shareholding has spread share-ownership across wide sections of the population. Millions of people may subscribe to one or more forms of institutionally managed savings or investment funds (e.g. through their insurance policies or pension schemes), and thus have a stake of one kind or another in the nation's productive assets, and, indeed, in foreign companies and enterprises. The spatial webs of individual share-ownership involved in these managed institutional, or 'collective', funds are thus extraordinarily complex and diffuse. However, in this form, individual share-ownership is also very indirect, as the buying and selling of the shares that make up pension and similar investment funds is undertaken by the fund managers, and most individual investors in those funds have little or no idea of – or control over – the companies in which those shares are held. However, the size of these institutional shareholders is such that they can wield considerable

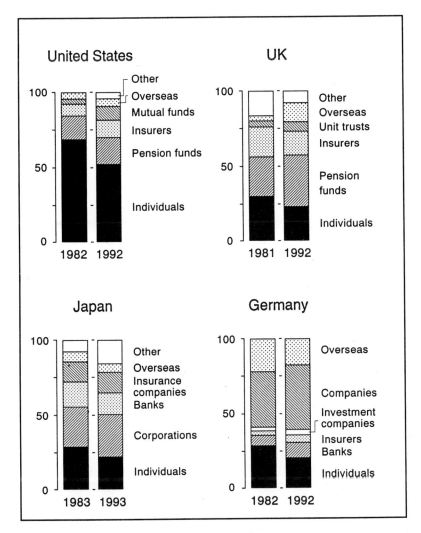

Figure 1 Share-ownership Structures: Four Different National Systems

influence over equity markets and company stock valuations. Thus, they constitute a key component of what Minsky (1989), in a different context, has labelled the rise of 'money manager capitalism'.

In the UK, institutional share-ownership expanded rapidly after the early 1960s. The growth of company pension schemes and private insurance plans, aided by tax reliefs granted by successive governments in favour of institutionalised insurance and pension provision, led to a marked shift in the pattern of shareholding away from the private individual to large financial institutions based mainly in London, the nation's financial centre. Whereas in the late 1950s, two-thirds of the company shares registered in the UK were owned by individuals (Chapman, 1990), by the end of the 1960s this had fallen to 50%, while the proportion of shares held by institutions had risen to nearly 40% (Table 1). By the end of the 1970s, the proportion of the equity market owned by private individuals had slipped further, to less than 30%, while institutions controlled more than two-thirds (see Figure 1). A similar development also took place in the United States, although there the erosion of individual shareholding by institutional shareholders – pension funds, mutual funds and insurers – progressed more slowly, so that even by the beginning of the 1980s households still owned more than half of the equity market. In Germany and Japan, the growth of institutional shareholding took a different form, centred not on pension funds but on banks and insurance companies (Table 1 and Figure 1).[4] Nevertheless, the result was somewhat similar to the UK case in that, by the late 1970s, the proportion of the equity market owned by private households had also declined.

During the post-war period the nationalisation of industries and services took place in many countries. In principle, nationalisation can be viewed – and indeed was sometimes justified politically – as a mechanism for securing a widespread 'public' ownership of capital. In practice, however, 'publicly-owned' meant *state*-owned. Ordinary members of the population were not issued with shares in the nationalised industries and public services. Their 'ownerships' could not be traded; and they did not carry any entitlements to dividends or profits. In practice, therefore, nationalisation did nothing to spead effective capital-ownership more widely through society. It did not return any control of key industries and enterprises to the local community; on the contrary, it concentrated financial control and management in the political centre.

However, according to some commentators, the wave of privatisation that has swept through much of the capitalist world (and, of course, much of the former socialist bloc) over the last 15 years has done what nationalisation actually failed to do, namely returned the ownership of industries and enterprises to the people by providing them with the opportunity to buy shares in the newly privatised activities. In the view of writers such as Moore (1992) and Saunders and Harris (1994), privatisation has promoted a new phase in the evolution of capital-ownership. Although privatisation sell-offs amongst advanced capitalist nations were not new – there had been examples in various countries during the 1960s and 1970s – the movement only really took off at the beginning of the 1980s. It began in the UK in 1979, when the Conservative Government embarked on the first, somewhat tentative, stages of what was to develop into an extensive programme of selling off nationalised industries and public utilities (Veljanovski, 1987; Bishop and Kay, 1989; Chapman, 1990). Other capitalist countries in the Organisation for Economic Cooperation and Development (OECD) group then followed with similar privatisation schemes, for example in Italy, Germany, France, Canada and Japan (Bös, 1993; Button and Weyman-Jones, 1994; Richardson, 1990; Vuylsteke, 1988). By the mid 1980s, the process had spread to the newly industrialised countries of South-East Asia, such as Thailand and Malaysia (Gouri, 1991; Milne, 1991 and 1992; Ng and Toh, 1992), and later that decade even developing countries had begun to transfer public enterprises into private ownership, though more so more in Latin and South America (e.g. Mexico and Chile) and India, than in African states (Bhaskar, 1992; Cowan, 1990). Most recently, as a result of the collapse and break-up of the socialist bloc and the efforts to establish a capitalist economic system there, privatisation has become a key policy in Eastern Europe, Central Asia and even in China (Bohm and Somenti, 1992; Ash, Hare and Canning, 1994). The precise scale of privatisation sales has, of course, varied widely from nation to nation, reflecting national differences in the pre-exisiting size of the state sector, as well as in economic, political and cultural conditions. Nevertheless, by the early 1990s, 'selling off the state' was not just confined to the leading industrial economies, but had become a global phenomenon and an accepted state economic policy in a diverse range of countries (Ramanadham, 1993; Goodrich, 1990). According to the World Bank, by the early 1990s the privatisation of state assets was averaging about US$50 billion a year worldwide, and reached nearly US$90 billion in 1995.[5]

Although by no means the primary political reason for selling off state assets, privatisation has often been seen as providing a mechanism for reversing the relative decline of individual shareholding that took place in the post 1945 period, for promoting a new 'shareholder capitalism'. There has been considerable debate about how that aim would best be achieved. One view is that instead of selling shares in privatised firms, states should give them away to the entire adult population, to whom – supposedly – those activities already belong. Alternatively, governments can bundle the shares of several privatised firms together into diversified portfolios run by investment trusts, in which individual members of the public are then given share entitlements (e.g. vouchers). Such schemes are thought likely to be much more successful in fostering widespread share-ownership, particularly where the domestic income base is low. With few exceptions, however, Western states have preferred to sell privatisation shares to the general public rather than give them away, no doubt mindful of the large sums of money that such sales generate for the state exchequer.[6] The question is whether this policy leads to a social and geographical extension of individual share-ownership on the scale hoped for. The UK case, as is shown in the rest of this chapter, suggests that it may not.

Privatising the UK State: The Great Sell-off of the Public Space Economy

Compared to other nations, the UK privatisation programme has been notable both in its scale and its high national and international profile. The origins of the programme resided in the Conservative Party's conversion, whilst in political opposition in the mid 1970s, to neoliberal free-market monetarist economics, with its aims of reducing state spending and lowering taxation, its anti-collectivism, anti-trade unionism (public sector unions not only enjoyed high levels of membership, but had also become very militant), and its espousal of private property and economic individualism (Thompson, 1990; Gilmour, 1992). This is not to say that when the Conservatives came to power under Mrs Thatcher in 1979 they had a carefully mapped out privatisation strategy. In the early years of Conservative Government, privatisation policy was rather *ad hoc*: 'sales were generally regarded within government as a series of individual market transactions best conducted within the traditional parameters that had served the city [of London] well over the years' (Grimstone, 1987: 23).[7] Although a number of public sector companies, or parts of them, were sold, most of these (e.g. British Petroleum in 1979 and 1983, Associated British Ports in 1981 and 1982, British Aerospace in 1981, Cable and Wireless in 1981 and 1983, Britoil in 1982 and Amersham International in 1982) were relatively small, and played no role in broader public policy objectives. By mid 1984 there had been nine such privatisation flotations totalling only £2.2 billion.

The growth of privatisation into a central feature of the UK Government's political programme began with the flotation of British Telecom, the state-owned telecommunications utility, in late 1984. The desire of British Telecom to undertake a substantial investment programme of more than £1 billion in new electronic switching technologies came at a time when the government was attempting to reduce the public sector borrowing requirement as part of its medium-term financial strategy. Not only was the treasury strongly opposed to any idea of increasing government borrowing by such an amount, it also refused to allow British Telecom to raise the necessary investment finance from the capital market. The only solution was to sell part of British Telecom. Some 38.5% of the share offer was sold to the UK public (including British Telecom employees). The share sale attracted 2.3 million applications for 3 billion shares, and was 9.7 times oversubscribed. More than half of the people who bought shares had been lured into shareholding for the first time. A total of £3.92 billion in equity proceeds was raised, which at the time represented the biggest market flotation of any kind anywhere in the world.[8]

The unexpected popularity of the British Telecom privatisation paved the way for future stock flotations of other state assets, and over subsequent years privatisation sales developed into one of the core policies of Thatcherism (Hamnett, 1998). In 1986 major sales of state activities included £5.43 billion of shares in British Gas and £1.36 billion in the Trustee Savings Bank. This was followed in 1987 by British Airways (£0.90 billion), Rolls Royce (£1.363 billion), British Airports Authority (£1.28 billion) and a £5.2 billion flotation of shares in British Petroleum. The end of the 1980s saw the further large sales of £2.5 billion of shares in British Steel in 1988 and £3.4 billion in the 10 regional water companies in 1989. Then, in 1990, the 12 regional electricity companies were also sold off, eventually bringing the state £5.5 billion of share proceeds. In the following year, both Scottish Power–Hydro-Electric (£2.90 billion) and National Power–Power-Gen (£2.1 billion) were also floated. By early 1991, more than 50% of the state sector had been sold off. Altogether, some £70 billion of state-owned activities, services and utilities have been transferred to the private sector since 1979 (see Figure 2).

Thus, although privatisation initially emerged without a pre-conceived plan, by the mid 1980s it had become a central plank in the Conservative Government's policy, and was justified in terms of three basic objectives (HM Treasury, 1995): (i) to promote efficiency and competition in order to benefit consumers; (ii) to obtain the best value for the taxpayer (in practice, the government) of each industry and

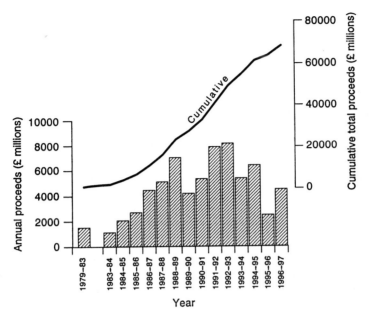

Figure 2 UK Privatisation Proceeds (Out-Turn), 1979–1997

service sold; and (iii) to spread share-ownership as widely as possible amongst the population.[9] As Gilmour (1992) points out, it became the Conservative equivalent of the Labour Party's historic Clause Four, with its fundamental objective to 'secure for the workers the common ownership of the means of production'.[10] Most Thatcherites and many others in the Conservative Party became as dogmatic in their zeal to sell off the state as the old left-wing Labour socialists had been in their desire to sweep all the UK's major industries and services into the state sector.[11] Arguably, the Conservatives proved to be far more successful in fulfilling their aim than Labour had been in achieving theirs.

By the late 1980s, this aim had become incorporated into an explicit vision of a society in which the mass of the population would own shares, with a new culture of capitalism emerging as a result. Just as it was government policy to create a 'nation of home-owners', so its long-term ambition was now 'to make the British people a nation of shareholders, too, to create a popular capitalism in which more and more men and women have a share in Britain's industry and business' (Chancellor Nigel Lawson, Budget Speech, 1986). This vision was spelled out in the Conservative Party's 1987 election manifesto, which promised to extend the privatisation programme in order to bring about an 'historic transformation' in society in which share-ownership 'would become the expectation of the many'. As John Moore, one of the ministers responsible for the initial stages of the UK privatisation programme put it:

> Our aim is to build upon our property-owning democracy and to establish a people's capital market, to bring capitalism to the place of work, to the high street, and even to

the home. As we dispose of state-owned assets, so more and more people have the opportunity to become owners. . . . (Quoted in Aharoni, 1988: 41)

It was a quest also enthusiastically voiced by the leaders of the business world, including the Chairman of the Confederation of British Industry:

The vision we share is of a country where the majority of its citizens are part owners of the large and small companies that create the wealth that we all enjoy. . . . They will thus become aware of and feel part of the system by which wealth is created. (Confederation of British Industry, 1990: 7)

Certainly, the success of the major privatisations in the late 1980s and early 1990s was such as to suggest that this vision would become reality. If the privatisation of British Telecom in 1984 had been enormously popular, the sale of British Gas in 1986 was even more so. British Gas at that time was the largest integrated gas supply monopoly in the Western world. The British Gas flotation attracted an unprecedented 4.6 million applications, twice the number received for British Telecom. The flotation of the electricity industry in 1990 generated even more individual share subscribers. Given the huge size of the industry, the government decided to divide it up geographically into 12 English and Welsh regional supply companies, two Scottish generating and supply companies, three power generating companies and a national grid company. Some 5.7 million people made more than 12 million applications for shares in one or more of these enterprises. Even the sale of the 10 regional water companies in 1989, which initially attracted considerable public criticism – on the grounds that it was unwise and unethical to transfer such a basic industry into the private sector – received applications from 2.5 million people.

In order to promote interest from the public, the government explicitly allocated a significant proportion of privatisation shares to individual purchasers: in many of the privatisation sales half or more of the share allocations were reserved for the general public (Table 2).[12] The government was also keen to boost employee shareholding in the privatised activities, and offered various incentives such as free shares, discounts and priority in allocation. In most cases, a high proportion of employees bought shares in the company in which they worked.[13] Perhaps above all, the public responded in such large numbers because the shares were offered at substantially undervalued prices, and therefore were perceived as 'no-lose' opportunities. Largely as a result of these privatisations, the number of individual shareholders in Britain increased dramatically. Whereas in 1979 fewer than 3 million individuals (only 7% of the adult population) held shares, by 1991 it had risen to 11 million (representing 25% of Britain's adult population) (Figure 3).

'POPULAR CAPITALISM'? THE GEOGRAPHICAL LIMITS OF PRIVATE SHARE-OWNERSHIP

However, despite the fact that the privatisation programme continued apace throughout the first half of the 1990s, by 1995 the number of individuals owning

Table 2 *The Public and Institutional Allocations of Some Major UK Privatisation Flotations*

Privatisation	Total shares offered	General public allocation (%)	Institutional allocation (%)
British Telecom (1984)	3 012 000 000	38.5	61.5
British Telecom (1991)	1 597 500 000	65.7	34.3
British Telecom (1993)	1 311 500 000	55.8	44.2
British Gas (1986)	4 025 500 000	62.0	38.0
British Airport Authority (1987)	500 000 000	46.0	54.0
Regional Water Companies (1989)	2 185 000 000	44.0	56.0
Regional Electricity Companies (1990)	2 311 614 000	50.6	49.4
Northern Ireland Electricity (1991)	164 600 000	67.0	33.0
Scottish Electricity (1991)	1 198 259 000	58.0	42.0
RailTrack (1995–1997)	500 000 000	42.0	58.0
GENCO (1991)	1 253 565 706	49.4	51.6
GENCO (1995)	759 691 098	51.3	48.7
British Energy (1991)	700 000 000	42.6	57.4

Source: HM Treasury (1996) *Guide to the Privatisation Programme*

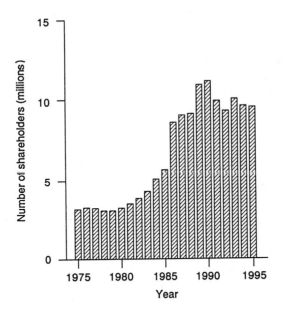

Figure 3 *The Growth in the Number of Private Shareholders, UK, 1975–1995*

shares had retreated somewhat to 9 million. For many individuals, privatisation has not encouraged a general shift to a long-term shareholding culture. As a result of their initial undervaluation, privatisation share prices have tended to rise very sharply after flotation, and this has encouraged many shareholders to cash in their shares to make a quick windfall profit.[14] For example, in the case of the 1984

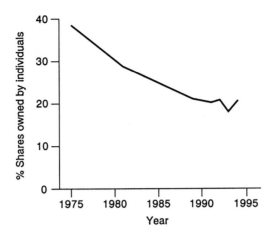

Figure 4 The Declining Presence of the Individual Shareholder in the UK Equity Market, 1975–1995

British Telecom and 1986 British Gas privatisations, 30% of the original shareholders had sold their shares within one year of the flotation; in the case of British Associated Ports, British Airport Authorities and the 12 Regional Electricity Companies, 60% or more no longer held their shares after one year. At the same time, privatisation sales do not seem to have inspired many individuals to invest in the stock market more generally. Of those who first bought shares as part of a privatisation offer, a mere 2% have gone on to buy shares in a non-privatised company (*The Economist*, 1996). As buyers of privatised shares have not ventured into other equity investment, holdings are typically small and narrow: half of all individual shareholders own shares in only one company. The fact is that although the numbers of individual shareholders has trebled since 1979, the percentage of the equity market now owned by private households has actually continued to fall, from 30% to about 20% (Figure 4), and the domination of the equity market by the large institutional shareholders has continued to increase. Thus, despite Britain's massive privatisation programme, the ownership structure of the capital market has continued to shift away from individuals, in the same way that it has done over the same period in other major advanced economies, such as the United States, Germany, France and Japan, where the scale of privatisation has been considerably less (see Table 1 and Figure 1). Hence, although share-ownership in the UK is certainly wider than it was, it is not deeper.

However, the widening of individual shareholding has not been a uniform, nationwide process either. Information on the geographical incidence of shareholding across the UK is surprisingly thin.[15] There are no regular government or stock market statistics on the geographical incidence of privatisation shareholding by the public, or indeed on private shareholding in general, and little is known about the spatial distribution of this aspect of the UK's financial landscape. However, unpublished surveys of household finances undertaken by the private market research company NOP Financial Research can be used to throw some light

Table 3 The Regional Distribution of Share-ownership in the UK, 1994–1966
(Regional Proportions of National Total Number of Households Owning Shares)

	Total shares	Number of companies in which shares held					Proportion of UK population
		One	Two	Three	Four–ten	11 or more	
South East	43	42	45	42	43	45	31.4
East Anglia	5	5	3	1	2	3	3.8
South West	10	10	8	14	11	0	8.4
East Midlands	6	6	5	2	7	6	7.4
West Midlands	7	7	9	15	4	7	9.6
Yorkshire/Humberside	5	5	9	7	7	4	9.1
North West	8	6	7	9	10	5	11.4
North	4	5	3	3	5	8	5.7
Wales	4	5	4	2	3	3	4.0
Scotland	8	9	7	5	8	9	9.2
UK	100	100	100	100	100	100	100.0

Source: Based on data supplied by NOP Financial

on the geography of 'popular capitalism' in the UK.[16] The regional incidence of private share-ownership in 1994–6 is shown in Table 3. The feature that stands out is that the south-east region of the UK contains well in excess of 40% of the nation's private shareholders, by far the highest percentage of any region (Hamnett, 1992). The remaining shareholders are spread thinly across the rest of the country, the south-western region being the only other area containing any sizeable proportion (10%). The northern regions of the country – the North, Wales and Scotland – account for only small proportions. Of course, the regional distribution of share-owning households will reflect regional variations in the distribution of the adult population, and as the South East contains the largest proportion of the country's population, it would be expected to dominate the geography of share-ownership. However, it is clear from Table 3 that the concentration of private shareholders in the South East (43% of the national total) is considerably greater than would be expected on the basis of the region's share of the country's population (31%).

This suggests that there are major differences between the regions in the *propensity* of the population to own shares. Table 4 shows the proportion of each region's households that owns shares of different types. Several features are evident. First, the three southern regions – the South East, South West and East Anglia – have significantly higher *rates* of shareholding than the rest of the UK: 24–28% of households in these areas directly own shares of one kind or another, compared to 16–18% in northern England. Second, in almost every region about 60% of share-owning households hold privatisation shares. Third, the rate of privatisation share-ownership is about 50% higher in the South East and South West than elsewhere. Fourth, the degree of employee share-ownership is low everywhere (only about 3–5% of households). It would seem, therefore, that – with the exception of employee share-ownership – private shareholding is indeed substantially more extensive in the South East than in other regions of the UK.

Table 4 The Share-ownership Densities in the UK Regions, by Broad Type of Share, 1994–1996 (% of Households)

	SE	EA	SW	EM	WM	YH	NW	N	Wa	Sc
All shares	28	24	26	17	21	17	18	16	18	18
Privatisation shares	17	10	16	11	10	11	11	11	10	11
Employee shares	4	4	3	3	3	3	5	2	3	3
PEPs and unit trusts	11	12	12	7	12	8	6	7	10	10

Note: Based on data from NOP Financial
Key: SE – South East YH – Yorkshire/Humberside
  EA – East Anglia NW – North West
  SW – South West N – North
  EM – East Midlands Wa – Wales
  WM – West Midlands Sc – Scotland

Even in the case of privatisation flotations, where perhaps the expectation might be – and certainly the political intention was – that the take up of shares would be more evenly spread across the country,[17] the South East emerges as the region with the highest share-ownership rate. To the extent that privatisation has promoted 'popular capitalism', it has clearly been much more 'popular' in the South East than in other regions of the UK.

This pattern almost certainly reflects variations in household incomes between the regions. Individuals or households with high average incomes are able to save more, and to be less risk-averse in their savings and investment, than those households with lower incomes. Thus, households with higher incomes are more likely to include shares as part of their financial portfolios. The evidence on household savings and incomes across the UK tends to bear this out (Figure 5 and Table 5). Thus, at one extreme, the South East, with the highest share-ownership rate, also has by far the highest average household income (nearly 20% above the national figure). At the other extreme, the North and Wales with the lowest share-ownership rates also have the lowest household incomes (more than 20% below the national average). However, probably of more importance than regional differences in average household income are variations in household income distributions between regions. Thus, in the South East the income distribution is skewed towards the top income bands, whereas in the North, North West, Wales and Scotland it is skewed instead towards the bottom income groups (Martin, 1995). Significantly, the South East has twice the proportion of high-income households (gross weekly incomes above £550) than has the rest of the country.

The patterns of share-ownership associated with specific major privatisations reveal even higher concentrations in the South East. The British Telecom, British Gas and Electricity Company sell-offs have been the three most popular privatisations; nationally, these companies account for about two-thirds of privatisation shareholdings. In the case of British Gas, the South East accounts for 44% of the households owning shares in this company, but in the case of British Telecom and the Regional Electricity Companies more than half of the private shareholders are to be found in this region (Table 6). In its privatisation publicity, the government argued that shares sold in major privatisation offers should be distributed as widely as possible to private individuals in order to increase the

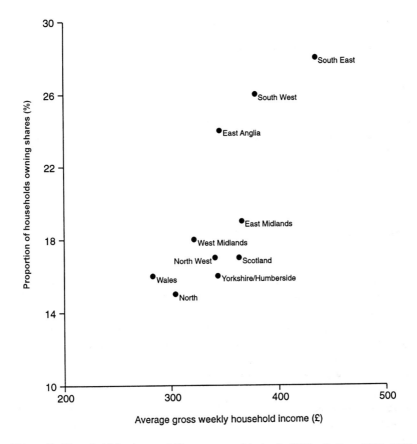

Figure 5 Household Income and Share-ownership in the UK by Region, 1995–1996

Table 5 Household Income and Savings by UK Region, 1994–1995 (% of Households with Particular Types of Savings)

	Average household income (£)	Building society savings account	Tax exempt special savings account	National savings	Stocks and shares
South East	435.4	65	13	10	26
East Anglia	345.6	59	12	9	23
South West	378.9	63	11	10	23
East Midlands	365.8	61	12	8	19
West Midlands	321.8	57	9	7	18
Yorkshire/Humberside	343.4	53	9	7	16
North West	340.9	52	10	7	17
North	304.0	50	7	6	15
Wales	282.7	48	8	7	16
Scotland	363.0	41	8	6	17

Sources: Family Expenditure Survey, Office for National Statistics and Family Resources Survey

Table 6 The Regional Distribution of Share-ownership in some Major UK Privatisations, 1995–1996

	British Telecom	British Gas	Electricity
South East	52	44	52
East Anglia	4	4	1
South West	4	10	13
East Midlands	4	5	8
West Midlands	6	7	1
Yorkshire/Humberside	9	7	5
North West	4	7	9
North	4	4	5
Wales	4	5	3
Scotland	9	7	3
Great Britain	100	100	100

Source: Based on data supplied by NOP Financial

Note: Proportions of households owning privatisation shares

proportion of economic assets owned by individual, rather than institutional, investors, and thereby make privatised industries accountable to a wider constituency of shareholders (HM Treasury, 1995). The privatisation of the gas, water and electricity distribution industries provided the opportunity to introduce an explicit geographical dimension into these objectives. Under state-ownership, all three sectors had been organised on a regional basis (as regional boards). Although in the event British Gas was sold off as a single concern, the water and electricity industries were sold off as regional distribution companies (in the case of electricity, there were also national generating and grid companies). Here were public industries – local monopolies – with a substantial local concern and captive customer base. Some attempt was made to 'regionalise' the public flotations by giving local customers incentives to invest in their own local regional distribution companies, but it would appear that significant levels of non-local individual investment occurred in certain regions.[18] Moreover, about half of the shares (54% of water and 49% of electricity) were in any case reserved for institutions. In the years since privatisation both the regional water and elecricity companies have witnessed major foreign involvement. Thus, French firms have taken significant holdings in two of the regional water companies, while (as of mid 1997) no less than seven of the 12 regional electricity companies have been bought out by US electricity corporations – further reducing the limited 'local' character of these privatised regional utilities and shifting their share-ownership structures overseas.[19] These takeovers have typically involved generous share-price offers to private shareholders in the UK, whose ownership in these privatised utilities has fallen as a consequence.[20]

That share-ownership would grow more in the South East could have been predicted. Not only is this the wealthiest region in the country, it was also the region that politically was most supportive of the Conservative Governments of the 1980s. It benefited most from the Conservative's economic and employment boom in the second half of the decade, and was also the main beneficiary of the cuts in personal taxation championed by the Conservative Government in the late 1980s.[21] The southern electorate has been much more politically and ideologically inclined towards both the policy of privatisation and the idea of private shareholding. The combination and interaction of these factors has almost certainly contributed to the higher propensity of the adult population in this area to purchase and hold on to privatisation shares. A further factor, however, may well have been the peculiar institutional landscape of Britain's financial sector, concentrated as it is in London and the South East, an issue discussed below.

Although the absolute number of shareholders has certainly risen substantially because of privatisation, the available evidence indicates that rather than spreading share-ownership widely, 'popular capitalism' has only really achieved some measure of success in southern Britain, and even there it is hardly a generalised phenomenon. Further, the retrenchment in the numbers of private shareholders between 1990 and 1995 suggests that privatisation did not succeed in stimulating a societal shift towards a more general private share-ownership culture. The bulk of the population prefers to save in less risky cash-based investments, such as building society savings accounts and tax-free savings schemes (Table 5).[22] In a survey of the barriers to private share-ownership, the Stock Exchange (1996) argued that a key problem is that the role and benefits of equities as a form of long-term personal saving and investment are simply not understood by the general population. In the case of the privatisations, however, there have also been other problems. The large profits being made by the privatised utilities, and the opening up of different pricing regimes in different regions for the same utility, are seen by many as running counter to the whole ethos of universal public utility provision at common standards and prices regardless of location.[23] Privatisation clearly raises important tensions between the interests of consumers and shareholders (groups that may overlap but which are certainly not identical), tensions which a partial, and socially and geographically uneven, 'popular capitalism' serves to accentuate rather than ameliorate.

The Institutional Geography of the Equity Market

In addition to variations in wealth and socio-economic culture, a further factor shaping the scale and geographical pattern of private share-ownership is the institutional 'thickness' of the capital market, that is whether there is a well-developed and easily accessible system of financial institutions and agents through which individuals can purchase and sell shares. In one sense, the UK's financial system is based on an extremely sophisticated, diverse and 'thick' institutional network, with a long history of accumulated expertise in almost every type of monetary and financial transaction. Yet, according to the Stock Exchange (1996), one of the reasons for the limited growth of shareholder capitalism in Britain has been the difficulty private individuals experience in accessing the equity market, which is characterised by deficiencies in both the distribution network and the quality of the institutions of which that network is composed. A major problem is that the British financial system is not well geared to promoting and marketing share-ownership to the small investor, instead being much more orientated towards the large corporate and institutional players in the market. Partly as a result of this, the public has an adverse perception of the financial market, as 'mysterious', 'élitist' and dominated by 'the City' (of London).

The high degree of geographical concentration of Britain's financial system in London is well documented. Based around the 'City' – the institutional nexus consisting of the Bank of England, the Stock Exchange, Lloyds and the Treasury – London is the centre of the nation's banking, insurance, equity and currency markets. Although recent years have seen a growth in provincial financial centres, it is in London that the overwhelming majority of banks, insurance companies, stockbrokers, pension fund managers, venture capitalists and accountancy firms are located, and where major financial decisions and policies are made (Leyshon and Thrift, 1989; Martin, 1989; Martin and Minns, 1995). Through its role as a national and, indeed, a world monetary centre, the agglomeration of financial institutions and activity in London far exceeds anything found in other major cities in Britain. Whether this agglomeration in London makes for an efficient and accessible market regardless of an investor's location (as argued by Clark, 1997b), or whether it benefits the South East over the rest of the UK (as argued by Martin, 1989; Martin and Minns, 1994; Hutton, 1995), is a subject of some debate.

This issue is of direct relevance to the privatisation programme. All of the share flotations associated with privatisation sales have been been undertaken on behalf of the government by London-based banks and lawyers.[24] The argument has been that the big London institutions were best placed – professionally, financially and geographically – to perform this task, and that they alone were capable of underwriting the privatisations. Further, not only is London at the natural apex of the global financial triangle, an advantage widely recognised by the international financial community, it is the world's largest market for trading foreign equities, which makes it the most liquid international equities market. This means that London banks are well positioned to distribute shares across the globe. Many of the major UK privatisation flotations have included an explicit 'international offer' (on the grounds that this helps to maximise the sale proceeds). The use of firms with strong international distribution capabilities has thus been seen as essential to the success of these international offers. In most of the privatisations, syndicates of London-based UK banks have dominated share allocations in the United States, Europe, Canada, Japan and other global regions. Such has been the success of London-based institutions in executing the privatisation sales for the UK government, that by the early 1990s they had also become the leading players in privatising state activities around the world (Marshall and Harding, 1993; HM Treasury, 1995). As the UK's privatisation programme began to slacken in the mid 1990s, so these London institutions increasingly shifted their attention to the rising tide of privatisations overseas.

Despite their role in marketing and underwriting the privatisation sell-offs for the UK government, these big London-based institutions have shown relatively little interest in the UK's small shareholders. Indeed, what the privatisation programme has highlighted is a major gap in the small investor retail layer of the equity market. This need has been particularly acute since the City of London 'Big-Bang' of 1986 when major changes were introduced to the workings of the Stock Exchange, and the interests of private investors were further subordinated to those of the large institutions. A number of institutional initiatives have been introduced in an attempt to remedy this problem. The Share Shop scheme, introduced by the government in its 1991 and 1993 sales of British Telecom, was intended to provide a geographically dispersed and specially designated retail service (selected stock brokers, banks and building societies) which would give private individuals easy access to privatisation shares. A second insitutional innovation was the formation in 1992 of ProShare, an independent, London-based non-profit-making company, funded jointly by the Treasury, the Stock Market and a number of private sector companies, to represent the views of individual shareholders to government and provide its members with a range of information and advisory services, both on privatisation share offers and the equity market in general. The impact of both of these institutional developments appears to have been limited.

More recently, in an attempt to balance the powerful influence of the major institutional investors and the London-based brokers, a nationwide Association of Private Client Investment Managers and Stockbrokers (APCIMS) has been established, consisting of more than 350 members (stockbrokers and investment managers) who specialise in services for personal (non-institutional) shareholders. As several writers have argued (O'Brien, 1992), in principle the new on-line information technologies and electronic trading systems that have revolutionised financial markets allow institutions to disperse from the major financial centres such as London and to locate near to potential customers wherever they may be whilst having direct and immediate access to the central markets for buying and selling shares. Although APCIMS members are to some extent spatially dispersed across the country, with notable presences in the major provincial cities (Bristol, Birmingham, Manchester, Leeds, Edinburgh and Glasgow), 40% of these organisations are found in the South East (see Figure 6). London alone accounts for 26%, and significantly, many of those outside the South East are in fact branches of London-based institutions.

In any case, recent and on-going reforms to the finanical system and the City of London threaten to squeeze the private investor out of the market. The use of 'nominee accounts' in association with the new electronic paperless share trading and settlement systems is likely to limit further the ability of private investors to participate in the equity market. The latest change to the system – the move to computer-matched 'order-driven trading' – also militates against the individual small investor, as deals worth less than £4000 are excluded (Prosser, 1997). Although the Stock Exchange recognises the individual investor's lack of direct access to the equity market, it has done little to remedy the problem, and thus far its new Private Investors Advisory Committee has had little positive impact. Not only is the Stock Exchange extraordinarily myopic when it comes to individual shareholding in the regions, by virtue of being 'self-regulating' there is little internal impetus for radical institutional change. There are, then, still no signs of a 'people's capital market' emerging. This will remain unlikely with an equity market

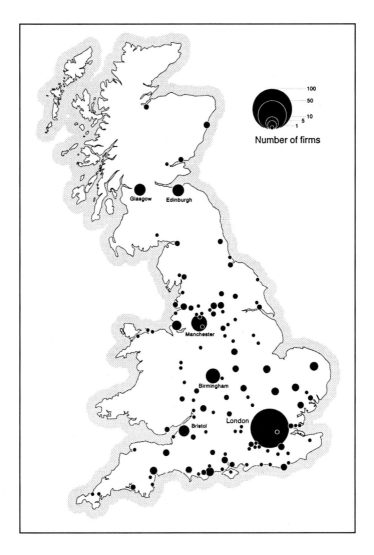

Figure 6 Private Client Investment Managers and Brokers, UK 1997. Source: APCIMS

dominated by the large, London-based and London-oriented financial institutions, with no real interests in the small investor, and the lack of any radical changes to the institutional structure of the market.

CONCLUSIONS AND IMPLICATIONS

As the UK case demonstrates, economic, socio-cultural and institutional factors may all limit the overall scale and the geographical spread of private share-ownership. In the UK, privatisation has not stimulated the 'historic transformation' towards shareholder or 'popular' capitalism that was hoped for. Although privatisation helped to treble the number of private shareholders between 1979 and 1995, it would seem that 'popular capitalism' has been largely confined to the South East, where much of the nation's wealth was already concentrated, and where socio-cultural and political attitudes were more supportive of privatisation policy and more receptive to the idea of investing in shares. Of course, it could be argued whether this limited, spatially biased success matters. After all, shares are a risky form of investment, especially for lower-income households, who are better advised to invest in more secure cash-based forms of savings. Given that there is an upper limit on average household savings ratios – of about 20% of disposable income – there are also limits in any case on how far private share-ownership might be expected to permeate the population. On the other hand, given that share values have consistently and substantially outperformed cash-based savings over the long term, the uneven social and spatial distribution of shareholding has served to reinforce social and geographical inequalities in wealth and income. Spreading share-ownership more widely across the population might then be considered a desirable goal if it helps to increase wealth accumulation amongst middle- and lower-income groups. Arguably, however, this is best achieved by encouraging greater participation in less risky 'collective' investment funds such as unit trusts (e.g. the British tax-exempt personal equity plans or PEPs) than via direct private share-ownership.[25]

By the late 1990s building society de-mutualisation had replaced privatisation as the main mechanism for creating mass share-ownership in the UK. Over recent years, several building societies (e.g. Abbey National, Alliance and Leicester, Halifax, Woolwich and Northern Rock) have 'floated' and become joint stock companies quoted on the stock market. By giving free shares to all their savers, these demutualised nationwide building societies created very large numbers of 'instant' shareholders, distributed much more evenly across the country. The Halifax flotation, for example, created 7 million new shareholders overnight. However, as in the case of the state privatisations, many of these new 'instant' shareholders have cashed in their free shares for windfall cash sums. For example, when the Alliance and Leicester Building Society converted to a public company in April 1997, shares were issued to 2.321 million individual members. Some 0.596 million of these sold their shares through the free dealing service offered on day one. Within six months of flotation, more than a million (45%) had sold their shares. Similarly, nearly half of the 0.88 million members of the Northern Rock Building Society immediately sold their free shares when the society demutualised in late 1997. Like privatisation, building society demutualisation may have helped to widen share-ownership, but much less than was expected.

In addition, there are certain aspects of contemporary, finance-dominated capitalism that limit, and even militate against, any sustained growth in individual shareholding. Despite the fact that, in the past, equities have provided a much higher return than cash-based forms of savings, equity markets have become increasingly subject to dramatic swings in fortune. The globalisation and deregulation of securities markets, combined with large-scale speculation and global economic uncertainty, have introduced a latent instability in share prices. The worldwide stock market crashes of October 1987, November 1997 and August–September 1998 have made direct individual shareholding (and even collective funds) a potentially high-risk investment for the average household. Yet, as states increasingly retreat from expensive pension and welfare schemes, and households have to make their own provision (as seems inevitable across much of Western Europe), so these institutional funds are likely to expand further (see Chapter 12). Under these circumstances, a major growth in individual private shareholding would seem to be unlikely.

Furthermore, there may be inherent structural contradictions between contemporary systems of corporate governance and dispersed small-scale private share-ownership. As Modigliani and Perotti (1996) point out, the development of equity markets depends on how well laws protect the property rights of minority shareholders. The more dispersed the shareholder base of corporate equity, the more important those minority property rights (voting rights, powers of veto over management, etc.) are if there are to be external checks and balances on corporations and their activities. On the other hand, the more dispersed that share-owership is, the less efficient corporate governance may be. In the UK, the hope that privatisation would disperse corporate accountability to the population at large was unrealistic, even if private shareholding had progressed much further than it actually has. The securities market is the major mechanism for monitoring and disciplining firms, and the major aim of shareholding is portfolio performance (dividends and capital growth) rather than corporate control; shareholders largely leave company managers to their own devices, monitoring them through their trading in the stock market, with share prices acting as a sort of 'approval rating'. However, transaction costs and incomplete information limit trading for small shareholders, and in any event, individual shareholders rarely own sufficient proportions of a company's stocks to be able to exercise significant influence over its policies and management, either through transactions on the stock market or at general shareholders' meetings. These issues have unfavourable implications for privatisation, where the emphasis has been on encouraging dispersed small shareholdings. Ironically, dispersal of share-ownership could in fact impair the capital market's monitoring of the behaviour and performance of the privatised enterprises (McDonald, 1993: 49). In Britain, even big institutional shareholders, such as the pension funds, have tended not to use their substantial voting rights in the companies in which they invest.[26]

Thus, there may well be structural limits to the spread of private shareholding arising from the nature of contemporary, 'money-manager capitalism'. To the extent that the creation of 'popular capitalism' is one of the hopes of other West European states' on-going privatisation schemes, the limits and weaknesses of the UK experience provide some salutary lessons. In the former centrally planned socialist states of Eastern and Central Europe, China, and many developing countries, the obstacles and limits to spreading popular capital-ownership through the mass privatisation of the state are even more formidable, especially given the low disposable income base in these societies (Ash, Hare and Canning, 1994). Although these countries have been pushing ahead with free share schemes, voucher systems and mutual share funds, unless at the same time they develop the appropriate financial infrastructures, systems of property rights, financial incentives and corporate governance structures needed to promote and support the individual minority shareholder, not only is the overall scale of individual capital ownership likely to prove disappointing, but regional and institutional biases in private shareholding seem inevitable.[27]

ACKNOWLEDGEMENTS

An earlier version of this chapter was given as a paper at the Regional Studies Association conference, 'Regional Frontiers', held in Frankfurt-Oder, Germany, in September, 1997. The comments made on that occasion were helpful in revising the arguments in this chapter.

The research for the paper forms part of a larger project on 'The Geographies of Private Shareholding', funded by the ESRC (contract R000222519).

ENDNOTES

CHAPTER 1

1. For example, Lösch recorded how US bank interest rates during the period 1915–25 in general rose with increasing distance from the liquid credit markets of New York city. This reflected the tendency for uncertainty, risk and transaction costs to grow the greater the distance between lenders and borrowers.
2. The main focus of this work was to propose a hierarchy of money forms, from local money to global money, and to theorise their implications for credit creation. I am indebted to Christian Berndt of the Department of Geography, Katholische Universität, Eichstätt, Germany, for assistance in translating this unknown paper.
3. Jean Labasse was president of a large private bank in Paris and a professor of urban planning in the University of Paris. Many years earlier, he had published a study in regional finance (*Les Capitaux et la Region*, 1955; Paris: Colin), which analysed the banking system and money flows of the Lyons area in France.
4. Although the rise of the credit-card economy, of telephone banking, electronic fund transfers, call centres, and related developments suggests that for these and certain other retail financial services (such as insurance) the need for locally based institutions may be declining.
5. The pre-eminence of London, New York and Tokyo as global financial centres is intimately bound up with the historical international political-economic roles of their nation-states as former imperial powers (in the case of the UK and Japan) or as geopolitical economic hegemons (as in the case of the United States).
6. The same was true of the stock market. During the nineteenth century, there were active stock markets in a number of provincial cities (including Birmingham, Leeds, Bradford, Bristol, Manchester and Cardiff) as well as in London. Again by the early part of the twentieth century, the regional markets had either closed down or had become regional offices of the London stock market.
7. The geography of the Gold Standard, for example, is vividly elucidated in Eichengreen and Flandreau (1996). Currency boards have been mainly a feature of developing countries. The Basle Banking Accord was created among the G10 countries, led by the United States and the UK.
8. The share of government in GDP has consistently been highest in European countries, especially Sweden, Germany, France and the UK, and lowest in the United States and Japan.
9. These include post-industrial society, the information economy, post-modernity, post-Fordism, flexible accumulation and even post-capitalism.
10. Most large stores and retail chains now offer their own credit cards, and an increasing number loan, insurance, investment and savings services. In effect, many retail companies have become quasi-banks.
11. Indeed, the degree of 'time-space shrinkage' claimed by many writers on globalisation has been greatly exaggerated.
12. Explanations of the wave of bank failures usually place primary blame on deteriorating local economic conditions. However, the increase in failures in the 1980s also coincides with the deregulation of the US banking system, which has allowed greater risk-taking by banks (Hook, 1994).
13. The stock markets of Indonesia, Thailand, South Korea, Malaysia and the Philippines sustained losses of 60% or more during 1997, and share prices in Hong Kong and

Singapore also declined severely. In total some $600 billion, the equivalent of two-fifths of the region's GDP, was wiped off the stock markets' value.

14 The merger movement is not confined to banks and insurance companies – stock exchanges are also merging. In Australia, for example, the six regional exchanges merged into the Sydney-based Australian Stock Exchange in 1987; in Germany, the regional exchanges in Frankfurt, Düsseldorf and Munich merged with Deutsche Termin Börse into Deutsche Börse in 1993; and in Switzerland, in 1995, the Zurich, Geneva and Basle exanges merged with Soffex, the screen-based derivatives exchange, into the Swiss Exchange.

15 Notable examples include the merger of the Bavarian banks Bayerische Vereinsbank and Bayerische Hypo-Bank, and the merger of the Berlin banks Nord Landesbank and Bankgesellschaft Berlin.

16 Examples include the merger of Barnett Banks of Florida with NationsBank of North Carolina, of Wells Fargo with First Interstate, Chemical with Chase, and First Union in Charlotte with First Fidelity of New Jersey.

17 Local or community currencies schemes are proliferating in the United States, Canada, Europe and Australia. Legal and easily manageable, local currencies enable people to gain access to goods and services outside the formal national banking system, which in any case is often abandoning them. These community currencies may be true currencies, having a physical form and traded for goods and services alongside the national currency; or they may be local barter and exchange networks, with no physical currency being used (hence the term 'local exchange trading systems'). The 'rules' for these different forms of local money vary from community to community.

CHAPTER 2

1 In this and other respects, it will be evident that the theory is more compatible with post-Keynesian, institutionalist and evolutionary economics than with the standard, mainstream neoclassical tradition. The idea that 'history matters' in the organisational evolution – and as will be seen, the spatial structure – of the financial system is another example of the 'path dependence' that dominates much of the economy (see David, 1994; Arthur, 1994).

2 Of course, there may be switches in the location of the financial centre within a country, but such switches usually arise because of exceptional events or changed circumstances. Once the locational shift has occurred, cumulative path dependence once again becomes crucial for the new centre to succeed over the old (see Chapter 5).

3 Before 1975, *intra*-state restrictions on branching by commercial banks were commonplace: only 17 states allowed banks to branch statewide. Since then, and mainly during the 1980s, restrictions on intra-state branching have been removed or relaxed in all states. Until the late 1970s, no state permitted *inter*-state branching. State barriers to inter-state branching began to fall in 1978, and during the 1980s and early 1990s every state (except Hawaii) dismantled at least some of its restrictions on inter-state banking.

4 In fact, a major restructuring of the German regional banking system has begun, involving alliances between the Landesbanks, mergers between public and private sector banks (e.g. in Berlin), and mergers between neighbouring banks (Bayerische Vereinsbank and Bayerische Hypo-Bank in Munich) to form 'super-regional' banks capable of competing across the EU market as a whole.

5 Indeed, a worrying trend is that the reorganisation of the banking system that is taking place in some countries appears to involve spatially selective rationalisation, in that it is precisely the poorer (less 'credit-worthy') localities – such as certain inner-city and depressed urban neighbourhoods – that seem to be bearing the brunt of branch closures (as in the UK and the United States) (Leyshon and Thrift, 1995). The withdrawal of

banks from such areas makes them vulnerable to high-interest charging non-bank and individual money-lenders. Alternatively, such communities may be forced to develop their own alternative 'local monies' such as local exchange and trading systems (LETS).

6 This, arguably, has been a traditional strength of the German and Italian regional banking systems. Both of these systems are now in the process of being restructured, so their future is by no means assured (for an analysis of the Italian case, see Chapter 4).

CHAPTER 3

1 This material is drawn from a wider study of the banking industry in Los Angeles (Pollard, 1995a).
2 While Transamerica was building up a chain of banks in seven western states, the Bank Holding Company Act of 1956 was being drafted, with provisions designed to halt the spread of interstate banking and to force holding companies such as Transamerica to divest itself of either its bank or non-bank holdings.
3 During the 1980s, California was the recipient of 20% of federal defence expenditure dollars, a ratio it has maintained since the 1950s (State of California, 1995).
4 In the mid 1980s, Barclays and Midland sold their California banks to Wells Fargo, Standard Chartered sold Union Bank to the Bank of Tokyo and Sanwa Bank took over Lloyds Bank of California.
5 By one estimate, 40% of all Japanese FDI in the late 1980s went to California (Anonymous, 1988).
6 Foreign banks entering California have been aided by US banking and anti-trust regulations which have hindered US out-of-state banks' efforts to expand into California (Doti and Schweikart, 1990).
7 By 1989, California's gross product of US$703 billion was surpassed only by that of the United States, Japan, Britain, Germany, France and Italy (State of California, 1996).
8 By 1988, 582 S&Ls (out of 3200 nationwide) had been closed and another 800 were being supported by regulators (Pizzo, Fricker and Muolo, 1989). The FSLIC, the federal overseer of deposit insurance for S&Ls, was insolvent by 1987 and had its deposit insurance function transferred to the Federal Deposit Insurance Corporation (FDIC) in the Financial Institutions Reform, Recovery and Enforcement Act of 1989.
9 The S&L bail out has a distinct geography, with north-eastern and midwest states losing the most and south-western states such as Texas receiving up to US$4775 *per capita* in federal transfers (Warf, 1994).
10 These are the equivalent of unit trusts in Britain.
11 In 1988, Brazilian debt was trading in secondary markets at 29 cents on the dollar; and Argentinean debt at 18 cents on the dollar (Johnston, 1990).
12 These countries together with Italy, France, Germany, Belgium, Sweden, Luxembourg, the Netherlands and Switzerland formed the Basle Committee on Banking Regulations and Supervisory Practices which reported in 1988.
13 Two of the large retailers offer discounts to customers using only telephone and ATM services; if such customers perform routine transactions in a branch, however, they are subject to a transaction fee (Bornemeier, 1995).
14 Centralisation also gave banks more control over loan quality – an important consideration after their losses on LDC debt – while automated, at-a-distance credit scoring was also a defence against law suits for racial discrimination in lending.
15 Banks have lobbied hard for deregulation to expand their product base and now sell annuities and unit trusts, and, courtesy of a Federal Appeals Court decision to overturn a Federal Reserve Board ruling in 1991, can underwrite and sell insurance nationwide (Hilder and Lambert, 1991).
16 Banks are required, under the auspices of the Community Reinvestment Act of 1977, to help meet the credit needs of low- and moderate-income communities in which they are

chartered. Federal regulators can assess banks' Community Reinvestment Act records when examining applications for branch openings and closures, new charters and so on.

17 In 1979, bank workers in Los Angeles received US$8.74 an hour (in 1989 $US), compared with US$10.14 an hour for all industries.
18 One of the largest retailers in California, recognizing the growing market power of computing professionals, has created a separate pay scale for computing workers, pitched 27% higher than that for their other job grades (Pollard, 1995a).
19 By 1989, for example, women's median hourly earnings were 80.2% of those of men, up from 69.9% in 1980 (Pollard, 1995a).
20 Census data probably understates the growth of these sales occupations as many bank workers acquire 'managerial'/'officer' job titles once they attain supervisory status.
21 The big four, together with Abbey National and TSB, wrote off £23.4 billion against profits between 1990 and 1994 (Mintel, 1995).

CHAPTER 4

1 Among others, see Chick and Dow (1988), Dow (1990), Vives (1991), Gentle and Marshall (1992), Greenwald, Levinson and Stiglitz (1993), Corbridge, Martin and Thrift (1994), Chiapporri, Perez-Castrillo and Verdier (1995), Jayaratne and Strahan (1996), Zazzaro (1997) and with particular reference to the Italian case, Faini, Galli and Giannini (1993), Alessandrini (1992 and 1996a), Messori (1995).
2 See Alessandrini (1994) and Dei Ottati (1994).
3 For a discussion of the main institutional innovations introduced in the Italian banking system, see De Cecco (1994).
4 The administrative regions considered here correspond to the NUTS I definition.
5 See Pettenati (1990) and Canullo and Pettenati (1994).
6 See Crivellini and Pettenati (1994) for a broad analysis of the different phases of Italian regional development over the four decades up to the 1980s.
7 See Graziani (1986) and Del Monte and Giannola (1997).
8 Among others, see Fuà (1983 and 1988), Becattini (1990), Brusco (1986 and 1989) and Garofoli (1989).
9 See Minsky (1986), Moore (1988) and Stiglitz and Weiss (1988).
10 On financial centres, see Kindleberger (1978), Amin and Thrift (1992). On the organization of the financial system on a hierarchical order from centre to periphery, and on the differences between international and national levels, see Alessandrini (1989).
11 The regional classification of the Italian banking system in Figure 1 is based on a cluster analysis conducted by P. Alessandrini (1996b). He used more than 30 indicators, ranging from traditional indicators of banking structure (such as the ratio of the percentage share of branches operating in an area to the percentage share of resident population or the amount of loans for each branch) to indicators of financial innovation (for example the ratios of portfolio managements and deposit certificates to population).
12 See Faini, Galli and Giannini (1993).
13 This transfer of costs on interest rates is a sign of a higher level of monopoly power of the bank operating in the Mezzogiorno. That the southern credit market is still less competitive is confirmed by the higher values of the Herfindhal index (see De Bonis and Ferrando, 1996).
14 It should be pointed out, however, that these are average figures, which are heavily influenced by the negative results of the Banco di Napoli and the Banco di Sicilia.
15 See Alessandrini (1996a).
16 For this view, see Chick and Dow (1988).
17 See Castelli, Martiny and Marullo Reedtz (1995). Indirectly, this is also evident from the data on the loans/deposits ratio shown in Figure.4.

18 Becattini and Rullani (1993) and Sabel (1996) give a stimulating general outline of this approach, although they do not refer directly to financial issues.
19 See Alessandrini (1989).
20 The introduction of a single European currency (the 'Euro'), a European central bank, etc.
21 On this point, see Conigliani (1990) and Favero and Papi (1995).
22 As shown by Ferri (1997), in the Italian banking system branch manager turnover is higher for national banks than for local ones.
23 See Zazzaro (1997).
24 As Clemenz (1991: 337–8) notes, 'credit markets are quite different from market for ordinary goods, such as bread. . . . The baker does not lose money on the bread he has already sold if he sells more bread. A lender, in contrast, may (adversely) affect his expected returns on existing loans by granting additional ones.'
25 Since 1995, the Bank of Italy has provided statistics classifying banks for territorial diffusion, in addition to the traditional classification by size. With regard to this new classification, banks are distinguished as national, inter-regional, regional, interprovincial, provincial or local on the basis of the geographical coverage of their distributive network.
26 The only southern bank operating on a national scale is the Banco di Napoli, which, due to its serious financial distress, is about to be absorbed by an outside bank (the Banca Nazionale del Lavoro). The Banco di Sicilia, the most important inter-regional bank in the Mezzogiorno, was acquired by MedioCredito Centrale, from the Centre North, after a crisis.
27 See Niccoli and Papi (1993).
28 Obviously, in the case of small firms the banks could find it feasible to issue asset backed securities only against a pool of small loans, on a declaration of their reliability.

Chapter 5

1 See Krugman's (1993) simple but elegant general equilibrium model of city location which illustrates these principles.
2 There are 60 'mini-states' functioning as OFCs, but only a handful are important on a worldwide scale (Laulajainen, 1998: chap. 7). Geographically, everything which is separated by a body of water is 'off-shore'. However, within financial economics, the term has a wider connotation, namely financial structures which follow different rules from the rest of the economy and are separated from it. Accepting the latter definition, euro markets are off-shore; as are countries and territories which are part of a mainland, but whose economy is largely orientated towards the needs of the international financial community. Therefore, Luxembourg, Liechtenstein and even Switzerland can be seen in these terms.

Chapter 6

1 A regime of accumulation can be defined as 'a systematic organisation of production, income distribution, exchange of the social product and consumption' (Dunford, 1990: 305). In other words, it refers to the organisation of the labour process within a particular macroeconomic environment in which the demands of the accumulation of capital are balanced by providing means of sustaining the social needs of labour. The 'mode of regulation' refers to the institutional–social rules of the game which sustain the corresponding regime of accumulation.

2. Urbanisation economies comprise a number of benefits which accrue to an individual firm from locating alongside firms in different activities. Examples include good transportation and communications facilities and the availability of specialised services which are not particular to a specific activity, as well as access to a large pool of labour. Localisation economies refer to benefits which firms in the same activity gain from joint location. They include cross-referral of business, existence of specialised labour, the availability of specialised business services relevant to the particular activity as well as access to high-quality information. In the case of financial markets, the latter is often on an informal basis, generated by physical proximity. Agglomeration economies are often referred to as a class of external economies of scale which only materialise if the spatial constraint is satisfied. These external economies can either be pecuniary, in the form of reduction in the cost of firms' inputs, or technological in the form of information spillovers.

3. The accounting concept of the balance sheet is central to the regulation of commercial banks, through the fractional reserve system. That is, banks can generate mulitiple liabilities from a single liquid asset base. A similar logic can be extended to the relationship between investment banks' assets and their exposure to market trading opportunities.

4. This was the era of petro-dollars, so called because oil prices are denominated in dollars. The four- or five-fold increase in oil prices in 1973 increased the dollar wealth of the oil-producing companies at a time of falling worldwide aggregate demand and international trade. The outcome was the classic 'too much money chasing too few goods' type of inflation.

5. Arbitrage is a process of taking advantage of price differences between two different markets. There are risk and riskless variants. The first type arises out of either bank-ruptcy or merger and acquisitions activity, where investors expect to trade in existing assets for those generating a potential greater return. The second is associated with common stocks traded in different locales which lead to price discrepancies. Both types of arbitrage correspond to the 'law of one price'. That is, assets with the same underlying risk should have the same expected rate of return. In other words, the prices of assets in capital markets should adjust until equivalent risk assets have identical expected returns. This 'law' is the basic postulate of finance theory.

6. The Euro dollar market was the first Euro market, established in the late 1950s and 1960s mainly among branches of US banks in Europe, particularly in London. They did so in response to regulations in the United States which restricted returns to foreign investors. Regulations in London were more liberal, giving rise to the concept of regulatory asymmetry. The prefix 'Euro' means off-shore. The Asian dollar market, based in Hong Kong and Singapore, is formally part of the Euro dollar market.

7. Prices of financial assets are the rates of return generated on an investment weighted by the amount of risk engaged in.

8. Time preference rates refer to the rate at which individuals forgo present consumption in order to gain a larger amount of consumption in the future, by engaging in current investment.

9. If all claims on all goods could be made in a full range of markets, financial asset markets would not be necessary. For example, if I contract to buy three cups of coffee on a cold day for the same price as one cup on a hot day, I could gain an advantage from the supplier if I claimed that I was telephoning from a cold place, although it was actually a hot day. This problem of moral hazard can be overcome if there is a centralised market for this type of claim. Financial asset markets allow the possibility of these type of claims being negotiated over time and space. Essentially, the value of financial assets are only sustainable if they are backed by tradable commodities. So long as financial instruments have a range of maturities and they are deemed liquid, the relationship between the size of asset markets and contingent commodity markets will not be crucial, at least in the short term.

10. Quote taken from the Presidential Task Force on Market Mechanisms *Report*, January 1988, Washington DC: Government Printer's Office, 10.

ENDNOTES

11 Tobin (1969) proposed a theory that makes the economy-wide rate of investment depend on the ratio of the aggregate market value of the outstanding stock of equities plus bonds to the aggregate stock of capital – the physical stock of plant and equipment valued at current costs of production or acquisition. This ratio is known as the q ratio. This ratio essentially reduces to the return on investment relative to the cost of capital.

12 Metallgessellschaft is one of the largest mining companies in the world. In 1994, its treasury department speculated on the price of oil futures. The losses on these transactions were so large that the federal government spent large sums of public money in order to prevent the company going bankrupt. The Orange County debacle in the United States was caused by the county treasury department speculating on derivatives based on US government bonds to boost its investment portfolio. Legally, the portfolio was restricted to investment in fixed-income government stock, but the legislation did not include any reference to derivatives. The Baring case is possibly the most infamous. The trading strategy of one individual led to the collapse of one of the oldest and most famous British merchant banks. By attempting to take advantage of very small price differences in the derivatives contracts, based on a rise in Japan's Nikkei stock exchange index, traded on the Singapore SIMEX exchange and Osaka futures and options exchange, and committing large amounts of the bank's capital, Nick Leason hoped to make considerable arbitrage profits. After the Kobe earthquake the Nikkei declined rapidly leaving Barings to cover losses which were much larger than the bank's total capitalization. In the Metallgessellschaft and Orange County cases, the transactions were OTC. In the Baring case, the transactions were made on a designated exchange.

13 Changes in financial intermediation have occurred because of disintermediation. That is, that investors can access capital markets directly without reference to traditional banks. This process of disintermediation has been aided by securitisation – the issue of a financial instrument secured against the value of a traded asset.

14 The term 'regulatory space' denotes not only the formal regulation of financial markets, through bodies such as the Securities Investment Board in Britain and the Securities and Exchange Commission in the United States, but the economic and social environment in which the various markets operate in each financial centre. Thus, the original meaning of '*régulation*' in the French Regulation School of macroeconomic dynamic adjustment is combined with the Anglo-Saxon meaning of 'regulation' as formal rule-making.

15 The interest parity theorem is closely connected to arbitrage conditions. It states that the interest differential between two currencies is equal to the expected rate of change in the exchange rate between them:

$$r_F = r_G + d_{F/G}$$

where F and G are France and Germany; r_F and r_G are interest rates on similar maturity, similar risk instruments, denominated respectively in franc and deutschmark; and $d_{F/G}$ is the exacted rate of depreciation against the deutschmark.

16 'Dual capacity' refers to the demarcation, before 1986, between brokers and jobbers. Jobbers were responsible for undertaking the trading of shares on the floor of the stock exchange. Brokers acted as the intermediary between clients and the actual trading of shares, advising them on their investment portfolios and acting as their agents in the market. The 'Big Bang' reforms abolished this distinction.

17 The recent East-Asian crisis tends to support the view that there is a strong relationship between asset markets and real estate markets. The problems faced by Japanese banks stem mainly from badly performing property loans in the region's capital cities. The region's currencies have rapidly devalued against the dollar and the yen and output has also decreased considerably. Thus, rental values have become expensive in local terms and capital values have fallen dramatically, with the result that Japanese banks' balance sheets have been seriously weakened and with dire consequences for the Japanese financial system. In essence, this episode appears to replicate regionally the bursting of

the Japanese 'bubble economy' during which the intimate relationship between real estate and asset values unwound in 1991.
18 The 'yield gap' refers simply to the difference in yields of two assets. The yield is the regular income stream divided by the current capital value of the asset.

CHAPTER 7

1 By 1964 the value of foreign holdings of US dollars exceeded the value of US gold reserves (Volcker and Gyohten, 1992).
2 By current standards, the balance of payments of around US$2 billion per year may seem insignificant, but at the time they were large and a threat to the stability of the international monetary system and the position of the United States within it (Volcker and Gyohten, 1992).
3 This simplifies things somewhat. US banks, US companies who received inward investment via the OFCs, and US citizens who make use of them are clearly not opposed to their existence. However, the US Government and its agencies have certainly sought to restrict the development of OFCs (Hudson, 1996).
4 The development of cyber-space banking may well provide an interesting arena in which to explore this idea further.
5 Alternatively, we might opt to totally deconstruct 'the economic', lumping everything together as 'the social'. However, although it is impossible to draw clear and fixed boundaries around aspects of social life, temporary conceptual boundaries seem to the author to be important for the practice of critical social science (compare Thrift and Olds, 1996).

CHAPTER 8

1 'Bootstrapping' can be defined as 'highly creative ways of acquiring the use of resources without borrowing money or raising equity financing from traditional sources' (Freear, Sohl and Wetzel, 1995a: 395). Examples include working out of the home, reduced, forgone or delayed salary, advances from customers and free or subsidised access to machinery and equipment.
2 The internal rate of return (IRR) is the compound rate of return over the lifetime of the investment, including realised capital gain, dividends and income.
3 A tenfold return in five years yields an annual rate of return of 58%; a fivefold return yields 39% (Bygrave, 1997b).
4 'Gearing' (or 'leverage' in US usage) refers to the proportion of debt and equity finance in the structure of the firm: the higher the proportion the more highly geared the firm is said to be. In general, the higher the debt/equity ratio, the higher the risk of bankruptcy (Edwards, 1989; Jensen and Meckling, 1976).
5 The pooled IRR for independent venture capital funds in the UK from 1980 to 1991 were: 4.3% for funds specialising in start-ups, 6.9% for funds specialising in development capital, 16.2% for mid-MBO funds, 23.8% for large MBO funds and 9.7% for generalist funds. Considering just the top quartile of funds with the highest IRRs, the pooled average amongst start-up funds was 16.1%, but here again this compared unfavourably with the equivalent figure for MBO funds (British Venture Capital Association, 1996).
6 That is, the proportion of total venture capital investments in region i, divided by its share of the total business stock. A location quotient greater than 1 indicates that region i has more than its expected share of venture capital investments; a location quotient of less than 1 indicates that region i has less than its expected share of venture capital investments.

7 Probably also East Anglia, but because the data on MBO/MBI investments are not disaggregated down to the level of individual regions, it is not possible to say for sure.
8 The data used in Figure 8 and Table 4 have been taken from the Price Waterhouse Technology Industry Group Quarterly Survey of Venture Capital, undertaken in conjunction with the National Venture Capital Association (NVCA) and the National Association of Small Business Investment Companies. For 1996, 550 venture capital funds responded to the survey, and on average 375 funds respond per quarter. The response rate per quarter ranges from 31% to 61%. However, given the small size of some venture capital funds, and the consequent 'lumpy' investment patterns over time, the effective response rate is likely to be higher than these figures suggest. In 1996 Price Waterhouse data covered US$9.5 billion invested (including SBIC investments), compared with US$10.5 billion (plus US$1.5 billion invested by SBICs) recorded by the more comprehensive VentureOne/NVCA data. Price Waterhouse data are used in this analysis of the geography of venture capital investments in the United States because they are available at a more appropriate level of spatial disaggregation.
9 High technology industries are defined as comprising biotechnology, communications, electronics and instrumentation, environmental equipment/services, computer and peripherals, medical instruments and devices, semi-conductors and information.
10 Martin and Minns (1995) provide a general discussion of the spatial organisation of the UK pension fund system.

CHAPTER 9

1 By focusing on this particular example, it is not the intention to suggest that the processes were confined to the United States. Indeed, the leveraged restructuring wave, and the use of high-yield ('junk') bonds (see Note 6) which effectively opened capital markets to new, high-risk firms and served to eliminate mere size of a company as an effective deterrent against take-over, quickly spread through most Western economies. Elsewhere (Wrigley, 1995), the author considers the equivalent UK industry during this period. In particular, there is an examination of the LBO of Gateway (the fifth largest UK food retailer of the 1980s) by the Isosceles consortium led by the US investment bank Wasserstein Perella, opposed in a battle for control by a group led by the US LBO specialist Kohlberg, Kravis and Roberts.
2 Agency costs arise, as Jensen and Meckling (1976) demonstrated, because of conflicts of interest between managers and shareholders, and shareholders and debt-holders. Managers have incentives to overexpand the size and scope of the firm to satisfy their own interests over those of shareholders and will enjoy privileged knowledge of the company's operations. Therefore, managers may use debt capital which they raise for purposes other than originally stated – in effect, they can shift it to higher-risk investment. Suppliers of debt capital could avoid such substitution by detailed monitoring, but that is likely to be an extremely costly exercise. Debt-holders, recognising these potential problems, may attempt to impose restrictive convenants on the debt contract, but are also likely to seek compensation by imposing a premium, that is by raising the cost of debt financing.
3 This can be placed in context by noting that Houston and James (1996: Table 1) report that, in a random sample of 250 NASDAQ-traded firms in the United States during the 1980s, the leverage level of the average firm was 35%.
4 A leveraged recapitalisation is a transaction in which a firm borrows in order to pay a large dividend to shareholders (at least 50% of the former equity value of the firm), whereas an LBO occurs when managers of the firm (or others) offer to pay a premium over the prevailing market price of the firm. As the debt-to-capitalisation (or market value) levels assumed in both transactions tend to be similar, they are often treated together for the purposes of analysis (Chevalier, 1995a and 1995b). However, there are

important differences between the two in terms of the resulting ownership/governance structure. LBOs are characterised by larger increases in managerial ownership and are sponsored by active investors (such as Kohlberg, Kravis and Roberts) who represent a majority of the equity and gain control of a majority of seats on the board of directors (Denis and Denis, 1995).

5 In 1983 and 1984, the FTC offered no challenge to the major mergers of the food retailers Kroger and Dillon, and American Stores and Jewel. The FTC and Department of Justice then issued revised merger guidelines which removed any doubts about the anti-trust agencies' new attitude to the legality of such mergers, and which represented 'explicit decisions by the anti-trust enforcement agencies not to enforce the existing anti-trust laws' (Müller and Paterson, 1986: 403).

6 Issues of junk bonds by US firms rose sharply in the 1980s, from just 25 in 1981 (with a face value of US$1.3 billion) to over 150 with a value exceeding US$30 billion in each of the years 1986, 1987 and 1988. In 1990, regulatory tightening reduced the total to just nine new issues, but by 1992 the high yield market had recovered and had returned to late 1980s levels in terms of numbers and value of issues. Gilson and Warner (1997) provide insight into how firms choose between junk bonds and bank debt. Their analysis suggests that junk bonds offer important advantages of financial flexibility over bank debt (fewer and looser contractual restrictions, such as financial ratio covenants), even when bank debt is structured as a revolving credit agreement ('revolver').

7 However, Hutchinson (Personal Correspondence, 1998) makes the valuable point that gains may only partly have been at the expense of unionised labour. His argument relates to the discussion above that during the period of high-leveraged transactions, projected future profits were capitalised and built into the financial base of the firm through increased debt. He argues that, given that financial markets are reasonably efficient, they are likely to have anticipated that leveraged restructuring, by increasing local market concentration/regional monopoly power, was likely to enhance future profit streams and thus their present value. This, in turn, would partly explain the acceptance by financial markets of the premiums and special dividends paid, on the basis of the expected increase in future net cash flows arising from *both* a reduction in costs (as a result of wage and work practice concessions) and an increase in monopoly rents.

8 Denis (1994) draws a very interesting comparison between Safeway and Kroger in this respect. Safeway's strategy was to offer parts of its store/distribution network intact to potential buyers that were unionised with already agreed labour cost concessions in place (in this way, Safeway obtained a higher price for those assets). Those concessions (wage cuts of between 20% and 30%) were obtained by facing the unions with the credible threat of either sale to non-unionised buyers or total shutdown. However, Safeway did not renegotiate wages in the parts of its store network which it kept. In contrast, Kroger obtained labour cost reductions in the stores it did not sell – the average hourly wage rate of Kroger workers being reduced by approximately 13%.

9 Defined as earnings before interest, taxes, depreciation and amortisation (EBITDA) divided by interest expense.

10 Theoretical articles date back to Brander and Lewis (1986) and Bolton and Scharfstein (1990). Recent articles by Chevalier (1995a and 1995b), Phillips (1995) and Kovenock and Phillips (1995 and 1997) attempt to reconcile theory and empirical evidence.

CHAPTER 10

1 Although the use of local money as a form of community association raises a further series of questions surrounding the complex intersections of, and indivisibilities between, the economic and non-economic dimensions of social life (see Barnes, 1996: Chapter 1).

2 In elections held during late April 1998 in Saxony-Anhalt, the poorest of eastern Germany's five states, the right-wing German People's Union (DVU) gained almost 13% of the vote and 16 seats in the state parliament after years of persistently high unemployment, social malaise and increasing hostility towards those regarded as 'foreigners' in the region following the political and monetary reunification of Germany.
3 See the series of papers published in the special issue of *Environment and Planning*, A, 28, 1996.
4 Unpublished research conducted by the author.

CHAPTER 11

1 Whether the deep financial crisis that hit East Asia in late 1997 has brought the meteoric rise of these economies to an end is an open question. Between 1965 and 1996, the annual average growth rate of GDP in these countries was almost three times that of the rich advanced countries of the Organisation for Economic Cooperation and Development (OECD). Plunging currencies and stock markets in 1997 put the East Asian economic miracle in the deep freeze, and minds are now concentrated simply on survival. Until the mess in East Asian banking has been cleaned up, new lending will remain severely curbed, so that in many of these economies growth may remain weak for some time to come.
2 Many commentators agree that a new social regulation model is emerging, based on 'workfare' rather than the welfare of the post-war era. The United States and the UK are leading this move to what Jessop (1993 and 1994b) and others have called the 'Schumpeterian workfare state'.
3 Arguably, these regional blocs might be seen as a way in which states can combine to achieve some degree of power in the global economy whilst being able to liberalise and deregulate their own domestic economic spaces.
4 This is akin to arguing that path dependence – the legacy of the past, or inertia – has ensured the continued supremacy of New York, rather than any contemporary absolute dynamic competitive advantage.
5 In fact, figures suggest that there are more foreign banks in the Cayman Islands than in any other major financial centre, including London and New York.
6 Economics has always subscribed to the idea of a (downward, negatively sloping) trade-off between equity and efficiency, and to the notion that policy-makers are only able to move up or down this trade-off curve (i.e. more equity means less efficiency and *vice versa*). Even if the existence of such a trade-off is accepted (and it is a dubious assumption), there seems to have been remarkably little discussion about how to shift the whole curve outwards (i.e. the possibility of securing increased efficiency with no reduction in equity.
7 Local economic development by local governments has become much more diverse since the early 1980s. Although inter-area competition for footloose (including foreign) capital still takes place, more emphasis is being put on promoting 'indigenous' activity (e.g. new small firms). The assumption is that such indigenous activity avoids the zero-sum game of inter-area competition for capital, and that such enterprises are much more committed to and rooted in the locality. It would be a mistake, however, to assume that local indigenous businesses are somehow isolated from the pressures, competition and imperatives of the global economy.
8 There is a paradox in the arguments of writers such as Ohmae (1995b) who proclaim the 'end of the nation-state' and the rise of more localised 'regional-states' in their place. Applying the same globalisation logic, these regional-states would presumably be subject to the same process of emasculation as nation-states.
9 This pro-market ideology has not been the preserve of traditional right-wing political parties, as the Labour Government in New Zealand and the recently elected 'New' Labour Government in the UK demonstrate.

CHAPTER 12

1. Making the connection between the state and global finance is an urgent theoretical and practical task. There are many issues to be considered. Although the author is not convinced that mobile capital is the ultimate 'enemy' of nation-state capacity, he does agree with Gill (1997: 21) and others who state that a profound 'shift is occurring away from the socialisation of risk provision . . . towards a privatised system of self-help'. Pension funds are deeply implicated in this process.
2. See Drache and Gertler (1991) on the recent performance of the Canadian economy and the threats to domestic policy occasioned by the emergence of meta-regional trading blocs.
3. Froud et al. define the pauper state in the following terms (emphasis removed): it is a state 'whose revenue problems and liberal economic ideology encourage it to run down nationalised industries, privatise utilities, squeeze provision of high-quality public services and, through deregulation, strip out the floors under competition even in sheltered sectors' (1997: 366).
4. In this respect, the author is not convinced that the state has been, or is being, 'hollowed-out' if that is interpreted to mean fewer financial resources (contra Jessop, 1994a, who tends to see the hollowing-out process as a result of increased constraints on nation-state functional capacity due to 'internationalisation' and who is also more optimistic than the author with respect to the opportunities for other tiers of the state to grasp the 'powers' given up by the nation-state).
5. Clark and Dear (1984) provides a summary of the literature referencing both left-of-centre and right-of-centre perspectives before developing a neomarxist account of the capitalist state in relation to democratic politics.
6. See also Rosecrance (1996) on the 'virtual state', namely that state left as a consequence of the demands of mobile capital, and the new sites of productive value found in knowledge and information (not land). According to Rosencrane, the state 'no longer commands resources as it did in mercantilist yester-year' (1996: 46).
7. At a conference in Oxford held in June 1997 and hosted by the Institute of Fiscal Studies (a research institute funded by the UK Economic and Social Research Council), speakers from the Institute and from the Social Market Foundation noted three related trends in taxpayer attitudes: (i) an unwillingness to pay more taxes; (ii) distrust of the state with respect to its ability to spend efficiently the available revenue; and (iii) an increasing demand for higher quality services. Tyrie (1996) provides a useful overview of UK public expenditure and a social market perspective on the political forces which shape spending priorities.
8. See Wood (1997) on the dilemmas posed by identity politics for social and community mobilisation and compare that argument with Peterson (1997), who believes that identity politics are a necessary step towards emancipation from the state.
9. Analytically, a useful treatment of the issue of fragmented voter solidarity is to be found in Ordeshook (1986: Chapter 5). There the game theory is used to demonstrate how and why the lack of voter solidarity may make it so difficult to get public agreement on increased public spending, and why it may lead to outcomes that are undesirable for all voters, even if on an issue-by-issue basis clusters of voters are advantaged by non-cooperation. This is the logic, of course, of the prisoners' dilemma.
10. See Hollingsworth and Boyer (1997) on Fordism, and Hirst and Zeitlin (1997) on post-Fordism and related notions of flexible accumulation. There is now a massive literature on this issue, overlapping and intersecting with geography, economics, political science and sociology. See Amin (1994).
11. Data for the UK comes from the Office on National Statistics (ONS), and refers to the personal sector which is made up of individuals, unincorporated businesses and other organisations relying on particular individuals. No direct measures are taken to estimate the data recorded in Table 1. The ONS makes estimates based on information provided

from other sectors and assumptions made about the allocation of financial assets between various sectors. In general, data on pension coverage, individuals' assets and liabilities and net worth are not as reliable in the UK as they are in the United States.

12. The rapid growth of mutual funds over the last 20 years in the United States has been remarkable (compared to other Anglo-American countries) (Blume, 1997). Although institutional investors dominate US and global securities' markets by virtue of their command of large tranches of assets, individual investors through their mutual funds have maintained a significant presence in these markets. Much of their growth can be traced to the success of companies such as Fidelity and Massachusetts Financial Services in the retail market for investment and retirement planning. It might also be argued that the systematic misrepresentation of mutual funds' performance has led to consumers over-estimating long-term returns (on the issue of survivorship bias in estimating mutual fund performance, see Carhart, 1997; for a cross-country comparison of shareholding, see Martin, Chapter 13).

13. The data on European pension fund assets are less reliable than one might hope. This chapter uses industry sources (Deutsche Bank AG, 1997). Other rather different estimates can be found in Davis (1995) and Harrison (1995).

14. Disney (1996) contains a very useful treatment of the demographics and economics of retirement in the twenty-first century. Of course, the issue is more complex than simple dependency ratios would suggest. We should also take into account the likely financial implications of social security pension values, recognising that some countries' benefits are more lucrative than others. Disney suggests that the UK Government's decision to sever the link between social security benefits and real wage increases once a person is retired has, in effect, significantly discounted the long-term financial burden of social security in the UK (compared, for example, with Italy and Germany).

15. Vittas (1996) and Vittas and Michelitsch (1995) provide useful overviews of the current and evolving situation in Eastern Europe. Ploug and Kvist (1996) provide an overview of the role and status of social security in Europe, placing the welfare state provision of pensions in the context of the fiscal crisis of the state.

16. Spence (1992) provides a useful introduction to the relevant issues concerning infrastructure and urban and regional development.

17. An essential feature of pension fund investment decision-making in the Anglo-American world is its decentralised nature: trustees are ultimately responsible for the allocation of assets, the choice of investment products and the management of the investment process. Over the last few decades, Anglo-American governments have systematically eliminated restrictions on trustees' investment options, and have been concerned to narrow responsibility to trustees rather than any other related or unrelated group (Langbein, 1995).

18. For a survey of the rapid growth in private equity deals compared to the situation in the mid 1990s see Goldman Sachs (1995 and 1997).

CHAPTER 13

1. Similarly, under a Keynesian interpretation, state-ownership funded by general taxation not only offered a mechanism for income redistribution, but also gave governments the leverage over the economy required by Keynesian techniques of demand management. Of course, the form of the Fordist–Keynesian mixed economy varied between individual countries (Hall, 1989; Esping-Andersen, 1990). However, in most cases, nationalisation formed an integral part of the state's approach to regulating the economy.

2. This is not only in the sense that rich regions provide the bulk of the tax revenues used to provide nationwide services and utilities, but also that the inherently higher unit costs of providing such services to remote and thinly populated regions are subsidised by the lower unit costs of provision in more central and densely populated areas.

3   Until the mid nineteenth century the UK had a well-established local (county) banking system, but by the beginning of the twentieth century this had been more or less replaced by a national system of branch banking, dominated by a handful of major banks most of which had become head-quartered in London. In other countries, such as the United States, Germany and Italy, local or regional banks still form a central component of the financial system, although increasing pressures for concentration and centralisation raise questions about the future of this institutional layer of the financial system.

4   Neither of these two countries has a specialised pension fund industry. In Germany, generous state pensions have reduced the need for and development of private pension plans and funds. In Japan, the pension system is largely company-based (see Chapter 12).

5   Telecommunications has been by far the most frequently privatised sector, accounting for more than one-quarter of the world total.

6   Another argument against giving privatisation shares away to the general public is that although it may well be popular, it is not capitalism. Thus, Moore (1992) asserts that giving rather than selling shares to the public is unlikely to inculcate the property-owning responsibilities that are essential to capitalism as an economic system.

7   Initially, Mrs Thatcher seemed rather reticent about 'selling the family silver' (as the traditional Tory, Harold Macmillan, put it), in part because major state assets embodied the cumulated investments of tax-payers' money, and in part because she believed such a policy would not be politically popular (Jenkins, 1995).

8   Before British Telecom, the world's largest equity sales had been a secondary offering of an already existing company, AT&T, in the United States, at just over US$1 billion.

9   The notion of a 'share economy' was also extolled by certain economists about this time. For example, Weitzman (1984) put forward a sophisticated argument that introducing employee share-ownership schemes would help to solve the problem of stagflation, as workers with a share in their companies' profits would be much more realistic about pushing for wage claims. In effect, the argument was that by making workers 'mini-capitalists', the basic capital-labour conflict over the distribution of incomes would be largely removed.

10 In 1996, Mr Blair, the new leader of the Labour Party succeeded in modernising this clause and in effect removed its historic commitment to large-scale nationalisation.

11 In Gilmour's view, privatisation and reform of the trade unions, represent the two success stories of the Thatcher era. In contrast, the two other flagship policies of Thatcherism – monetarism and the poll tax – proved singularly unsuccessful.

12 The massive scale of the public's response was also boosted by the various incentives that the government offered, for example the issuing of 'bonus shares', allowing payment by instalments and giving shareholders discounts on their bills with the privatised companies.

13 Although many of these same workers have since lost their jobs as the privatised companies have restructured and reduced employment. Some estimates suggest that because of this employment shake-out process, privatisation has resulted in the net loss to the economy of as many as 250 000 jobs.

14 In many cases, the values of shares increased by up to one-third by the end of the first week after the flotation.

15 The situation is similar in other countries. We know very little about about the geography of shareholding (for one attempt to investigate this issue in the United States, see Green, 1993).

16 The Family Resources Surveys by the Department of Social Security, based on 26 000 households, provide some information, but allow little disaggregation by types of shares held. The annual Financial Research Surveys conducted by NOP Financial Research involve a stratified random sample of 60 000 households. They cover virtually all types of financial product. The author is indebted to NOP Financial Research for providing various regional breakdowns of their survey results. The following interpretation of

these data is of course the author's own, and should not be attributed in any way to NOP Financial Research. Note that the NOP Financial Research survey data do not include Northern Ireland, which is therefore excluded from the following discussion.

17 Each major privatisation has included a maximum size of allocation per applicant, so this in itself should have encouraged a relatively even spread of shareholders.

18 There is evidence to suggest that significant proportions of the individual shareholders of some of the regional water and electricity boards live outside those regions. For example, when Northern Ireland Electricity was floated in June 1993, it had 300 000 private shareholders, but only 109 000 were resident in Northern Ireland.

19 The recent Americanisation of the UK regional electricity industry is somewhat ironic. In the 1920s, two Chicago businessmen, Samuel Insull and Harley Clarke, secretly bought up much of southern Britain's (then private) electricity supply industry. When, in 1936, the siphoning of profits back to the United States caused a public outcry, the British Government supported a buy-out of the Americans. Sixty years on, the bulk of Britain's privatised electricity distribution sector is once again in US hands.

20 It proved impossible to obtain information on the share-ownership structures of these utilities taken over by US concerns, although the general impression conveyed during personal discussions with UK representatives of these US-based companies was that 'substantial numbers' of private shareholders sold their shares at the time of takeover.

21 The region contains the largest proportion of higher-rate tax payers.

22 Specifically, the Tax Exempt Special Savings Account (TESSA). Nationally, about 10% of households have savings in TESSAs, and more than 50% have building society savings accounts. However, as with virtually all forms of savings, the proportion of households with TESSAs and building society savings accounts is half as high again in the South East as it is in northern and peripheral regions of the country.

23 With only one or two exceptions, the pre-tax profits of privatised companies have risen dramatically after privatisation (e.g. those of British Telecom, the regional electricity companies, and the water companies all doubled within just four years of their flotations, and those of Cable and Wireless and National Freight Corporation (NFC) quadrupled). Government moves to enforce price reductions through the utility regulators, and more recently to tax the windfall profits of privatised companies (in order to fund new state-run workfare programmes), have been attacked by the privatised concerns for undermining investment, long-term profitability and returns to shareholders. Estimates of the windfall tax range from £4.7 billion to £5.2 billion. The cost of the Labour Government's 'welfare into work' programme to reduce unemployment is estimated to be about £3.2 billion. The latter figure ignores the on-going costs of the unemployed associated with the job-cutting undertaken by the privatised industries themselves.

24 The names have been the familiar ones, for example, Kleinwort Benson, S.G. Warburg, N.M. Rothschild, Price Waterhouse, Hill Samuel, Schroder Wagg, Barclays de Zoete Wedd and Samuel Montagu.

25 PEPs were introduced in 1987 by the Conservative Government and allow individuals to invest up to £6000 per annum, tax free, in managed share funds; by the mid 1990s about 10% of the adult population had one or more PEPs as can be seen in Table 4. The New Labour Government (1997) plans to introduce a new Individual Savings Account, to come into operation in 1999, which is intended to promote greater saving by middle- and lower-income households. These ISAs will replace existing tax-exempt savings vehicles such as the cash-based TESSA and PEPs, and by offering tax credits on share earnings, it is hoped that the new scheme will encourage greater share-ownership across these income groups.

26 However, there are some signs that this may be changing. As the US example of Calpers, the interventionist California Public Employees Retirement System or pension fund, demonstrates, a large (and in this case a regionally based) institutional shareholder can indeed influence corporate boards in the interests of its individual shareholder members. Calpers has led numerous campaigns, both in the United States and in

Europe, directed at companies in which it has investments with the aim of improving their management, performance and accountability in the interests of its workers' retirement funds. As Calpers' average holding in a firm is about US$35 million, it is able to excercise considerable influence.

27 A number of the former centrally planned socialist economies of East and Central Europe (Hungary, Poland, the Czech Republic and Russia) are developing investment trusts to act as holding companies or mutual trusts for shares in privatised state assets. Citizens are able to redeem vouchers against shares in these trusts. However, as the voucher privatisation in the Czech Republic from 1992 onwards testifies, such schemes can be problematic. Some 6.1 million people (83% of the adult population) took up their share vouchers in the first and second waves of privatisation, but by 1996 more than 2 million, almost one-third, had sold all their shares (Blazek, 1997). Social, regional and institutional biases in shareholding have already emerged in these countries (Willer, 1997).

BIBLIOGRAPHY

Aaronovitch, S. (1983) 'A Marxist View', *Money Talks*. Eds. A. Horrox and G. McCredie. London: Methuen. 34–53.

Advisory Council on Science and Technology (1990) *The Enterprise Challenge: Overcoming Barriers to Growth in Small Firms*. London: HMSO.

Aglietta, M. (1979) *A Theory of Capitalist Regulation: The Experience of the United States*. London: New Left Review Editions.

Agnew, J. (1994) 'The Territorial Trap: The Geographical Assumptions of International Relations Theory', *Review of International Political Economy*, 1, 3–80.

Agnew, J. and Corbridge, S. (1995) *Mastering Space: Hegemony, Territory and International Political Economy*. London: Routledge.

Aharoni, Y. (1988) 'The United Kingdom: Transforming Attitudes'. *The Promise of Privatization*. Ed. R. Vernon. New York: Council on Foreign Relations.

Alessandrini, P. (1989) 'I Flussi Finanziari Interregionali: Interdipendenze Funzionali ed Indizi Empirici sulla Realta Italiana'. *Credito e Sviluppo. Evoluzione delle Strutture Finanziarie e Squilibri Territoriali*. Ed. A. Niccoli. Milan: Giuffre Editore.

Alessandrini, P. (1992) 'Squilibri Regionali e Dualismo Finanzario in Italia: Alcune Riflessioni', *Moneta e Credito*, 177, 6–81.

Alessandrini, P. Ed. (1994) *La Banca in un Sistema Locale di Piccole e Medie Imprese*. Bologna: Il Mulino.

Alessandrini, P. (1996a) 'I Sistemi Locali del Credito in Regioni a Diverso Stadio di Sviluppo', *Moneta e Credito*, 196, 567–600.

Alessandrini, P. (1996b) 'L'Evoluzione degli Squilibri Finanziari Territoriali in Italia. Un'Analisi Statistica', *Economia Marche*, 2, 277–97.

Alexander, A.F. and Pollard, J.S. (1998) 'Sleeping with the Enemy? Banks, Grocers and the Retailing of Financial Services', *Mimeo*, Birmingham: School of Geography, University of Birmingham.

Aliber, R. (1979) 'Monetary Aspects of Off-shore Markets', *Columbia Journal of World Business*, 14, 8–16.

Allen, J. (1997) 'Economies of Power and Space'. *Geographies of Economies*. Eds. R. Lee and J. Wills. London: Arnold.

Allen, J. and Hamnett, C. Eds. (1995) *A Shrinking World? Global Unevenness and Inequality*. Oxford: Oxford University Press.

Allen, F. and Gale, D. (1994) 'A Welfare Comparison of the German and US Financial Systems', *Working Paper*, 13-94. Philadelphia: Rodney White Center for Financial Research, Wharton School, University of Pennsylvania.

Amin, A. Ed. (1994) *Post-Fordism: A Reader*. Oxford: Blackwell.

Amin, A. (1997) 'Placing Globalisation', *Theory, Culture and Society*, 14, 123–37.

Amin, A. and Thrift, N.J. (1992) 'Neo-Marshallian Nodes in Global Networks', *International Journal of Urban and Regional Research*, 16, 571–87.

Amin, A. and Thrift, N. (1994) Living the Global. *Globalization, Institutions and Regional Development in Europe*. Eds. A. Amin and N.J. Thrift. Oxford: Oxford University Press.

Amos, O.M. (1992) 'The Regional Distribution of Bank Closings in the United States from 1982 to 1988', *Southern Economic Journal*, 58, 805–15.

Amos, O.M. and Wingender, J.R. (1993) 'A Model of the Interaction Between Regional Financial Markets and Regional Growth', *Regional Science and Urban Economics*, 23, 85–110.

Anderson, J. (1996) 'The Shifting Stage of Politics: New Medieval and Post-modern Territorialities', *Environment and Planning, D: Society and Space*, 14, 133–53.

Anonymous (1988) 'This Takeover Could Open California to All Comers', *Business Week*, 29 February, 31.
Anonymous (1993) 'Clinging on', *The Economist*, 20 November, 88–93.
Arbomeit, H. (1986) 'Privatization in Great Britain', *Annals of Public and Cooperative Economy*, 57, 153–79.
Arrow, K. and Debreu, G. (1954) 'Existence of Equilibrium in a Competitive Economy', *Econometrica*, 22, 265–90.
Arthur, W.B. (1988) 'Urban Systems and Historical Path Dependence'. *Cities and their Vital Systems*. Eds. J.H. Ausubel and R. Herman. Washington DC: National Academy Press.
Arthur, W.B. (1989) 'Competing Technologies, Increasing Returns and Lock-in by Historical Events', *Economic Journal*, 99, 116–31.
Arthur, W.B. (1994) *Increasing Returns and Path Dependence in the Economy*. Michigan: Michigan University Press.
Ash, T., Hare, P. and Canning, A. (1994) 'Privatisation in the Former Centrally Planned Economies'. *Privatisation and Regulation: A Review of the Issues*. Eds. P.M. Jackson and C.M. Price. London: Longman. 213–36.
Ashby, D. (1981) 'Will the Eurodollar Market Go Back Home?' *The Banker*, 131-1, 93–8.
Avery, R.B., Bostic, R.W., Calem, P.S. and Canner, G.B. (1997) 'Changes in the Distribution of Banking Offices', *Federal Reserve Bulletin*, 83, 707–25.
Balakrishnan, S. and Fox, I. (1993) 'Asset Specificity, Firm Heterogeneity and Capital Structure', *Strategic Management Journal*, 14, 3–16.
Bank for International Settlements (1993) *International Banking Statistics, 1977–91*. Basle: BIS.
Bank for International Settlements (1995) *Central Bank Survey of Derivatives Market Activity*. Basle: BIS.
Bank of England (1996) *The Financing of Technology-based Small Firms*. London: Bank of England.
Bank of England (1997) *Quarterly Report on Small Business Statistics*. December, London: Bank of England.
Banking, Insurance and Finance Union (1995) *Killing Communities*. London: BIFU.
Banuri, T. and Schor, J. Eds. (1992) *Financial Openness and National Autonomy*. Oxford: Clarenden Press.
Barnes, T. (1996) *Logics of Dislocation: Models, Metaphors and Meanings of Economic Space*. New York: Guilford Press.
Barnet, R. and Cavanagh, J. (1996) 'Electronic Money and the Casino Economy'. *The Case Against the Global Economy and for a Turn towards the Local*. Eds. J. Mander and E. Goldsmith. San Francisco: Sierra Books. 360–3.
Bayliss, V. (1998) *Redefining Work*. London: Royal Society for the Encouragement of Arts, Manufactures and Commerce.
BCR (1995) *Venture Capital in the UK: A Report and Guide to the Venture Capital Industry*. London: HMSO.
Beare, J.B. (1976) 'A Monetarist Model of Regional Business Cycles', *Journal of Regional Science*, 16, 57–63.
Becattini, G. (1990) 'The Marshallian Industrial District as a Socio-Economic Notion'. *Industrial Districts and Inter-Firm Co-operation*. Eds. F. Pyke, G. Becattini and W. Segerberger. Geneva: International Institute for Labour Studies. 52–74.
Becattini, G. and Rullani, E. (1993) 'Sistema locale e mercato globale', *Economia e Politica Industriale*, 80, 25–48.
Berger, S. and Dore, R. Eds. (1996) *National Diversity and Global Capitalism*, Ithaca: Cornell University Press.
Berglöf, E. (1997) 'Reforming Corporate Governance: Redirecting the European Agenda', *Economic Policy*, 2, 93–123.
Berle, A.A. and Means, G. (1932) *The Modern Corporation and Private Property*. New York: Macmillan.
Berman, M. (1982) *All that is Solid Melts into Air*. New York: Penguin.

Bertrand, O. and Noyelle, T.J. (1988) *Human Resources and Corporate Strategy: Technological Change in Banks and Insurance Companies: France, Germany, Japan, Sweden, United States*. Paris: Organization for Economic Co-operation and Development.

Bhaskar, V. (1992) 'Privatization and the Developing Countries: The Issues and the Evidence', *Discussion Paper*, 47, Geneva: UNCTAD.

Bhide, A. (1992) 'Bootstrap Finance: The Art of Start-ups', *Harvard Business Review*, 70(6), 109–17.

Bias, P.V. (1992) 'Regional Financial Segmentation in the United States', *Journal of Regional Science*, 32, 321–34.

Bienen, H. and Waterbury, J. (1989) 'The Political Economy of Privatization in Developing Countries', *World Development*, 17, 5, 617–32.

Binkley, J.K. and Connor, J.M. (1996) 'Market Competition and Metropolitan-area Grocery Prices', *Staff Paper*, 96-15, Dept. of Agricultural Economics, Purdue University.

Binks, M. and Ennew, C. (1996) 'Growing Firms and the Credit Constraint', *Small Business Economics*, 8, 17–25.

Bishop, M. and Kay, J. (1989) 'Privatization in the United Kingdom: Lessons from Experience', *World Development*, 17, 5, 643–57.

Blake, D. (1995) *Pension Schemes and Pension Funds in the United Kingdom*. Oxford: Clarendon Press.

Blanden, M. (1997) 'US Regionals: A State of Merger Mania', *The Banker*, October, 40–2.

Blazek, J. (1997) 'The Stock Market and Regional Development in the Czech Republic', *Mimeo*, Prague: Department of Social Geography and Regional Development, Charles University.

Block, F.L. (1996) *The Vampire State, and other Myths and Fallacies about the US Economy*. New York: The New Press.

Blum, R. (1984) *Offshore Haven Banks, Trusts and Companies: The Business of Crime in the Euromarket*. New York: Praeger Publishers.

Blume, M. (1997) 'An Anatomy of Morningstar Rating', *Working Paper*, 12-97, Philadelphia: Rodney White Center for Financial Research, Wharton School, University of Pennsylvania.

Boddy, M. (1976) 'The Structure of Mortgage Finance: Building Societies and the British Social Formation', *Transactions of the Institute of British Geographers*, NS, 1, 20–33.

Boddy, M. (1980) 'Finance Capital, Commodity Production and the Production of the Urban Built Form'. *Urbanization and Urban Planning in Capitalist Societies*. Eds. M. Dear and A.K.J. Scott. Chicago: Maaroufa Press.

Bohm, A. and Somenti, M. Eds. (1992) *Privatization in Central and Eastern Europe*. Ljubljana: CEEPN.

Bolton, P. and Scharfstein, S.D. (1990) 'A Theory of Predation Based on Agency Problems in Financial Contracting', *American Economic Review*, 80, 93–106.

Bonefeld, W. and Holloway, J. Eds. (1996) *Global Capital, Nation State and the Politics of Money*. London: Macmillan.

Boot, A.W. and Thakor, A.V. (1997) 'Financial System Architecture', *Review of Financial Studies*, 10, 693–733.

Bornemeier, J. (1995) 'First Chicago Bank's $3.00 Fee for Teller Use Draws Fire', *Los Angeles Times*, 4 May, D2.

Bös, D. (1991) *Privatization: A Theoretical Treatment*. Oxford: Clarendon Press.

Bös, D. (1993) 'Privatization in Europe: A Comparison of Approaches', *Oxford Review of Economic Policy*, 9, 1, 95–111.

Boycko, M., Shleifer, A. and Vishny, R. (1995) *Privatizing Russia*. Cambridge, Mass: MIT Press.

Boyd, J.H. and Gertler, M. (1994) 'The Role of Large Banks in the Recent US Banking Crisis', *Federal Reserve Bank of Minneapolis Quarterly Review*, 18, 1, 2–21.

Boyer, R. and Drache, D. Eds. (1996) *States Against Markets: The Limits of Globalization*. London: Routledge.

Brander, J.A. and Lewis, T.R. (1986) 'Oligopoly and Financial Structure', *American Economic Review*, 76, 956–70.

Braudel, F. (1985) *Civilisation and Capitalism 15th–18th Century, Volume I The Structures of Everyday Life*. London: Fontana Press.

British Venture Capital Association (1996) *Performance Measurement Survey 1994*. London: BVCA.

British Venture Capital Association (1997a) *Sources of Business Angel Capital 1997/98*. London: BVCA.

British Venture Capital Association (1997b) *Directory 1997/8*. London: BVCA.

Brohman, J. (1996) 'Post-war Development in the Asian NICs: Does the Neoliberal Model Fit Reality?' *Economic Geography*, 72, 107–30.

Brophy, D.J. (1997) 'Financing the Growth of Entrepreneurial Firms'. *Entrepreneurship 2000*. Eds. D.L. Sexton and R.W. Smilor. Chicago: Upstart. 5–28.

Bruchesi, J. et al. (1943) *Montreal Economique*. Montreal: Editions Fides.

Brusco, S. (1986) 'Small Firms and Industrial Districts: The Experience of Italy'. *New Firms and Regional Development*. Eds. D. Keeble and E. Wever. London: Croom Helm. 184–202.

Brusco, S. (1989) *Piccole Imprese e Distretti Industriali*. Turin: Rosenberg & Sellier.

Bryan, L.L. (1991) *Bankrupt: Restoring the Health and Profitability of our Banking System*. New York: Harper Business.

Budd, L. (1995) 'Globalisation, Territory and the Growth of Strategic Alliances in Different Financial Centres', *Urban Studies*, 32, 2, 345–60.

Budd, L. and Whimster, S. (1992) *Global Finance and Urban Living*. London: Routledge.

Burch, K. (1994) 'The "Properties" of the State System and Global Capitalism'. *The Global Economy as Political Space*. Ed. S. Rosow. London: Lynne Rienner.

Butlin, S.J., Hall, A.R. and White, R.C. (1971) Australian Banking and Monetary Statistics, 1817–1945. Reserve Bank of Australia, Occasional Publication 4A, Sydney.

Button, K.J. and Weyman-Jones, T. (1994) 'Impacts of Privatization Policy in Europe', *Contemporary Economic Policy*, 141, 4, 23–33.

Bygrave, W. (1997a) 'Calling on Family and Friends for Start-up Cash'. *Mastering Enterprise*. Eds. S. Birley and D. Muzyka. London: FT/Pitman. 70–2.

Bygrave, W. (1997b) 'How the Venture Capitalists Work Out the Financial Odds'. *Mastering Enterprise*. Eds. S. Birley and D. Muzyka. London: FT/Pitman.

Bygrave, W.D. and Timmons, J. (1992) *Venture Capital at the Crossroads*. Boston: Harvard Business School Press.

Bygrave, W.D. and Timmons, J.A. (1996) 'Venture Capital: Predictions and Outcomes', Paper to the 10th Anniversary Conference of the Centre for Management Buyout Research, University of Nottingham.

Cahn, E. and Rowe, J. (1992) *Time Dollars*. Emmaus, Penn: Rodale Press.

Cairncross, F. (1997) *The Death of Distance: How the Communications Revolution will Change our Lives*. London: Orion Business Books.

California Association of Realtors (1995) *California Existing Single Family Housing Markets: Historical Data Summaries*. San Francisco: California Association of Realtors.

Campbell, P. (1982) 'Impact and Implications of International Banking Facilities'. *Virginia Journal of International Law*, 22, 3, 521–54.

Campen, J. (1993) 'Banks, Communities and Public Policy'. *Transforming the US Financial System: Equity and Efficiency for the 21st Century*. Eds. G. Dymski, G. Epstein and R. Pollin. New York: ME Sharpe.

Canadian Bankers' Association (1991) *Bank Facts 1991*. Toronto: CBA.

Canullo, G. and Pettenati, P. (1994) 'Regional Convergence in the European Community'. *European Challenges and Hungarian Responses in Regional Policies*. Eds. H. Zoltan and H. Gyula. Centre for Regional Studies, Hungarian Academy of Science, 38–48.

Çapoglu, G. (1991) *Prices, Profits and Financial Structures*. Aldershot: Edward Elgar.

Cardew, R.V., Langdale, J.V. and Rich, D.C. (1982) *Why Cities Change – Urban Development and Economic Change in Sydney*. Sydney: Allen and Unwin.

Carhart, M. (1997) 'Mutual Fund Survivorship', *Mimeo*, Los Angeles: Marshall School of Business, University of Southern California.

Carlino, G. and Lang, R. (1989) 'Interregional Flows of Funds as a Measure of Economic Integration in the United States', *Journal of Urban Economics*, 26, 20–9.

Carnevali, F. (1996) 'Between Markets and Networks: Regional Banks in Italy', *Business History*, 38, 3, 84–100.

Carosso, V. (1976) 'Proposals for a Regional Financial History of the Middle Atlantic States'. *Regional Economic History: The Middle Atlantic States since 1700*. Ed. G. Porter. Wilmington, Delaware: Eleutherian Mills-Hagley Foundation.

Carrington, M., Langguth, P. and Steiner, T. (1997) *The Banking Revolution: How Technology is Creating Winners and Losers*. London: FT/Pitman.

Case, K., Schiller, R. and Weiss A. (1991) 'Index-based Futures and Options Markets in Real Estate', *Cowles Foundation Working Paper No. 1006*. New Haven: Cowles Foundation for Research in Economics at Yale University.

Castelli, F., Martiny, M. and Marullo Reedtz, P. (1995) 'La redditività degli sportelli bancari dopo la liberalizzazione', *Banca d'Italia, Temi di discussione*, n. 259.

Castells, M. (1989) *The Informational City*. Oxford: Blackwell.

Castells, M. (1996) *The Rise of the Network Society*. Oxford: Blackwell.

Castells, M. (1997) *The Power of Identity*. Oxford: Blackwell.

Caves, R.E. (1990) 'Lessons from Privatization in Britain: State Enterprise Behaviour, Public Choice and Corporate Governance', *Journal of Economic Behaviour and Organization*, 13, 2, 145–69.

Central Bank of the Bahamas and the Association of International Banks and Trust Companies (1986) *International Banking and Trust Activities in the Bahamas*. Nassau, Bahamas: Central Bank of the Bahamas.

Centre for Management Buyout Research (1996) *Quarterly Review of Management Buyouts*. Spring, CMBOR. Nottingham: University of Nottingham.

Centre for Real Estate and Urban Economics (1991) *California Real Estate Opportunities in the 1990s*. University of California, Berkeley.

Cerny, P.G. (1993) *The American Financial System: From Free Banking to Global Competition*. Working Paper 16. Department of Politics and International Studies, University of Warwick.

Chapman, C. (1990) *Selling the State: Has Privatization Worked?* London: Hutchinson.

Chevalier, J.A. (1995a) 'Capital Structure and Product-market Competition: Empirical Evidence from the Supermarket Industry', *American Economic Review*, 85, 115–35.

Chevalier, J.A. (1995b) 'Do LBO Supermarkets Charge More? An Empirical Analysis of the Effects of LBOs on Supermarket Pricing', *Journal of Finance*, 50, 1095–112.

Chiapporri, P.-A., Perez-Castrillo, D. and Verdier, T. (1995) 'Spatial Competition in the Banking System: Localization, Cross-Subsidies and the Regulation of Deposit Rates', *European Economic Review*, 39, 889–918.

Chick, V. (1986) 'The Evolution of the Banking System and the Theory of Saving, Investment and Interest', *Economies et Societes*, 20, *Monnaie et Production*, 3, 193–205. Reprinted in Arestis, P. and Dow, S.C. Eds. (1992) *On Money, Method and Keynes: Selected Papers of Victoria Chick*. London: Macmillan.

Chick, V. (1993a) 'Some Scenarios for Money and Banking in the EC and their Regional Implications'. *The Political Economy of Global Restructuring, Vol. 2: Trade and Finance*. Ed. I.H. Rima. Cheltenham: Edward Elgar.

Chick, V. (1993b) 'The Evolution of the Banking System and the Theory of Monetary Policy'. *Monetary Theory and Monetary Policy: New Tracks for the 1990s*. Ed. S.F. Frowen. London: Macmillan.

Chick, V. and Dow, S.C. (1988) 'A Post-Keynesian Perspective on the Relation Between Banking and Regional Development'. *Post-Keynesian Monetary Economics*. Ed. P. Arestis. Aldershot: Edward Elgar. 219–50.

Christiansen, R.E. Ed. (1989) 'Privatisation', *World Development, Special Issue*, 17, 5.

Clark, G.L. (1989) 'Remaking the Map of Corporate Capitalism: The Arbitrage Economy of the 1990s', *Environment and Planning*, A, 21, 997-1000.

Clark, G.L. (1993a) 'Global Interdependence and Regional Development: Business Linkages and Corporate Governance in a World of Financial Risk', *Transactions of the Institute of British Geographers*, 18, 309-25.

Clark, G.L. (1993b) *Pensions and Corporate Restructuring in American Industry: A Crisis of Regulation*. Baltimore: Johns Hopkins University Press.

Clark, G.L. (1997a) 'Pension Funds and Urban Investment: Four Models of Financial Intermediation', *Environment and Planning*, A, 29, 1297-316.

Clark, G.L. (1997b) 'The Functional and Spatial Structure of the Investment Management Industry', *Working Paper*, 97-7, Oxford: School of Geography, University of Oxford.

Clark, G.L. (1997c) 'Rogues and Regulation in Global Finance: Maxwell, Leeson and the City of London', *Regional Studies*, 31, 221-36.

Clark, G.L. (1998a) 'Pension Fund Capitalism: A Causal Analysis', *Working Paper*, 98-1, Oxford: School of Geography, University of Oxford.

Clark, G.L. (1998b) 'Why Convention Dominates Pension Fund Trustee Investment Decision-Making', *Environment and Planning*, A, in press.

Clark, G.L. (1998c) 'Stylized Facts and Close Dialogue: Methodology in Economic Geography', *Annals of the Association of American Geographers*, 88, in press.

Clark, G.L. and Dear, M.J. (1984) *State Apparatus: Structures and Language of Legitimacy*. London and Boston: Allen and Unwin.

Clark, G.L. and Evans, J. (1998) 'The Private Provision of Urban Infrastructure: Long-term Contracts as Financial Intermediation', *Urban Studies*, 35, 301-19.

Clark, G.L. and Wrigley, N. (1997a) 'The Spatial Configuration of the Firm and the Management of Sunk Costs', *Economic Geography*, 73, 285-304.

Clark, G.L. and Wrigley, N. (1997b) 'Exit, the Firm, and Sunk Costs: Reconceptualising the Corporate Geography of Disinvestment and Plant Closure', *Progress in Human Geography*, 21, 338-58.

Clarke, T. and Pitelis, C. Eds. (1993) *The Political Economy of Privatization*. London: Routledge.

Clemenz, G. (1991) 'Loans for Projects With Interdepent Returns', *Journal of Institutional and Theoretical Economics*, 337-53.

Coakley, J. (1994) 'The Integration of Property and Financial Markets', *Environment and Planning*, A, 26, 697-713.

Coakley, J. and Harris, L. (1983) *The City of London: London's Role as a Financial Centre*. Oxford: Blackwell.

Cochrane, J.H. (1997) 'Where is the Market Going? Uncertain Facts and Novel Theories', *Economic Perspectives*, 21(6), 3-37. Chicago: Research Department, Federal Reserve Bank of Chicago.

Code, W.R. (1971) 'The Spatial Dynamics of Financial Intermediation: An Interpretation of the Distribution of Financial Decision Making in Canada'. Unpublished Ph.D. Dissertation, U.C. Berkeley.

Code, W.R. (1991) 'Information Flows and the Processes of Attachment and Projection: The Case of Financial Intermediaries'. *Collapsing Space and Time*. Eds. S.D. Brunn and T.R. Leinbach. London: HarperCollins Academic.

Cohen, R. (1979) 'The Changing Transactions Economy and its Spatial Implications', *Ekistics*, 274, 7-15.

Cohen, J. and Maeshiro, A. (1977) 'The Significance of Money at the State Level', *Journal of Money, Credit and Banking*, 9, 672-8.

Cohen, S.S., Garcia, C.E. and Loureiro, O. (1993) *From Boom to Bust in the Golden State: The Structural Dimensions of California's Prolonged Recession*. Working Paper 64. Berkeley Roundtable on the International Economy, University of California, Berkeley.

Commission for the European Communities (1990) 'One Market, One Money: An Evaluation of the Potential Benefits and Costs of Forming an Economic and Monetary Union', *European Economy*, 44 October, 1-351.

Confederation of British Industry (1990) *A Nation of Shareholders*. London: CBI.
Confederation of British Industry (1997) *Tech Stars: Breaking the Growth Barriers for Technology-based SMEs*. London: CBI.
Conigliani, C. (1990) *La concetrazione Bancaria in Italia*. Bologna: Il Mulino.
Connor, J.M. (1996) 'Did the Competitive Regime Switch in the 1980s?', *American Journal of Agricultural Economics*, 78, 1192–7.
Connor, J.M. (1997) 'Concentration and Mergers in US Wholesale Grocery Markets', *Staff Paper*, 97-09, Dept. of Agricultural Economics, Purdue University.
Conzen, M.P. (1977) 'The Maturing Urban System in the United States', *Annals of the Association of American Geographers*, 67, 1, 88–108.
Coopers & Lybrand (1996) *The Economic Impact of Venture Capital in the UK*. London: British Venture Capital Association.
Coopers & Lybrand L.L.P./VentureOne (1996) *Seventh Annual Economic Impact of Venture Capital Study*. Coopers and Lybrand on behalf of the National Venture Capital Association.
Corbridge, S. (1994) 'Bretton Woods Revisited: Hegemony, Stability and Territory', *Environment and Planning*, A, 26, 1829–59.
Corbridge, S., Martin, R.L. and Thrift, N. Eds. (1994) *Money, Power and Space*. Oxford: Blackwell.
Cotterill, R.W. (1990) 'Food Mergers: Implications for Performance and Policy', *Review of Industrial Organization*, 5, 189–202.
Cotterill, R.W. (1993) 'Food Retailing: Mergers, Leveraged Buyouts and Performance'. *Industry Studies*. Ed. L. Duetsch. Englewood Cliffs: Prentice-Hall.
Cotterill, R.W. and Haller, L.E. (1992) 'Barrier and Queue Effects: A Study of Leading US Supermarket Chain Entry Patterns', *Journal of Industrial Economics*, 40, 427–40.
Courlet, C. and Soulage, B. (1995) 'Industrial Dynamics and Territorial Space', *Entrepreneurship and Regional Development*, 7, 287–308.
Covill, L. (1997) 'Germany: A Regional Outlook', *The Banker*, October, 33–4.
Cowan, L. (1990) *Privatization in the Developing World*. New York: Greenwood Publishing.
Cowen, T. and Kroszner, R. (1994) *Explorations in the New Monetary Economics*. Oxford: Blackwell.
Cox, K. Ed. (1997) *Spaces of Globalization: Reasserting the Power of the Local*. New York: Guilford Press.
Cox, K. and Mair, A. (1988) 'Locality and Community in the Politics of Local Economic Development', *Annals of the Association of American Geographers*, 78, 307–25.
Cressey, P. and Scott, P. (1992) 'Employment, Technology and Industrial Relations in the UK Clearing Banks: Is the Honeymoon Over?' *New Technology, Work and Employment*, 3, 83–96.
Crivellini, M. and Pettenati, P. (1994) 'Patterns of Regional Development and the Italian Case'. *Italian Economic Papers, Vol. II*. Ed. L. Pasinetti. Bologna: Il Mulino; Oxford: Oxford University Press, 151–87.
Croall, J. (1997) *LETS Act Locally: The Growth of Local Exchange Trading Systems*. London: Calouste Gulbenkian Foundation.
Cronon, W. (1991) *Nature's Metropolis*. New York: W.W. Norton and Co.
Cross, I.B. (1927) *Financing an Empire: History of Banking in California, Vols. I–IV*. Los Angeles: SJ Clarke Publishing Company.
Daly, M.T. (1984) 'The Revolution in International Capital Markets: Urban Growth and Australian Cities', *Environment and Planning*, A, 16, 1003–20.
Daly, M. and Logan, M. (1989) *The Brittle Rim: Finance, Business and the Pacific Region*. Harmondsworth: Penguin.
David, P. (1988) 'Path Dependence: Putting the Past into the Future of Economics', *The Economic Series Technical Report*, 533. Institute for Mathematical Studies in the Social Sciences, Stanford: Stanford University.
David, P. (1994) 'Why are Institutions the "Carriers of History"? Path Dependence and the

Evolution of Conventions, Organisations and Institutions', *Structural Change and Economic Dynamics*, 5, 205–20.

Davidson, J. Ed. (1986) *The Sydney–Melbourne Book*. Sydney: Allen and Unwin.

Davies, G. (1994) *A History of Money: From Ancient Times to the Present Day*. Cardiff: University of Wales Press.

Davis, E.P. (1995) *Pension Funds: Retirement Income Security and Capital Markets: An International Perspective*. Oxford: Clarendon Press.

De Bonis, R. and Ferrando, A. (1996) 'Le Determinanti dei Tassi di Interesse sui Prestiti nei Mercati Locali'. Mimeo. Bank of Italy.

Debreu, G. (1959) *Theory of Value*. New Haven: Yale University Press.

De Cecco, M. (1994) 'The Italian Banking System to a Historical Turning Point', *Review of Economic Conditions in Italy*, n. 1, 51–67.

Dei Ottati, G. (1994) 'Trust Interlinking Transactions and Credit in the Industrial District', *Cambridge Journal of Economics*, 18, 529–46.

Del Monte, A. and Giannola, A. (1997) *Istituzioni Economiche e Mezzogiorno. Analisi delle Politiche di Sviluppo*. Rome: NIS.

Denis, D.J. (1994) 'Organizational Form and the Consequences of Highly Leveraged Transactions: Kroger's Recapitalization and Safeway's LBO', *Journal of Financial Economics*, 36, 193–224.

Denis, D.J. and Denis, D.K. (1995) 'Causes of Financial Distress Following Leveraged Recapitalizations', *Journal of Financial Economics*, 37, 129–57.

Deutsche Bank AG. (1997) *Aspiring Equity Culture in Germany*. Frankfurt am Main: Deutsche Morgan Grenfell European Equity Research, Mainzer Landstrasse 16.

Diamond, P. (1994) 'Privatisation of Social Security: Lessons from Chile', *Revista de Analisis Economico*, 9, 21–53.

Dicken, P. (1976) 'The Multi-Plant Business Enterprise and Geographical Space: Some Issues in the Study of External Control', *Regional Studies*, 10, 401–12.

Dicken, P. (1986) *Global Shift*. London: Harper and Row.

Dicken, P. (1992) *Global Shift: The Internationalisation of Economic Experience*, 2nd Edition. New York: Guilford Press.

Dicken, P. and Lloyd, P. (1991) *Location in Space: A Theoretical Approach to Economic Geography*, 3rd Edition. New York: Harper & Row.

Disney, R. (1996) *Can We Afford to Grow Older? A Perspective on the Economics of Ageing*. Cambridge, MA: MIT Press.

Dixon, R. (1991) 'Venture Capital and the Appraisal of Investments', *Omega*, 5, 333–44.

Dodd, N. (1994) *The Sociology of Money: Economics, Reason and Contemporary Society*. Cambridge: Polity Press.

Donovan, P. (1996) 'NatWest to Cut More Jobs', *The Guardian*, 1 May, p. 18.

Doti, L.P. and Schweikart, L. (1990) *Banking in the American West: From the Gold Rush to Deregulation*. Norman: University of Oklahoma Press.

Dow, S.C. (1987) 'Money and Regional Development', *Studies in Political Economy*, 23, 732–94.

Dow, S.C. (1988) 'Incorporating Money in Regional Economic Models'. *Recent Advances in Regional Economic Modelling*. Eds. F. Harrigan and P. McGregor. Papers in Regional Science 19. London: Pion.

Dow, S.C. (1990) *Financial Markets and Regional Economic Development: The Canadian Experience*. Aldershot: Avebury.

Dow, S.C. (1993) *Money and the Economic Process*. Aldershot: Edward Elgar.

Dow, S.C. (1996a) 'European Monetary Integration, Endogenous Credit Creation and Regional Economic Development'. *Wealth from Diversity: Innovation, Structural Change and Finance for Regional Development in Europe*. Eds. X. Vence-Deza and J.S. Metcalfe. Boston: Kluwer.

Dow, S.C. (1996b) 'Horizontalism: A Critique', *Cambridge Journal of Economics*, 4, 497–508.

Dow, S.C. and Earl, P.E. (1982) *Money Matters: A Keynesian Approach to Monetary Economics*. Oxford: Martin Robertson.
Dow, S.C. and Rodriguez-Fuentes, C.J. (1997) 'Regional Finance: A Survey', *Regional Studies*, 31, 903–20.
Dow, S.C. and Smithin, J. (1992) 'Free Banking in Scotland, 1695–1845', *Scottish Journal of Political Economy*, 39, 374–90.
Drache, D. and Gertler, M. Eds. (1991) *The New Era of Global Competition: State Policy and Market Power*. Kingston and Montreal: McGill-Queens University Press.
Drake, P.J. (1980) *Money, Finance and Development*. Oxford: Martin Robertson.
Drucker, P. (1969) *The Age of Discontinuity: Guidelines to Our Changing Society*. New York: Heineman.
Dunford, M. (1990) 'Theories of Regulation', *Environment and Planning*, D, 8, 310–19.
Dunleavy, P. (1986) 'Explaining the Privatization Boom: Public Choice versus Radical Approaches', *Public Administration*, 64, 2, 13–34.
Dunn, J. (1990) 'Introduction'. *The Economic Limits of Modern Politics*. Ed. J. Dunn. Cambridge: Cambridge University Press.
Dymski, G., Veitch, J. and White, M. (1991) *Taking it to the Bank: Poverty, Race and Credit in Los Angeles*. Los Angeles: Western Centre on Law and Poverty.
Economist, The (1992) 'Can the Centre Hold? A Survey of Financial Centres', *The Economist Magazine*, Survey, 27 June.
Economist, The (1995) 'The Death of Distance, A Survey of Telecommunications', *The Economist Magazine*, Survey, 30 September.
Economist, The (1996) 'Unpopular Capitalism', 18 May, 32–5.
Edwards, F. (1981) 'The New "International Banking Facility": A Study in Regulatory Frustration', *Columbia Journal of World Business*, 16, 6–18.
Edwards, J.S.S. (1989) 'Gearing'. *The New Palgrave: Finance*. Eds. J. Eatwell, M. Milgate and P. Newman. London: Macmillan. 159–63.
Eichengreen, B. (1990) 'One Money for Europe? Lessons from the US Currency Union', *Economic Policy*, 10, 117–87.
Eichengreen, B. (1996) *Globalizing Capital: A History of the International Monetary System*. Princeton: Princeton University Press.
Eichengreen, B. and Flandreau, M. (1996) 'The Geography of the Gold Standard'. *Currency Convertibility: The Gold Standard and Beyond*. Eds. J. Braga de Macedo, B. Eichengreen and J. Reis. London: Routledge.
Elango, B., Fried, V.H., Hisrich, R.D. and Polonchek, A. (1995) 'How Venture Capital Firms Differ', *Journal of Business Venturing*, 10, 159–79.
Esping-Anderson, G. (1990) *The Three Worlds of Welfare Capitalism*. Cambridge: Polity.
European Venture Capital Association (1993) *Venture Capital: Policy Paper*. Zaventem, Belgium: EVCA.
European Venture Capital Association (1995) *Boosting Europe's Growing Companies*. Zaventem, Belgium: EVCA.
European Venture Capital Association (1997) *Yearbook 1997*. Zaventem, Belgium: EVCA.
Faini, R., Galli, G. and Giannini, C. (1993) 'Finance and Development: the Case of Southern Italy'. *Finance and Development: Issues and Experience*. Ed. A. Giovannini. Cambridge; Cambridge University Press. 158–214.
Fainstein, S. (1991) 'Promoting Economic Development: Urban Planning in the United States and Great Britain', *Journal of the American Planning Association*, 57, 22–33.
Favero, C.A. and Papi, L. (1995) 'Technical and Scale Efficiency in the Italian Banking Sector. A Non-parametric Approach', *Applied Economics*, n. 4, 385–95.
Fazzari, S.M., Hubbard, R.G. and Peterson, B.C. (1988) 'Financing Constraints and Corporate Investment', *Brookings Papers on Economics Activity*, 2, 141–95.
Featherstone, M. Ed. (1990) *Global Culture: Nationalism, Globalization and Modernity*. London: Sage.
Federal Deposit Insurance Corporation (1996) *Historical Statistics on Banking*. Washington DC: FDIC.

Ferri, G. (1997) 'Branch Manager Turnover and Lending Efficiency: Local vs National Banks', *BNL Quarterly Review, Special Issue*, 229–47.

Fiet, J.O. (1995) 'Risk Avoidance Strategies in Venture Capital Markets', *Journal of Management Studies*, 32, 551–74.

Financial Times (1983a) 'Nassau Steels Itself for an Exodus of Eurodollar Business', 17 March.

Financial Times (1983b) 'Carrot and Stick', 28 November.

Financial Times (1984) 'Business Picks Up after IBF Blow', 29 May.

Financial Times (1997) 'Property Backed Term Loan Well Met', *Financial Times*, 30 April.

Fischer, G.C. (1968) *American Banking Structure*. New York: Columbia University Press.

Florida, R.L. and Kenney, M. (1988a) 'Venture Capital, High Technology and Regional Development', *Regional Studies*, 22, 33–48.

Florida, R.L. and Kenney, M. (1988b) 'Venture Capital and High Technology Entrepreneurship', *Journal of Business Venturing*, 3, 301–19.

Florida, R.L. and Smith, D.F. (1990) 'Venture Capital, Innovation and Economic Development', *Economic Development Quarterly*, 4, 345–60.

Florida, R.L. and Smith, D.F. (1993) 'Venture Capital Formation, Investment and Regional Industrialisation', *Annals of the Association of American Geographers*, 83, 434–51.

Florida, R.L., Smith, D.F. and Sechoka, E. (1991) 'Regional Patterns of Venture Capital Investment'. *Venture Capital: International Comparisons*. Ed. M. Green. London: Routledge.

Foster, C.D. (1992) *Privatization, Public Ownership and the Regulation of Natural Monopolies*. Oxford: Blackwell.

Franklin, A.W. and Cotterill, R.W. (1993) 'An Analysis of Local Market Concentration Levels and Trends in the US Grocery Retailing Industry', *Food Marketing Policy Center Research Report*, 19, Dept. of Agricultural and Resources Economics, University of Connecticut.

Freear, J. and Wetzel, W.E. (1990) 'Who Bankrolls High-tech Entrepreneurs?' *Journal of Business Venturing*, 5, 77–89.

Freear, J., Sohl, J.E. and Wetzel, W.E. (1994) 'Angels and Non-angels: Are There Differences?' *Journal of Business Venturing*, 9, 109–23.

Freear, J., Sohl, J.E. and Wetzel, W.E. (1995a) 'Who Bankrolls Software Entrepreneurs?' *Frontiers of Entrepreneurship Research*. Eds. W.D. Bygrave, B.J. Bird, S. Birley, N.C. Churchill, M. Hay, R.H. Keeley and W.E. Wetzel. Babson Park, MA: Babson College.

Freear, J., Sohl, J.E. and Wetzel, W.E. (1995b) 'Angels: Personal Investors in the Venture Capital Market', *Entrepreneurship and Regional Development*, 7, 85–94.

Fried, V.H. and Hisrich, R.D. (1994) 'Toward a Model of Venture Capital Investment Decision-making', *Financial Management*, 23, 3, 28–37.

Fried, V.H. and Hisrich, R.D. (1995) 'The Venture Capitalist: A Relational Investor', *California Management Review*, 37, 101–13.

Friedman, J.J. (1995) 'The Effects of Industrial Structure and Resources upon the Distribution of Fast-growing Small Firms Among US Urbanised Areas', *Urban Studies*, 32, 863–83.

Froud, J., Haslam, C., Johal, S., Williams, J. and Williams, K. (1997) 'From Social Settlement to Household Lottery', *Economy and Society*, 26, 340–72.

Fuà, G. (1983) 'Rural Industrialization in Later Developed Countries: The Case of Northeast and Central Italy', *BNL Quarterly Review*, 147, 351–77.

Fuà, G. (1988) 'Small-scale Industry in Rural Areas: The Italian Experience'. *The Balance between Industry and Agriculture in Economic Development, Vol. I, Basic Issue*. Ed. K. Arrow. London: Macmillan. 259–79.

Fujita. K. (1991) 'A World City and Flexible Specialization: Restructuring the Tokyo Metropolis', *International Journal of Urban and Regional Research*, 15: 269–84.

Furnham, A. and Argyle, M. (1998) *The Psychology of Money*. London: Routledge.

Futures and Options World (1995) June Edition.

Futures and Options World (1996) June Edition.

Futures and Options World (1997) June Edition.

Gaffikin, F. and Morrissey, M. (1990) *Northern Ireland: The Thatcher Years*. London: Zed Books.

Gaffikin, F. and Warf, B. (1993) 'Urban Policy and the Post-Keynesian State in the United Kingdom and the United States', *International Journal of Urban and Regional Research*, 17, 67–84.

Galbraith, J.K. (1975) *Money: Whence it Came, Where it Went*. Boston: Houghton Mifflin.

Gardener, E.P.M. (1988) 'Innovations and New Structural Frontiers in Banking'. *Contemporary Issues in Money and Banking*. Ed. P. Arestis. London: Macmillan. 7–29.

Gardener, E. and Molyneux, E. (1990) *Changes in Western European Banking*. London: Unwin Hyman.

Garofoli, G. (1989) 'Modelli Locali di Sviluppo: i Sistemi di Piccola Impresa'. *Modelli locali di sviluppo*. Ed. G. Becattini. Bologna: Il Mulino. 75–90.

Gaston, R.J. (1989) *Finding Private Venture Capital For Your Firm: A Complete Guide*. New York: Wiley.

Gentle, C.J.S. (1993) *The Financial Services Industry: The Impact of Corporate Reorganisation on Regional Eonomic Development*. Aldershot: Avebury.

Gentle, C.J.S. and Marshall, N. (1992) 'The Deregulation of the Financial Services Industry and the Polarization of Regional Economic Prosperity', *Regional Studies*, 26, 581–6.

Gentle, C.J.S. and Marshall, N. (1993) 'Corporate Restructuring in the Financial Services Industry: Converging Markets and Divergent Organisational Structures'. *Mimeo*. CURDS, University of Newcastle-upon-Tyne.

Gibb, R. and Michalak, W. Eds. (1994) *Continental Trading Blocs: The Growth of Regionalism in the World Economy*. New York: John Wiley.

Gibbs, D. (1991) 'Venture Capital and Regional Development: The Operation of the Venture Capital Industry in Manchester', *Tijdschrift voor Economische en Sociale Geografie*, 82, 242–53.

Giddens, A. (1981) *A Contemporary Critique of Historical Materialism*. Berkeley: University of California Press.

Giddens, A. (1985) *The Nation-State and Violence*. Cambridge: Polity Press.

Gill, S. (1989) 'Global Hegemony and the Structural Power of Capital', *International Studies Quarterly*, 33, 475–99.

Gill, S. (1997) 'Transformation and Innovation in the Study of World Order'. *Innovation and Transformation in International Studies*. Eds. S. Gill and J.H. Mittelman. Cambridge: Cambridge University Press. 5–23.

Gillispie, A. and Williams, H. (1988) 'Telecommunications and the Reconstruction of Comparative Advantage', *Environment and Planning*, A, 20, 1311–21.

Gilmour, I. (1992) *Dancing with Dogma: Britain Under Thatcherism*. London: Simon and Schuster.

Gilson, S.C. and Warner, J.B. (1997) 'Junk Bonds, Bank Debt, and Financing Corporate Growth'. *Mimeo*. Harvard Business School.

Glynn, A. (1996) 'Taxing and Spending', *Imprints*, 1, 48–58.

Goldman Sachs (1995) *Executive Summary: Survey of Alternative Investments by Pension Funds*. New York: Goldman Sachs.

Goldman Sachs (1997) *Report on Alternative Investing by Tax Exempt Organisations*. New York: Goldman Sachs.

Goldsmith, R.W. (1969) *Financial Structure and Development*. New York: Yale University Press.

Gompers, P.A. (1994) 'The Rise and Fall of Venture Capital, *Business and Economic History*, 23(2), 1–26.

Gompers, P.A. (1995) 'Optimal Investing: Monitoring and the Staging of Venture Capital', *Journal of Finance*, 50, 1461–89.

Goodrich, J. Ed. (1990) *Privatization in Global Perspective*. London: Pinter.

Gorman, M. and Sahlman, W.A. (1989) 'What do Venture Capitalists do?' *Journal of Business Venturing*, 4, 231–48.

Gouri, G. (1991) *Privatization and Public Enterprise: The Asian-Pacific Experience*. Oxford: IBH Publishing.

Gowa, J. (1983) *Closing the Gold Window: Domestic Politics and the End of Bretton Woods*. London: Cornell University Press.

Graziani, A. (1986) 'Il Mezzogiorno e l'Economia Italiana'. *L'economia e il Mezzogiorno. Sviluppo Imprese e Territorio*. Ed. A. Giannola. Milan: F. Angeli. 15–45.

Greco, T. (1994) *New Money for Healthy Communities*. Tuscon, Arizona: T.H. Greco.

Green, M.B. (1991) 'Preferences for US Venture Capital Investment 1970–1988'. *Venture Capital: International Comparisons*. Ed. M. Green. London: Routledge.

Green, M.B. (1993) 'A Geography of Institutional Stock Ownership in the United States', *Annals of the Association of American Geographers*, 83, 66–89.

Greenwald, B.C., Levinson, A. and Stiglitz, J.E. (1993) 'Capital Market Imperfections and Regional Economic Development'. *Finance and Development: Issues and Experience*. Ed. A. Giovani. Cambridge: Cambridge University Press. 65–93.

Greenwald, B.C., Stiglitz, J. and Weiss, A. (1984) 'Information Imperfections in the Capital Market and Macroeconomic Fluctuations', *American Economic Review Papers and Proceedings*, 74, 194–7.

Greenwich Associates (1996) *New Paradigm, New Potency*. Greenwich, CT: Greenwich Associates.

Greider, W. (1987) *Secrets of the Temple: How the Federal Reserve Runs the Country*. New York: Simon and Schuster.

Grimstone, G. (1987) 'Privatization: The Unexpected Crusade', *Contemporary Record*, 1, 1, 20–30.

Groh, K. (1986) 'The Evolution of Telephone Banking', *Magazine of Bank Administration*, 62, 9, 14–16.

Grossman, S. (1995) 'Dynamic Asset Allocation and the Informational Efficiency of Markets', *Journal of Finance*, 50, 773–87.

Gupta, A.K. and Sapienza, H.J. (1992) 'Determinants of Venture Capital Firms' Preferences Regarding the Industry Diversity and Geographic Scope of their Investments', *Journal of Business Venturing*, 7, 347–62.

Gurley, J.G. and Shaw, E.S. (1967) 'Financial Structure and Development', *Economic Development and Cultural Change*, 15, April, 257–68.

Haldon, J. (1994) 'Invisible Cities, Hidden Agendas', *The Journal of Peasant Studies*, 21, 2, 308–21.

Hall, P. (1989) *Governing the Economy*. Cambridge: Polity

Hall, J. and Hofer, C.W. (1993) 'Venture Capitalists' Decision Criteria in New Venture Evaluation', *Journal of Business Venturing*, 8, 25–42.

Hall, P. and Preston, P. (1988) *The Carrier Wave: New Information Technology and the Geography of Innovation, 1846–2003*. London: Unwin Hyman.

Hamnett, C. (1992) 'Share-ownership still Reflects Social Divide', *The Independent*, 3 January.

Hamnett, C. (1998) 'The UK: From Pragmatic to Systematic Privatization'. *Shrinking the State: The Political Underpinnings of Privatization*. Eds. Feijenbaum, Henig and C. Hamnett. Cambridge: Cambridge University Press.

Harper, R. (1987) 'A Functional Classification of Management Centers of the United States', *Urban Geography*, 540–99.

Harris, L. (1995) 'International Financial Markets and National Transmission Mechanisms'. *Managing the Global Economy*. Eds. J. Michie and J. Grieve-Smith. Oxford: Oxford University Press.

Harris, S. and Bovaird, C. (1996) *Enterprising Capital*. Aldershot: Avebury.

Harrison, B. and Bluestone, B. (1990) 'Wage Polarisation in the US and the "Flexibility" Debate', *Cambridge Journal of Economics*, 14, 351–73.

Harrison, D. (1995) *Pension Fund Investment in Europe*. London: Financial Times Publishing.

Harrison, R.T and Mason, C.M. (1992) 'The Roles of Investors in Entrepreneurial Companies:

A Comparison of Informal Investors and Venture Capitalists'. *Frontiers of Entrepreneurship Research*. Eds. N.C. Churchill, S. Birley, W.D. Bygrave, D.F. Muzyka, C. Wahlbin and W.E. Wetzel. Babson Park, MA: Babson College.
Harrison, R.T. and Mason, C.M. Eds. (1996) *Informal Venture Capital: Evaluating the Impact of Business Introduction Services*. Hemel Hempstead: Prentice-Hall.
Harrison, R.T., McIntyre, P. and Mason, C.M. (1997) 'Informal Venture Capital and Early Stage Equity Investment: The Case of Northern Ireland'. Paper to the ISBA Small Firms Policy and Research Conference, Belfast, 19–21 November.
Harvey, D. (1974) 'Class Monopoly Rent, Finance Capitals and the Urban Revolution', *Regional Studies*, 8, 239–55.
Harvey, D. (1977) 'Government Policies, Financial Institutions and Neighbourhood Change in United States Cities', *Captive Cities*. Ed. M. Harloe. London: Wiley.
Harvey, D. (1982) *The Limits to Capital*. Oxford: Blackwell.
Harvey, D. (1985) *Consciousness and the Urban Experience*. Oxford: Blackwell.
Harvey, D. (1989a) 'From Managerialism to Entrepreneurialism: The Transformation in Urban Governance in Late Capitalism', *Geografiska Annaler*, 71, 3–17.
Harvey, D. (1989b) *The Condition of Postmodernity: An Enquiry into the Origins of Cultural Change*. Oxford: Blackwell.
Harvey, D. (1990) 'Between Space and Time: Reflections on the Geographical Imagination', *Annals of the Association of American Geographers*, 80, 418–34.
Harvey, D. and Chatterjee, L. (1974) 'Absolute Rent and the Restructuring of Space by Government and Financial Institutions', *Antipode*, 6, 22–36.
Hawawini, G. and Jacquillat, B. (1990) 'European Equity Markets in the 1990s'. *European Banking in the 1990s*. Ed. J. Dermine. Oxford: Blackwell. 69–104.
Hawley, J. (1986) *Dollars and Borders: US Government Attempts to Restrict Capital Outflows, 1960–1980*. New York: M.E. Sharpe.
Heinsohn, G. and Steiger, O. (1987) 'Private Ownership and the Foundations of Monetary Theory', *Economies et Societes*, 9, 229–57.
Helleiner, E. (1994a) *States and the Re-emergence of Global Finance: From Bretton Woods to the 1990s*. London: Cornell University Press.
Helleiner, E. (1994b) 'From Bretton Woods to Global Finance'. *Political Economy and the Changing Global Order*. Eds. R. Stubbs and G. Underhill. London: Macmillan.
Hepworth, M. (1986) 'The Geography of Technological Change in the Information Economy', *Regional Studies*, 20, 6, 407–24.
Hepworth, M. (1990) *Geography of the Information Economy*. London: Guilford Press.
Hepworth, M. (1991) 'Information Technology and the Global Restructuring of Capital Markets'. *Collapsing Space and Time*. Eds. S.D. Brunn and T.R. Leinbach. London: HarperCollins.
Higgins, B. (1986) *The Rise and Fall of Montreal*. Moncton: Canadian Institute for Research on Regional Development.
Hilder, D.B. and Lambert, W. (1991) 'Banks Cleared to Underwrite, Sell Insurance', *The Wall Street Journal*, 11 June, A3.
Hindle, K. and Wenban, R. (forthcoming) 'An Exploratory Study of Australia's Informal Venture Capitalists: A Predicate to Theory Building in the Context of International Angel Research', *Venture Capital: An International Journal of Entrepreneurial Finance*, 1.
Hines, C. (1997) 'Putting a Bloc on the Global Wrecking Gang', *The Guardian*, 17 November, 19.
Hirst, P. and Thompson, G. (1996) *Globalization in Question: The International Economy and the Possibilities of Governance*. Cambridge: Polity Press.
Hirst, P. and Zeitlin, J. (1997) 'Flexible Specialisation: Theory and Evidence in the Analysis of Industrial Change'. *Contemporary Capitalism*. Eds. J.R. Hollingsworth and R. Boyer. Cambridge: Cambridge University Press.
HM Treasury (1994) *Share Ownership*. London: HM Treasury.
HM Treasury (1995) *Privatization: Sharing the UK Experience*. London: HM Treasury.
HM Treasury (1996) *Guide to the Privatization Programme*. London: HM Treasury.

Hoare, A. (1997) 'Privatisation Comes to Town: National Policies and Local Responses – The Bristol Case', *Regional Studies*, 31, 3, 253–65.

Hobsbawm, E. (1994) *Age of Extremes: The Short Twentieth Century: 1914–1989*. London: Michael Joseph.

Hollingsworth, J.R. and Boyer, R. Eds. (1997) *Contemporary Capitalism: The Embeddeness of Institutions*. Cambridge: Cambridge University Press.

Holly, B. (1987) 'Regulation, Competition, and Technology: The Restructuring of the US Commercial Banking System', *Environment and Planning*, A, 19, 633–52.

Hook, L.M. (1994) *Bank Failures and Deregulation in the 1980s*. New York: Garland Publishers.

Houston, J. and James, C. (1996) 'Bank Information Monopolies and the Mix of Private and Public Debt Claims', *Journal of Finance*, 51, 1863–89.

Howe, K. (1993a) 'B of A Chief Defends Plan to Reduce Full-time Workers', *San Francisco Chronicle*, 17 February, D1 and D2.

Howe, K. (1993b) 'B of A Cut in Job Hours Follows Trend', *San Francisco Chronicle*, 11 February, D1 and D4.

Hudson, A. (1996) 'Globalisation, Regulation and Geography: The Development of the Bahamas and the Cayman Islands Offshore Financial Centres'. Unpublished PhD Dissertation, Department of Geography, University of Cambridge, UK.

Hudson, A. (forthcoming) 'Reshaping the Regulatory Landscape: Border Skirmishes Around the Bahamas and Cayman Off-shore Financial Centres', *Review of International Political Economy*.

Hustedde, R.J. and Pulver, G.C. (1992) 'Factors Affecting Equity Capital Acquisition: The Demand Side', *Journal of Business Venturing*, 7, 363–74.

Hutchinson, R.W. (1995) *Corporate Finance: Principles of Investment, Financing and Valuation*. Cheltenham: Stanley Thomas.

Hutchinson, R.W. and Hunter, R.L. (1995) 'Determinants of Capital Structure in the Retailing Sector in the UK', *International Review of Retail Distribution and Consumer Research*, 5, 63–78.

Hutchinson, R.W. and Mckillop, D.G. (1990) 'Regional Financial Sector Models: An Application to the Northern Ireland Financial Sector', *Regional Studies*, 24, 421–31.

Hutton, W. (1995) *The State We Are In*. London: Jonathan Cape.

Ingham, G. (1984) *Capitalism Divided: The City and Industry in Britain*. London: Macmillan.

International Monetary Fund (1993) *International Capital Markets*, August. Washington: IMF.

International Monetary Fund (1996) *International Capital Markets*, August. Washington: IMF.

Ireland, J. (1981) 'IBFs: the Bahamas Eyes its Future', *The Banker*, July.

Irwin, M. and Kasarda, J. (1991) 'Air Passenger Linkages and Employment Growth in US Metropolitan Areas', *American Sociological Review*, 56, 524–37.

Jackall, R. (1978) *Workers in a Labyrinth: Jobs and Survival in a Bank Bureaucracy*. New Jersey: Allenheld Osmun.

Jackson, P.M. and Price, C.M. Eds. (1994) *Privatisation and Regulation: A Review of the Issues*. London: Longman.

James, J. (1976) 'Banking Market Structure, Risk and the Pattern of Local Interest Rates in the US, 1893–1911', *Journal of Economic History*, 36, 112–130.

James. J. (1978) *Money and Capital Markets in Postbellum America*. Princeton: Princeton University Press.

James, M. and James, B.R. (1954) *Biography of a Bank: The Story of Bank of America*. New York: Harper and Brothers.

Jayaratne, J. and Strahan, P.E. (1996) 'The Finance-Growth Nexus: Evidence from Branch Deregulation', *Quarterly Journal of Economics*, 639–70.

Jenkins, R. (1991) 'The Political Economy of Industrialization: A Comparison of Latin

American and East Asian Newly Industrializing Countries', *Development and Change*, 22, 197–231.
Jenkins, S. (1995) *Accountable to None: The Tory Nationalization of Britain*. London: Hamish Hamilton.
Jensen, M.C. (1986) 'Agency Costs of Free Cash Flow, Corporate Finance and Takeovers', *American Economic Review*, 76, 323–9.
Jensen, M.C. (1989) 'The Eclipse of the Public Corporation', *Harvard Business Review*, 67(5), 61–74.
Jensen, M.C. (1991) 'Corporate Control and the Politics of Finance', *Journal of Applied Corporate Finance*, 4, 13–33.
Jensen, M.C. and Meckling, W. (1976) 'Theory of the Firm: Managerial Behaviour, Agency Costs and Ownership Structure', *Journal of Financial Economics* 3, 305–60.
Jessop, B. (1992) 'The Schumpeterian Workfare State: Or Japanism and Post-Fordism', *8th Conference of Council for Europeanists*. Palmer House, Chicago. 27–29 March.
Jessop, B. (1993) 'Towards a Schumpeterian Workfare State? Preliminary Remarks on Post-Fordist Political Economy', *Studies in Political Economy*, 40, 7–39.
Jessop, B. (1994a) 'Post-Fordism and the State'. *Post-Fordism: A Reader*. Ed. A. Amin. Oxford: Blackwell.
Jessop, B. (1994b) 'The Transition to Post-Fordism and the Schumpeterian Workfare State'. *Towards a Post-welfare State?* Eds. R. Burrows and B. Loader. London: Routledge.
Jessop, B., Bonnett, K., Bromley, S. and Ling, T. (1988) *Thatcherism: A Tale of Two Nations*. London: Polity Press.
Johns, R. (1983) *Tax Havens and Offshore Finance*. London: Pinter.
Johns, R. and Le Marchant, C. (1993) *Finance Centres. British Isle Off-shore Development Since 1979*. London: Pinter.
Johnson, H.G. (1976) 'Panama as a Regional Financial Center: A Preliminary Analysis of Development Contribution', *Economic Development and Cultural Change*, 24, 2, 261–86.
Johnson, R.B. (1983) *The Economics of the Euro-Market: History, Theory and Policy*. London: Macmillan.
Johnston, M. (1990) *Roller Coaster: The Bank of America and the Future of American Banking*. New York: Ticknor and Fields.
Kambula, J., Keane, F. and Benadon, C. (1996) 'Price Risk Intermediation in the Over-the-Counter Derivatives Markets: Interpretation of a Global Survey', *Federal Reserve Bank of New York Economic Policy Review*, 2, 1, 1–16.
Kaplan, S. and Stein, J. (1993) 'The Evolution of Buyout Pricing and Financial Structure in the 1980s', *Quarterly Journal of Economics*, 108, 313–58.
Kapstein, E.B. (1994) *Governing the Global Economy: International Finance and the State*. Cambridge, Mass: Harvard University Press.
Karmel, R.S. (1993) 'Implications of the Stakeholder Model', *The George Washington Law Review*, 61, 1156–76.
Kay, J.A. and Thompson, D.J. (1986) 'Privatization: A Policy in Search of a Rationale', *Economic Journal*, 96, 18–32.
Keeble, D. (1994) 'Regional Influences and Policy in New Technology-based Firm Creation and Growth'. *New Technology-based Firms in the 1990s*. Ed. R. Oakey. London: Paul Chapman Publishing.
Keeble, D. (1997) 'Small Firms, Innovation and Regional Development in Britain in the 1990s', *Regional Studies*, 31, 281–93.
Keeble, D. and Walker, S. (1994) 'New Firms, Small Firms and Dead Firms: Spatial Patterns and Determinants in the United Kingdom', *Regional Studies*, 28, 411–27.
Keohane, R.O. (1984) 'The World Political Economy and the Crisis of Embedded Liberalism'. *Order and Conflict in Contemporary Capitalism: Studies in the Political Economy of Western European Nations*. Ed. J. Goldthorpe. Oxford: Oxford University Press.
Kerr, D. (1965) 'Some Aspects of the Geography of Finance in Canada', *Canadian Geographer*, 9, 4, 175–92.

Key, S. and Terrell, H. (1988) 'International Banking Facilities', *Board of Governors of the Federal Reserve System, International Finance Discussion Papers*, 333.

Kindleberger, C.P. (1974) *The Formation of Financial Centres: A Study in Comparative Economic History*. Princeton Studies in International Finance, 36, Princeton: Princeton University Press.

Kindleberger, C.P. (1978) 'The Formation of Financial Centers'. *Economic Response. Comparative Studies in Trade, Fianance and Growth*. Ed. C.P. Kindleberger. Cambridge, Mass: Harvard University Press.

Kindleberger, C.P. (1985) 'The Functioning of Financial Centres: Britain in the Nineteenth Century, the US since 1945'. *International Financial Markets and Capital Movements*. Eds. W. Ethier and R. Marston. Princeton: Essays in International Finance, 157.

Knudsen, D.C. Ed. (1996) *The Transition to Flexibility*. Boston: Kluwer.

Kobrin, S.J. (1997) 'Electronic Cash and the End of National Markets', *Foreign Policy*, Summer, 65–77.

Kolbenschlag, M., Taylor, C., Burton, J. and Darling, J. (1986) 'Los Angeles – the Number Two Financial Center', *Euromoney*, September, 2–14.

Kovenock, D. and Phillips, G.M. (1995) 'Increased Debt and Product-market Rivalry. How do we Reconcile Theory and Evidence?' *American Economic Review*, 85, 403–8.

Kovenock, D. and Phillips, G.M. (1997) 'Capital Structure and Product Market Behaviour: An Examination of Plant Exit and Investment Decisions', *Review of Financial Studies*, 20, 197–219.

KPMG Management Consulting (1992) *Investment Networking*. Glasgow: Scottish Enterprise.

Krugman, P. (1991a) *The Geography of Trade*. Cambridge, MA: MIT Press.

Krugman, P. (1991b) 'Increasing Returns and Economic Geography', *Journal of Political Economy*, 99, 484–99.

Krugman, P. (1993) 'First Nature, Second Nature and Metropolitan Location', *Journal of Regional Science*, 33, 2, 129–44.

Krugman, P. (1995) *Development, Geography and Economics*. Cambridge, MA: MIT Press.

Kurtzman, J. (1993) *The Death of Money*. Boston: Little, Brown and Co.

Kuttner, R. (1997) *Everything for Sale: The Virtues and Limits of Markets*. New York: Knopf.

Kynaston, D. (1994) *The City of London. Volume 1: A World of its Own, 1815–90*. London: Pimlico Press.

Labasse, J. (1974) *L'Espace Financier*. Paris: Colin.

Landström, H. (1993) 'Informal Risk Capital in Sweden and Some International Comparisons', *Journal of Business Venturing*, 8, 525–40.

Lang, W.W. and Nakamura, L.I. (1990) 'The Dynamics of Credit Markets in a Model with Learning', *Journal of Monetary Economics*, 26, 305–18.

Langbein, J. (1995) 'The Contractarian Basis of the Law of Torts', *Yale Law Journal*, 105, 625–75.

Langbein, J. (1997) 'The Secret Life of Trust: The Trust as an Instrument of Commerce', *Yale Law Journal*, 107, 165–89.

Langdale, J. (1989) 'The Geography of International Business Telecommunications: The Role of Leased Networks', *Annals of the Association of American Geographers*, 79, 501–22.

La Porta, R., Lopez-de-Silanes, F., Shleifer, A. and Vishny, R. (1997) 'Legal Determinants of External Finance', *Journal of Finance*, 52, 1131–50.

Lash, S. and Urry, J. (1987) *The End of Organized Capitalism*. Madison: University of Wisconsin Press.

Laulajainen, R. (1998) *Financial Geography*. Göteborg: School of Economics and Commercial Law.

Law, J. (1994) *Organizing Modernity*. Oxford: Blackwell.

Lee, R. (1989) 'Social Relations and the Geography of Material Life'. *Horizons in Human Geography*. Eds. D. Gregory and R. Walford. London: Macmillan. Chapter 2.4.

Lee, R. (1996) 'Moral Money? LETS and the Social Construction of Economic Geographies in South East England', *Environment and Planning*, A, 28, 1377–94.

Lee, R. (1998a) 'Access to the Gods? Social Relations and/or Regimes of Truth in the Social

Construction (the Geographies) of Material Life'. Paper given at the Annual Conference of the AAG, 27 March.
Lee, R. (1998b) 'Shelter from the Storm? Mutual Knowledge and the Construction of Moral Economic Geographies'. Paper given at the Annual Conference of RGS-IBG University of Surrey, Guildford, 6 January.
Lee, R. (forthcoming) *Access to the Gods? Social Relations and Geographies of Material Life*. London: Routledge.
Lee, R. and Schmidt-Marwede, U. (1993) 'Interurban Competition? Financial Centres and the Geography of Financial Production', *International Journal of Urban and Regional Research*, 17, 492–515.
Lee, R. and Wills, J. Eds. (1997) *Geographies of Economies*. London: Arnold.
Leibfritz, W., Roseveare, D., Fore, D. and Wurzel, E. (1995) 'Ageing Populations, Pension Systems and Government Budgets: How Do they Affect Savings?' *Working Paper*, 156, Paris: OECD Economics Department.
Leitner, H. (1990) 'Cities in Pursuit of Economic Growth: The Local State as Entrepreneur', *Political Geography Quarterly*, 9, 146–70.
Lerner, J. (1994) 'The Syndication of Venture Capital Investments', *Financial Management*, 23, 3, 16–27.
Levine, M.V. (1997) *The Feasibility of Economically Targeted Investing: A Wisconsin Case Study*. Brookfield: International Foundation of Employee Benefit Plans.
Leyshon, A. (1992) 'The Transformation of Regulatory Order: Regulating the Global Economy and Environment', *Geoforum*, 23, 249–67.
Leyshon, A. (1995a) 'Annihilating Space? The Speed Up of Communications'. *A Shrinking World?* Eds. J. Allen and C. Hamnett. Milton Keynes: Open University Press. Chapter 1.
Leyshon, A. (1995b) 'Geographies of Money and Finance: I', *Progress in Human Geography*, 19, 531–43.
Leyshon, A. (1996) 'Dissolving Difference? Money, Disembedding and the Creation of "Global Financial Space"'. *The Global Economy in Transition*. Eds. P.W. Daniels and W.F. Lever. Harlow: Longman. Chapter 5.
Leyshon, A. (1997) 'Geographies of Money and Finance: II', *Progress in Human Geography*, 21, 381–92.
Leyshon, A. and Pollard, J.S. (1998) 'Globalisation, Conventions of Restructuring and the Reorganisation of Retail Banking: Evidence from the United States and Britain'. *Mimeo*. Department of Geography, University of Bristol, UK.
Leyshon, A. and Thrift, N.J. (1989) 'South Goes North? The Rise of the British Provincial Financial Centre'. *The North–South Divide: Regional Change in Britain in the 1980s*. Eds. J. Lewis and A.R. Townsend. London: Paul Chapman.
Leyshon, A. and Thrift, N. (1992) 'Liberalisation and Consolidation: The Single European Market and the Remaking of European Financial Capital', *Environment and Planning*, A, 24, 49–81.
Leyshon, A. and Thrift, N. (1993) 'The Restructuring of the U.K. Financial Services Industry in the 1990s: A Reversal of Fortune?' *Journal of Rural Studies*, 9, 3, 223–41.
Leyshon, A. and Thrift, N. (1995) 'Geographies of Financial Exclusion: Financial Abandonment in Britain and the United States', *Transactions of the Institute of British Geographers*, NS, 20, 312–41.
Leyshon, A. and Thrift, N.J. (1997) *Money/Space: Geographies of Monetary Transformation*. London: Routledge.
Leyshon, A. and Tickell, A. (1994) 'Money Order? The Discursive Construction of Bretton Woods and the Making and Breaking of Regulatory Space', *Environment and Planning*, A, 26, 1861–90.
Leyshon, A., Thrift, N. and Pratt, J. (1998) 'Scales of Risk and Systems of Knowledge: Information, Decision-making and The Rise of Credit Scoring Systems in Retail Banking'. Paper given at the Annual Conference of the RGS-IBG, University of Surrey, Guildford, 8 January.
Lipietz, A. (1982) 'Towards Global Fordism?' *New Left Review*, 132, March–April, 33–47.

Lister, R.C. (1993) *Bank Behavior, Regulation and Economic Development: California, 1860–1910*. New York: Garland Publishing Inc.

Litan, R.E. (1990) 'Remedy for S&Ls: Operation "Clean Sweep"', *Challenge*, November–December, 26–32.

Logan, J. and Molotch, H. (1987) *Urban Fortunes: The Political Economy of Place*. Berkeley: University of California Press.

London Stock Exchange (1993) *Stock Market Quarterly*, Spring Edition.

London Stock Exchange (1994) *Stock Market Quarterly*, Spring Edition.

London Stock Exchange (1995a) *Stock Market Quarterly*, Spring Edition.

London Stock Exchange (1995b) *Quality of Markets Review*. Spring Edition.

London Stock Exchange (1996) *Quality of Markets Review*. Spring Edition.

Lord, J.D. (1987) 'Interstate Banking and the Relocation of Economic Control Points', *Urban Geography*, 8, 501–19.

Lord. J.D. (1992) 'Geographic Deregulation of the US Banking Industry and the Spatial Transfers of Corporate Control', *Urban Geography*, 13, 25–48.

Lösch, A. (1949) 'Theorie der Wahrung', *Weltwirtschaftliches Archive*, LXII, 35–88.

Lösch, A. (1954) *The Economics of Location*. New Haven: Yale University Press (Second English Edition; German Edition, 1939).

Lumme, A., Mason, C. and Suomi, M. (1998) *Informal Venture Capital: Investors, Investments and Policy Issues in Finland*. Dordrecht, NL: Kluwer.

Lyons, D. (1995) 'Agglomeration Economies among High Technology Firms in Advanced Production Areas: The Case of Denver/Boulder', *Regional Studies*, 29, 265–78.

Mackintosh, J. (1997) 'Another Way to Pay', *The Financial Times*, 6 September.

MacLeod, G. (1997) *From Mondragon to America: Experiments in Community Development*. Sydney: NS, University College of Cape Breton Press.

MacMillan, I.C., Siegal, R. and Subba Narasimba, P.N. (1985) 'Criteria Used by Venture Capitalists to Evaluate New Venture Proposals'. *Frontiers of Entrepreneurship Research 1985*. Eds. J.A. Hornaday, E.B. Shils, J.A. Timmons and K.H. Vesper. Babson Park, MA: Babson College.

MacMillan, I.C., Kulow, D.M. and Khoylian, R. (1989) 'Venture Capitalists' Involvement in their Investments: Extent and Performance', *Journal of Business Venturing*, 4, 27–47.

Magowan, P.A. (1989) 'The Case for LBOs: The Safeway Experience', *California Management Review*, 32, 9–11.

Maier, C. (1978) 'The Politics of Productivity: Foundations of American Economic Policy after World War II'. *Between Power and Plenty: Foreign Economic Policies of Advanced Industrial States*. Ed. P. Katzenstein. Madison: University of Wisconsin Press.

Mander, J. and Goldsmith, E. Eds. (1996) *The Case against the Global Economy and for a Turn towards the Local*. San Francisco: Sierra Books.

Marion, B.W. (1995) 'Changing Power Relationships in the US Food Industry: Brokerage Arrangements for Private Label Products'. Paper presented at Conference on Food Retailer–Manufacturer Competitive Relationships in the EU and USA, University of Reading, July.

Markowitz, H.M. (1952) 'Portfolio Selection', *Journal of Finance*, 7, 77–91.

Markusen, A. (1986) 'Defence Spending: A Successful Industrial Policy?' *International Journal of Urban and Regional Research*, 10, 105–22.

Markusen, A. and Gwiasda, V. (1994) 'Multipolarity and the Layering of Functions in World Cities: New York City's Struggle to Stay on Top', *International Journal of Urban and Regional Research*, 18, 167–93.

Markusen, A. and Yudken, J. (1992) *Dismantling the Cold War Economy*. New York: Basic Books.

Marshall, A. and Harding, C. (1993) 'How Britain Sets the World Afloat', *Independent on Sunday*, Business Section, 19 January, 12–13.

Martin, R.L. (1989) 'The Growth and Geographical Anatomy of Venture Capitalism in the United Kingdom', *Regional Studies*, 23, 389–403.

Martin, R.L. (1992) 'Financing Regional Enterprise: The Role of the Venture Capital Market'.

Regional Development in the 1990s: The British Isles in Transition. Eds. P. Townroe and R. Martin. London: Jessica Kingsley Publishers.

Martin, R.L. (1994) 'Stateless Monies, Global Financial Integration and National Autonomy: The End of Geography?' *Money, Power and Space*. Eds. S. Corbridge, R.L. Martin and N. Thrift. Oxford: Blackwell.

Martin, R.L. (1995) 'Income and Poverty Inequalities Across Regional Britain: The North–South Divide Lingers On'. *Off the Map: The Social Geography of Poverty in the UK*. Ed. C. Philo. London: CAPG.

Martin, R.L. and Minns, R. (1995) 'Undermining the Financial Basis of Regions: The Spatial Structure and Implications of the UK Pension Fund Industry', *Regional Studies*, 29, 125–44.

Martin, R.L. and Sunley, P. (1997) 'The Post-Keynesian State and the Space Economy'. *Geographies of Economies*. Eds. R. Lee and J. Wills. London: Arnold.

Martin, S. and Parker, D. (1997) *The Impact of Privatisation: Ownership and Corporate Performance in the UK*. London: Routledge.

Mason, C.M. and Harrison, R.T. (1989) 'The North–South Divide and Small Firms Policy in the UK: The Case of the Business Expansion Scheme', *Transactions of the Institute of British Geographers*, 14, 37–58.

Mason, C.M. and Harrison, R. (1991) 'Venture Capital, the Equity Gap and the North–South Divide in the UK. *Venture Capital: International Comparisons*. Ed. M. Green. London: Routledge.

Mason, C.M. and Harrison, R.T. (1994) 'The Informal Venture Capital Market in the UK'. *Financing Small Firms*. Eds. A. Hughes and D.J. Storey. London: Routledge.

Mason, C.M. and Harrison, R. (1995) 'Closing the Regional Equity Gap: The Role of Informal Venture Capital', *Small Business Economics*, 7, 153–72.

Mason, C.M. and Harrison, R.T. (1996a) 'Informal Venture Capital: A Study of the Investment Process and Post-investment Experience', *Entrepreneurship and Regional Development*, 8, 105–26.

Mason, C.M. and Harrison, R.T. (1996b) 'Why Business Angels Say No: A Case Study of Opportunities Rejected by an Informal Investor Syndicate', *International Small Business Journal*, 14, 2, 35–51.

Mason, C.M. and Harrison, R.T. (1997a) 'Business Angels – Heaven-sent or the Devil to Deal With?' *Mastering Enterprise: Your Single Source Guide to Becoming an Entrepreneur*. Eds. S. Birley and D.F. Muzyka. London: FT/Pitman Publishing.

Mason, C.M. and Harrison, R.T. (1997b) 'Business Angel Networks and the Development of the Informal Venture Capital Market in the UK: Is There Still a Role for the Public Sector?' *Small Business Economics*, 9, 111–23.

Mason, C.M. and Harrison, R.T. (1997c) 'Supporting the Informal Venture Capital Market: What Still Needs to Be Done? *Venture Finance Working Paper*, 15, Southampton: University of Southampton.

Mason, C. and Rogers, A. (1997) 'The Business Angel's Investment Decision: An Exploratory Analysis'. *Entrepreneurship in the 1990s*. Eds. D. Deakins, P. Jennings and C. Mason. London: Paul Chapman Publishing.

Massey, D. (1984) *Spatial Divisions of Labour: Social Structures and the Geography of Production*. London: Macmillan.

Massey, D. and Meegan, R. (1982) *The Anatomy of Job Loss*. London: Methuen.

Mayer, M. (1992) *The Greatest Ever Bank Robbery: The Collapse of the Savings and Loan Industry*. New York: Macmillan Publishing Company.

McCaffery, K., Hutchinson, R.W. and Jackson, R. (1997) 'Aspects of the Finance Function: A Review and Survey of the UK Retailing Sector', *International Review of Retail, Distribution and Consumer Research*, 7, 125–44.

McCrae, F. and Cairncross, A. (1991) *Capital City*. London: Methuen.

McDonald, K.R. (1993) 'Why Privatization is not Enough', *Harvard Business Review*, May–June, 423–42.

McDowell, L. (1997) *Capital Culture: Gender at Work in the City*. Oxford: Blackwell.

McGahey, R., Malloy, M., Kazanas, K. and Jacobs, M.P. (1990) *Financial Services, Financial Centres: Public Policy and the Competition for Markets, Firms and Jobs.* Boulder: Westview Press.

McIntyre, J. (1994) 'Competition in Japan's Information Technology Sector', *Competitiveness Review*, 4, 1–12.

McKillop, D.G. and Hutchinson, R.W. (1990) *Regional Financial Sectors in the British Isles.* Aldershot: Avebury.

McKillop, D. and Hutchinson, R. (1991) 'Financial Intermediaries and Financial Markets: A United Kingdom Perspective', *Regional Studies*, 25, 6, 543–54.

McMichael, P. (1996) 'Globalization: Myths and Realities', *Rural Sociology*, 61, 1, 25–55.

Meeker-Lowry, S. (1996) 'Community Money: The Potential of Local Currency'. *The Case against the Global Economy and for a Turn towards the Local.* Eds. J. Mander and E. Goldsmith. San Francisco: Sierra Books.

Messori, M. (1995) 'Banking and Finance in the Italian Mezzogiorno: Issues and Problems'. *Wealth from Diversity: Innovation, Structural Change and Finance for Regional Development in Europe.* Eds. J.S. Metcalfe and X. Vence Deza. Dordrecht: Kluwer Publishers.

Meyers, S.C. and Majluf, N.S. (1984) 'Corporate Financing and Investment Decisions When Firms Have Information that Other Investors Do Not Have', *Journal of Financial Economics*, 12, 187–222.

Michalak, W. and Gibb, R. (1997) 'Trading Blocs and Multilateralism in the World Economy', *Annals of the Association of American Geographers*, 87, 264–79.

Michie, J. and Greive-Smith, J. (1995) *Managing the Global Economy.* Oxford: Oxford University Press.

Milne, R.S. (1991) 'The Politics of Privatization in the ASEAN States', *ASEAN Economic Bulletin*, 7, 3.

Milne, R.S. (1992) 'Privatization in the ASEAN States: Who gets What, Why and with What Effect?' *Pacific Affairs*, 65, 1, 25–40.

Miller, R.J. (1978) *The Regional Impact of Money in the US.* Lexington, MA: Lexington Books.

Miller, R. (1981) 'The Caymans – Off-shore Paradise', *The Banker's Magazine*, 41.

Minsky, H. (1986) *Stabilizing an Unstable Economy.* New Haven: Yale University Press.

Minsky, H.P. (1989) 'Financial Crises and the Evolution of Capitalism: The Crash of 1987 – What Does it Mean?' *Capitalist Development and Crisis Theory: Accumulation, Regulation and Spatial Restructuring.* Eds. M. Gottdiener and N. Komninos. London: Macmillan.

Mintel (1995) *Personal Financial Reports.* London: Mintel.

Mitchell, K. (1995) 'Flexible Circulation in the Pacific Rim: Capitalisms in Cultural Context', *Economic Geography*, 71, 364–82.

Mitton, D.G. (1991) 'Tracking the Trends in Designer Genes: A Longitudinal Study of the Sources and Size of Financing in the Developing Biotech Industry in San Diego'. *Frontiers of Entrepreneurship Research 1991.* Eds. N.C. Churchill, W.D. Bygrave, J.G. Covin, D.L. Sexton, D.P. Slevin, K.H. Vesper and W.E. Wetzel. Babson Park, MA: Babson College.

Modigliani, F. and Miller, M. (1958) 'The Cost of Capital, Corporate Finance and the Theory of Investment', *American Economic Review*, 48, 261–97.

Modigliani, F. and Perotti, E. (1996) 'Protection of Minority Interest and the Development of Security Markets', *Mimeo*, London School of Economics.

Moffat, S. (1991) 'Huge Layoffs Expected after the Deal', *Los Angeles Times*, 13 August, D1 and D7.

Mollenkopf, J.H. and Castells, M. (1991) *Dual City: Restructuring New York.* New York: Russell Sage Foundation.

Moore, B.J. (1988) *Horizontalists and Verticalists. The Macroeconomics of Credit Money.* Cambridge: Cambridge University Press.

Moore, J. (1992) 'British Privatization – Taking Capitalism to the People', *Harvard Business Review*, 70, 115–24.

Moore, C.L. and Hill, J.M. (1982) 'Interregional Arbitrage and the Supply of Loanable Funds', *Journal of Regional Science*, 22, 499–512.
Moore, C.L., Karaska, G.J. and Hill, J.M. (1985) 'The Impact of the Banking System on Regional Analyses', *Regional Studies*, 19, 29–35.
Morgenson, G. (1990) 'The Buyout that Saved Safeway', *Forbes*, November, 88–92.
Motomura, A. (1994) 'The Best and Worst of Currencies: Seigniorage and Currency Policy in Spain, 1597–1650', *Journal of Economic History*, 54, 1, 104–27.
Mulgan, G. (1994) 'The End of Unemployment', *Demos Quarterly*, 2, 28–9.
Müller, W.F. and Paterson, T.W. (1986) 'Policies to Promote Competition'. *The Organization and Performance of the US Food Industry*. Ed. B.W. Marion. Lexington, Mass: Lexington Books.
Murray, G.C. (1994) 'The European Union's Support for New Technology-based Firms: An Assessment of the First Three Years of the European Seed Capital Fund Scheme', *European Planning Studies*, 2, 435–61.
Murray, G.C. (1996) 'Venture Capital'. *Small Business and Entrepreneurship*. Eds. P. Burns and J. Dewhurst. Basingstoke, Hampshire: Macmillan.
Murray, G.C. (1998) 'A Policy Response to Regional Disparities in the Supply of Risk Capital to New Technology-based Firms in the European Union: The European Seed Capital Fund Scheme', *Regional Studies*, 32, 405–19.
Murray, G.C. and Lott, J. (1995) 'Have UK Venture Capital Firms a Bias Against Investment in New Technology-based Firms?' *Research Policy*, 24, 283–99.
Murray, G.C. and Wright, M. (1996) 'Management's Search for Venture Capital in Smaller Buy-outs: The Role of Intermediaries and Marketing Implications', *International Journal of Bank Marketing*, 14, 2, 14–25.
Muzyka, D., Birley, S. and Leleux, B. (1996) 'Trade-offs in the Investment Decisions of European Venture Capitalists', *Journal of Business Venturing*, 11, 273–87.
Myers, S. (1984) 'The Capital Structure Puzzle', *Journal of Finance*, 29, 147–76.
Myrdal, G. (1957) *Economic Theory and Underdeveloped Regions*. London: Gerald Duckworth.
Myrdal, G. (1964) *Economic Theory and Under-developed Regions*. London: Methuen.
Nassau Guardian (1981) 'NY Banking Plan no "Real" Threat to Off-shore Business', 28 January.
Naylor, R. (1975) *The History of Canadian Business 1867–1914*. Toronto: James Lorimer and Co.
Neufeld, E.P. (1972) *The Financial System of Canada*. Toronto: Macmillan.
Ng, C.Y. and Toh, K.W. (1992) 'Privatization in the Asian-Pacific Region', *Asian-Pacific Economic Literature*, 6, 2, 168–90.
Niccoli, A. and Papi, L. (1993) *Debito Pubblico e Sistema Finanziario. Gli Effetti di Lungo Periodo*. Milan: Giuffrè.
North, P. (1998) 'Explorations in Heterotopia: Local Exchange Trading Schemes (LETS) and the Micro-politics of Money and Livelihood', *Society and Space*, in press.
Norton, W.E. and Kennedy, P.J. (1985) *Australian Economic Statistics 1949–50 to 1984–85*. Sydney: Reserve Bank of Australia Occasional Paper 8A.
Noyelle, T.J. (1987) *Beyond Industrial Dualism: Market and Job Segmentation in the New Economy*. Boulder: Westview Press.
Noyelle, T.J. Ed. (1989) *New York's Financial Markets: The Challenges of Globalisation*. Boulder: Westview Press.
O'Brien, R. (1992) *Global Financial Integration: The End of Geography*. London: Royal Institute of International Affairs.
O'Connor, J. (1973) *The Fiscal Crisis of the State*. New York: St Martin's Press.
Odell, K.A. (1992) *Capital Mobilization and Regional Financial Markets: The Pacific Coast States, 1850–1920*. New York: Garland Publishing.
Offe, C. and Henze, R.G. (1992) *Beyond Employment*. Cambridge: Polity Press.
Offer, A. (1997) 'Between the Gift and the Market: The Economy of Regard', *Economic History Review*, L, 3, 450–76.

Ogborn, M. (1998) 'Excise Geographies'. *Spaces of Modernity: London's Geographies 1680–1780*. London: Guilford Press. Chapter 5.

Ohmae, K. (1990) *The Borderless World*. London: HarperCollins.

Ohmae, K. Ed. (1995a) *The Evolving Global Economy*. Cambridge Mass: Harvard Business Review Books.

Ohmae, K. (1995b) *The End of the Nation-State: The Rise of Regional Economies*. London: HarperCollins.

O'hUallacháin, B. (1994) 'Foreign Banking in the American Urban System of Financial Organisation', *Economic Geography*, 70, 206–28.

Ordeshook, O. (1986) *Game Theory and Political Theory*. Cambridge: Cambridge University Press.

Organisation for Economic Cooperation and Development (1996) *Economic Outlook*. Paris: OECD.

Organisation for Economic Cooperation and Development (1997a) *Government Venture Capital for Technology-Based Firms*. Paris: OECD Committee for Scientific and Technological Policy.

Organisation for Economic Cooperation and Development (1997b) *Economic Outlook*, 61, Paris: OECD.

Organisation for Economic Cooperation and Development (various years) *OECD Economic Surveys – Australia*. Paris: OECD.

Organization for Economic Cooperation and Development (various years) *OECD Economic Surveys – Canada*. Paris: OECD.

O'Shea, M. (1995) *Venture Capital in OECD Countries*. Paris: OECD, Directorate for Financial, Fiscal and Enterprise Affairs.

Pain, N. and Young, G. (1996) 'The UK Public Finances: Past Experience and Future Prospects', *National Institute Economic Review*, 158 (Oct), 27–35.

Palan, R., Abbott, J. and Deans P. (1996) *State Strategies in the Global Political Economy*. London: Pinter.

Park, Y.S. and Essayyad, M. (1989) *International Banking and Financial Centres*. Norwood, MA: Kluwer.

Peck, J. and Tickell, A. (1994) 'Searching for a New Institutional Fix: The After-Fordist Crisis and the Global-Local Disorder'. *Post-Fordism*. Ed. A. Amin. Oxford: Blackwell.

Perlin, F. (1993) *The Invisible City: Administrative and Popular Infrastructures in Asia and Europe, 1500–1900*. Aldershot: Variorum.

Peterson, V.S. (1997) 'Whose Crisis? Early and Post-modern Masculinism'. *Innovation and Transformation in International Studies*. Eds. S. Gill and J.H. Mittelman. Cambridge: Cambridge University Press.

Petrella, R. (1996) 'Globalization and Internationalization: The Dynamics of the Emerging World Order'. *States Against Markets*. Eds. R. Boyer and D. Drache. London: Routledge.

Pettenati, P. (1990) 'Occupazione e Sviluppo Industriale: dalle Tre Italie al Neo-Dualismo'. *Piccola Impresa, Aree Depresse, Mercato del Lavoro*. Eds. R. Cafferata and C. Romagnoli. Milan: F. Angeli. 273–88.

Pfister, U. and Suter, C. (1987) 'International Financial Relations as Part of the World-System', *International Studies Quarterly*, 31, 239–72.

Phelps, N.A. (1993) 'Branch Plants and the Evolving Spatial Division of Labour: A Study of Material Linkage Change in the Northern Region of England', *Regional Studies*, 27, 87–101.

Phillips, G.M. (1995) 'Increased Debt and Industry Product Markets: An Empirical Analysis', *Journal of Financial Economics* 37, 189–238.

Pierson, C. (1994) *Dismantling the Welfare State? Reagan, Thatcher and the Politics of Retrenchment*. Cambridge: Cambridge University Press.

Pizzo, S., Fricker, M. and Muolo, P. (1989) *Inside Job: The Looting of America's Savings and Loans*. New York: McGraw-Hill Publishing Company.

Plender, J. and Wallace, P. (1985) *The Square Mile*. London: Century.

Ploug, N. and Kvist, J. (1996) *Social Security in Europe: Development or Dismantlement*. London: Kluwer Law International.

Pollard, J.S. (1995a) 'Industry Change and Labor Segmentation: The Banking Industry in Los Angeles, 1970–1990'. Unpublished Ph.D. Dissertation, Graduate School of Architecture and Urban Planning, University of California, Los Angeles.

Pollard, J.S. (1995b) 'The Contradictions of Flexibility: Labour Control and Resistance in the Los Angeles Banking Industry', *Geoforum*, 26, 2, 121–38.

Pollard J.S. (1996) 'Banking at the Margins: A Geography of Financial Exclusion in Los Angeles', *Environment and Planning*, A, 28, 7, 1209–32.

Porteous, D.J. (1995) *The Geography of Finance: Spatial Dimensions of Intermediary Behaviour*. Aldershot: Avebury.

Porter, R. (1994) *London: A Social History*. London: Hamish Hamilton.

Pratt, J. and Zeckhauser, R. Eds. (1985) *Principals and Agents: The Structure of Business*. Boston: Harvard Business School Press.

Pred, A.R. (1973) *Urban Growth and the Circulation of Information: The United States System of Cities 1790–1840*. Cambridge, MA: Harvard University Press.

Pred, A. (1977) *City Systems in Advanced Economies*. London: Hutchinson.

Price Waterhouse (1997) *Price Waterhouse National Venture Capital Survey*. Dallas: Price Waterhouse LLP Technology Industry Group.

Prosser, D. (1997) 'Time for Shareholders to Start a Revolution', *Investors Chronicle*, 19 September, 18–19.

Pryke, M. (1991) 'An International City Going Global: Spatial Change in the City of London', *Environment and Planning*, D, Society and Space, 9, 197–222.

Pryke, M. (1994) 'Urbanizing Capitals: Towards an Integration of Time, Space and Economic Calculation'. *Money, Power and Space*. Eds. S. Corbridge, R. Martin and N. Thrift. Oxford: Blackwell.

Ramanadham, V.V. (1993) *Privatization: A Global Perspective*. London: Routledge.

Ramanadham, V.V. (1995) *Privatization and Equity*. London: Routledge.

Remolona, E., Bassett, W. and Geoum, I.-S. (1996) 'Risk Management by Structured Derivative Products', *Federal Reserve Bank of New York Economic Policy Review*, 21, 17–38.

Richardson, H.W. (1972) *Regional Economics: Location Theory, Urban Structure and Regional Change*. London: World University.

Richardson, H.W. (1973) *Regional Growth Theory*. London: Macmillan.

Richardson, J. Ed. (1990) *Privatization and Deregulation in Canada and Britain*. Dartmouth: Institute for Research on Public Policy.

Riding, A. and Orser, B. (1997) *Beyond the Banks: Creative Financing for Canadian Entrepreneurs*. Toronto: Wiley.

Riley, B. (1996) 'Banking's Thriving Dinosaurs', *The Financial Times*, 2 November.

Robbie, K. and Wright, M. (1996) *Management Buy-Ins: Entrepreneurship, Active Investors and Corporate Restructuring*. Manchester: Manchester University Press.

Robbins, S. and Terleckyj, N. (1960) *Money Metropolis*. Cambridge, MA: Harvard University Press.

Roberts, E.B. (1991a) *Entrepreneurs in High Technology: Lessons from MIT and Beyond*. New York: Oxford University Press.

Roberts, E.B. (1991b) 'High Stakes for High-tech Entrepreneurs: Understanding Venture Capital Decision-making', *Sloan Management Review*, 32, 9–20.

Roberts, R.B. and Fishkind, H. (1979) 'The Role of Monetary Forces in Regional Economic Activity: An Econometric Analysis', *Journal of Regional Science*, 19, 1, 15–29.

Roberts, S. (1994) 'Fictitious Capital, Fictitious Spaces: The Geography of Offshore Financial Centres'. *Money, Power and Space*. Eds. S. Corbridge, R.L. Martin and N. Thrift. Oxford: Blackwell.

Roberts, S. (1995) 'Small Place, Big Money: The Cayman Islands and the International Financial System', *Economic Geography*, 71, 237–56.

Robinson, G. (1996) 'Property is Wealth', *RISK*, 9, 11, 12–15.

Rockoff, H. (1977) 'Regional Interest Rates and Bank Failures, 1870–1914', *Explorations in Economic History*, 14, 76–91.

Rodriguez, N. and Feagin, J. (1986) 'Urban Specialization in the World-System', *Urban Affairs Quarterly*, 22, 187–219.

Rogers, D. (1993) *The Future of American Banking: Managing for Change*. New York: McGraw-Hill.

Rosecrance, R. (1996) 'The Rise of the Virtual State', *Foreign Affairs*, 75(4), 45–61.

Rotstein, A. and Duncan, C.A.M. (1991) 'For a Second Economy'. *The New Era of Global Competition*. Eds. D. Drache and M. Gertler. Montreal: McGill-Queens University Press. 415–34.

Ruggie, J.G. (1993) 'Territoriality and Beyond: Problematizing Modernity in International Relations', *International Organization*, 47, 139–74.

Ruggie, J.M. (1982) 'International Regimes, Transactions and Change: Embedded Liberalism in the Post-War Economic Order', *International Organisation*, 36, 379–415.

Sabel, C. (1996) 'Intelligible Differences: on Deliberate Strategy and the Exploration of Possibility in Economic Life', *Rivista Italiana degli Economisti*, 1, 55–79.

Sahlman, W.A. (1988) 'Aspects of Financial Contracting in Venture Capital', *Journal of Applied Corporate Finance*, 1, 23–36.

Sahlman, W.A. (1990) 'The Structure and Governance of Venture-Capital Organisations', *Journal of Financial Economics*, 27, 473–521.

Sapienza, H.J. (1992) 'When Do Venture Capitalists Add Value?' *Journal of Business Venturing*, 7, 9–27.

Sassen, S. (1991) *The Global City*. Princeton, NJ: New York, London, Toyko. Princeton University Press.

Saunders, P. and Harris, C. (1994) *Privatization and Popular Capitalism*. Buckingham: Open University Press.

Saxenian, A. (1994) *Regional Advantage: Culture and Competition in Silicon Valley and Route 128*. Boston: Harvard University Press.

Scalera, D. and Zazzaro, A. (1997) 'Reputazione di gruppo e discriminazione nel mercato del credito: un modello dinamico con apprendimento', *Quaderni di Ricerca*, n. 97, Dipartimento di Economia, Università di Ancona.

Schamp, E.W., Linge, G.J.R. and Rogerson, C.M. Eds. (1993) *Finance, Institutions, and Industrial Change: Spatial Perspectives*. Berlin: W. de Gruyter.

Schneider-Lenné, E.R. (1994) 'Corporate Control in Germany', *Oxford Review of Economic Policy*, 8, 11–23.

Schor, J.B. (1992) 'Introduction'. *Financial Openness and National Autonomy and Constraints*. Eds. T. Banuri and J.B. Schor. Oxford: Clarendon Press.

Semple, K. (1985) 'Quaternary Place Theory', *Urban Geography*, 6, 4, 285–96.

Semple, R.K., Martz, D.J. and Green, M.B. (1985) 'Perspectives on Corporate Headquarters Relocation in the United States', *Urban Geography*, 6, 4, 370–91.

Shapiro, C. and Willig, R.D. (1990) 'Economic Rationales for the Scope of Privatization'. *The Political Economy of Public Sector Reform and Privatization*. Eds. E.N. Suleiman and J. Waterbury. Boulder: Westview Press.

Sharpe, W. (1964) 'Capital Asset Prices: A Theory of Market Equilibrium under Conditions of Risk', *Journal of Finance*, 19, 425–42.

Sharpe, S.A. (1990) 'Asymmetric Information, Bank Lending and Implicit Contracts: A Stylized Model of Customer Relationship', *Journal of Finance*, 45, 1069–87.

Shefter, M. (1993) *Capital of the American Century: The National and International Influence of New York City*. New York: Russell Sage Foundation.

Sheshunoff Information Services (1991) *Sheshunoff Banks of California*. Austin: Sheshunoff Information Services.

Shleifer, A. (1998) *Market Inefficiency*. Oxford: Clarendon Press.

Shonfield, A. (1965) *Modern Capitalism*. Oxford: Oxford University Press.

Short, J. and Nicholas, D.J. (1981) *Money Flows in the UK Regions*. Farnborough: Gower.

Short, D.M. and Riding, A.L. (1989) 'Informal Investors in the Ottawa–Carleton Region: Experiences and Expectations', *Entrepreneurship and Regional Development*, 1, 99–112.
Simmons, J. (1972) 'Interaction between the Cities of Ontario–Quebec'. *Urban Systems Development in Central Canada: Selected Papers*. Toronto: University of Toronto Press.
Smiley, G. (1975) 'Interest Rate Movements in the United States, 1888–1913', *Journal of Economic History*, 35, Supplement, 595–609.
Smith, A. (1776) *An Inquiry into the Nature and Causes of the Wealth of Nations*, Book I, Chapter IV. London: Routledge.
Smith, J. (1990) 'Address to the Eighth Bahamas International Financial Centre', *Mimeo*, 5 February.
Smith, M. (1998) 'EU Moves against Black Economy', *Financial Times*, 8 April.
Solomon, E.H. (1997) *Virtual Money: Understanding the Power and Risks of Money's High Speed Journey into Electronic Space*, Oxford: Oxford University Press.
Spence, N. (1992) 'Impact of Infrastructure Investment Policy'. *Regional Development in the 1990s*. Eds. P. Townroe and R.L. Martin. London: Jessica Kingsley.
Standeven, P. (1993) 'Financing the Early-stage Technology Firm in the 1990s: An International Perspective'. Discussion Paper for a Six Countries Programme meeting, Montreal.
Stanford Research Institute (1960) *The Savings and Loan Industry in California*. Report prepared for the Savings and Loan Commissioner, Division of Savings and Loan, State of California, SRI, Pasadena.
State of California (1990) *Economic Report of the Governor*. State of California, Sacramento.
State of California (1994) *Survey Estimates, California*. EDD, Los Angeles.
State of California (1995) *Statistical Abstract of California*. Department of Finance, Sacramento.
State of California (1996) *Statistical Abstract of California*. Department of Finance, Sacramento.
Statistics Canada (1983) *Historical Statistics of Canada, 2nd Edition*. Ottawa: Ministry of Supply & Services.
Stephens, J.D. and Holly, B.P. (1980) 'The Changing Pattern of Industrial Corporate Control in the Metropolitan United States'. *The American Metropolitan System: Present and Future*. New York: John Wiley.
Steier, L. and Greenwood, R. (1995) 'Venture Capitalist Relationships in the Deal Structuring and Post-investment Stages of New Firm Creation', *Journal of Management Studies*, 32, 337–57.
Stern, J. and Chew, D. (1992) *The Revolution in Corporate Finance*. Oxford: Blackwell.
Stiglitz, J.E. and Weiss, A. (1988) 'Banks as Social Accountants and Screening Device for the Allocation of Credit', NBER, *Working Paper*, 2710.
Stock Exchange (1996) *Report of the Committee on Private Share-ownership*. London: Gee Publishing.
Storey, D., Watson, R. and Wynarczyk, P. (1989) 'Fast Growth Small Businesses: Case Studies of 40 Small Firms in North East England', *Research Paper*, 67, Sheffield: Department of Employment.
Strange, S. (1986) *Casino Capitalism*. Oxford: Blackwell.
Strange, S. (1988) *States and Markets*. London: Pinter.
Strange, S. (1994) 'From Bretton Woods to the Casino Economy'. *Money, Power and Space*. Eds. S. Corbridge, R. Martin and N. Thrift. Oxford: Blackwell.
Sweeting, R.C. (1991) 'UK Venture Capital Funds and the Funding of New Technology-based Businesses: Processes and Relationships', *Journal of Management Studies*, 28, 601–22.
Swyngedouw, E. (1989) 'The Heart of the Place: The Resurrection of Locality in an Age of Hyperspace', *Geografiska Annaler*, 71, 31–42.
Swyngedouw, E. (1996) 'Producing Futures: Global Finance as a Geographical Project'. *The Global Economy in Transition*. Eds. P. Daniels and W. Lever. London: Longman. Chapter 8.

Taylor, M. Ed. (1984) *The Geography of Australian Corporate Power*. Sydney: Croom Helm Australia.

Taylor, M. and Thrift, N. (1981) 'The Changing Spatial Concentration of Large Company Ownership and Control in Australia 1953–1978', *Australian Geographer*, 15, 98–105.

Terrell, H. and Mills, R. (1983) 'International Banking Facilities and the Eurodollar Market', *Board of Governors of the Federal Reserve System, Staff Studies*, 124.

Thompson, C. (1989) 'The Geography of Venture Capital', *Progess in Human Geography*, 13, 62–98.

Thompson, G. (1990) *The Political Economy of the New Right*. London: Pinter.

Thomson Financial Publishing (1991) *Thomson Savings Directory, 1991*. Skokie, IL: Thomson Financial Publishing.

Thrift, N. (1986) 'The Geography of International Economic Disorder'. *A World in Crisis?* Eds. R. Johnston and P. Taylor. Oxford: Blackwell.

Thrift, N. (1994) 'On the Social and Cultural Determinants of International Financial Centres: The Case of the City of London'. *Money, Power and Space*. Eds. S. Corbridge, R.L. Martin and N. Thrift. Oxford: Blackwell.

Thrift, N. and Leyshon, A. (1994) 'A Phantom State? The De-traditionalisation of Money, the International Financial System and International Financial Centres', *Political Geography*, 13, 299–327.

Thrift, N. and Olds, K. (1996) 'Re-figuring the Economic in Economic Geography', *Progress in Human Geography*, 20, 311–37.

Thwaites, A.T. and Wynarczyk, P. (1996) 'The Economic Performance of Innovative Small Firms in the South East Region and Elsewhere in the UK', *Regional Studies*, 30, 135–49.

Tibbett, R. (1997) 'Alternative Currencies: A Challenge to Globalization?' *New Political Economy*, 2, 1, 127–35.

Tickell, A. (1994) 'Banking in Britain? The Changing Role and Geography of Japanese Banks in Britain', *Regional Studies*, 28, 291–304.

Tickell, A.T. (forthcoming) *Rogues, Regulation and the Culture of Finance*. New York: Guilford.

Timmons, J. and Bygrave, W.D. (1997) 'Venture Capital: Reflections and Projections'. *Entrepreneurship 2000*. Eds. D.L. Sexton and R.W. Smilor. Chicago: Upstart.

Tobin, J. (1969) 'A General Equilibrium Approach to Monetary Theory', *Journal of Money, Credit and Banking*, February, 15–29.

Towle, A.A. (1990) 'The Social and Spatial Effects of the Restructuring of US Commercial Banking'. Unpublished MA thesis, Department of City and Regional Planning, Cornell University, Ithaca.

Townsend, R.M. (1990) *Financial Structure and Economic Organization*. Oxford: Blackwell.

Triffin, R. (1960) *Gold and the Dollar Crisis*. New Haven: Yale University Press.

Tyebjee, T.T. and Bruno, A.V. (1984) 'A Model of Venture Capitalist Investment Activity', *Management Science*, 30, 1051–66.

Tyrie, A. (1996) *The Prospects for Public Spending*. London: Social Market Foundation.

UN (1998) *Human Development Report*. Geneva: United Nations.

Unger, R.M. (1996) *What Should Legal Analysis Become?* London: Verso.

US Bureau of the Census (1960) *Historical Statistics of the United States, Colonial Times to 1957*. Washington DC: USBC.

US Bureau of the Census (1980) *Census of Population and Housing 1980: Public Use Microdata Samples (5%)*. Washington DC: California USBC.

US Bureau of the Census (1990) *Census of Population and Housing 1990: Public Use Microdata Samples (5%)*. Washington DC: California, USBC.

Veljanovski, C. (1987) *Selling the State: Privatisation in Britain*. London: Weidenfeld and Nicolson.

VentureOne (1997) *National Venture Capital Association 1996 Annual Report*. San Francisco: VentureOne Corporation.

Vickers, J. and Yarrow, G. (1988) *Privatization: An Economic Analysis*. Cambridge, Mass: MIT Press.

Vittas, D. (1996) 'Private Pension Funds in Hungary: Early Performance and Regulatory Issues', *Policy Research Working Paper*, 1638, Washington, DC: World Bank.

Vittas, D. and Michelitsch, R. (1995) 'Pension Funds in Central Europe and Russia: their Prospects and Potential Role in Corporate Governance', *Policy Research Working Paper*, 1459, Washington, DC: World Bank.

Vives, X. (1991) 'Banking Competition and European Integration'. *European Financial Integration*. Eds. A. Giovannini and C. Mayer. Cambridge: Cambridge University Press. 9–34.

Volcker, P. and Gyohten, T. (1992) *Changing Fortunes: The Making of a Supra-National Economic Order*. London: Pluto.

Vuylsteke, C. (1988) 'Techniques of Privatization of State-owned Enterprises', *World Bank Technical Papers*, 88, Washington: World Bank.

Waite, S.R. (1991) 'The Eclipse of Growth Capital', *Journal of Applied Corporate Finance*, 4, 77–85.

Walmsley, J. (1983) 'A Tough Year for US Bankers', *The Banker*, February, 85–93.

Warf, B. (1989) 'Telecommunications and the Globalization of Financial Services', *Professional Geographer*, 41, 257–71.

Warf, B. (1994) 'Vicious Circle: Financial Markets and Commercial Real Estate in the United States'. *Money, Power and Space*. Eds. S. Corbridge, R.L. Martin and N.J. Thrift. Oxford: Blackwell.

Warf, B. (1995) 'Telecommunications and the Changing Geographies of Knowledge Transmission in the Late 20th Century', *Urban Studies*, 32, 361–78.

Warf, B. and Cox, J.C. (1995) 'US Bank Failures and Regional Economic Structure', *Professional Geographer*, 47, 3–16.

Webber, M.J. and Rigby, D. (1996) *The Golden Age Illusion: Rethinking Post-War Capitalism*. New York: Guilford Press.

Weitzman, M.L. (1984) *The Share Economy: Conquering Stagflation*. Cambridge, Mass: Harvard University Press.

Wetzel, W.E. (1994) 'Venture Capital'. *The Portable MBA in Entrepreneurship*. Ed. W.D. Bygrave. New York: Wiley.

Wheeler, J. and Mitchelson, R.L. (1989) 'Information Flows among Major Metropolitan Areas in the US', *Annals of the Association of American Geographers*, 79, 4, 522–43.

Willer, D. (1997) 'Corporate Governance and Shareholder Rights in Russia', *Discussion Paper* 343, London: London School of Economics, Centre for Economic Performance.

Williams, C.C. (1996) 'Local Exchange Trading Systems. A New Source of Work and Employment?' *Environment and Planning*, A, 28, 8, 1395–415.

Wilson, D. (1991) 'Urban Change, Circuits of Capital, and Uneven Development', *Professional Geographer*, 43, 403–16.

Wilson, N.C. (1964) *400 California Street: The Story of the Bank of California National Association and its First 100 Years in the Financial Development of the West Coast*. San Francisco: Bank of California.

Winborg, J. and Landström, H. (1997) 'Financial Bootstrapping in Small Businesses: A Resource-based View on Small Business Finance'. *Frontiers of Entrepreneurship Research 1997*. Eds. P.D. Reynolds, W.D. Bygrave, N.M. Carter, P. Davidsson, W.B. Gartner, C.M. Mason and P.P. McDougall. Babson Park, MA: Babson College.

Wise, D. (1982) 'International Banking Facilities and the Future of Off-shore Banking', *Fletcher Forum*, 6, 299–329.

Wolfson, M.H. (1993) 'The Evolution of the Financial System and the Possibilities for Reform'. *Transforming the US Financial System: Equity and Efficiency for the 21st Century*. Eds. G.A. Dymski, G. Epstein and R. Pollin. Armonk NY: ME Sharpe.

Wood, P. (1991) 'Flexible Accumulation and the Rise of Business Services', *Transactions of the Institute of British Geographers*, 16, 160–72.

Wood, E.M. (1997) *Democracy Against Capitalism: Renewing Historical Materialism*. Cambridge: Cambridge University Press.

World Bank (1997) *World Development Report, 1997: The State in a Changing World.* Oxford: Oxford University Press.

Wray, L.R. (1990) *Money and Credit in Capitalist Economies: The Endogenous Money Approach.* Aldershot: Edward Elgar.

Wright, M. and Robbie, K. (1997) 'Entrepreneurial Spirit Propels Buy-outs'. *Mastering Enterprise.* Eds. S. Birley and D. Muzyka. London: FT/Pitman.

Wright, M., Robbie, K. and Ward, M. (1994) 'Management Buyouts in the Regions: A Tale of Two Recessions', *Regional Studies,* 28, 319–25.

Wright, M. and Thompson, S. (1994) 'Divestiture of Public Sector Assets'. *Privatisation and Regulation: A Review of the Issues.* Eds. P.M. Jackson and C.M. Price. London: Longman. Chapter 2.

Wright, M., Thompson, S., Chiplin, S. and Robbie, K. (1991) *Buy-Ins and Buy-Outs: New Strategies in Corporate Management.* London: Graham and Trotman.

Wrigley, N. (1992) 'Antitrust Regulation and the Restructuring of Grocery Retailing in Britain and the USA', *Environment and Planning,* A, 24, 727–49.

Wrigley, N. (1997) 'British Food Retail Capital in the USA. Part 2: Giant Prospects?', *International Journal of Retail and Distribution Management,* 25, 48–58. Reprinted in *British Food Journal,* 99, 427–37.

Wrigley, N. (1998a) 'Market Rules and Spatial Outcomes: Insights from the Corporate Restructuring of US Food Retailing', *Regions, Regulations and Institutions: Towards a New Industrial Geography.* Eds. T. Barnes and M.S. Gertler. London: Routledge.

Wrigley, N. (1998b) 'European Retail Giants and the Post-LBO Reconfiguration of US Food Retailing', *International Review of Retail, Distribution and Consumer Research,* 8, 127–46.

Wrigley, N. (1998c) 'Understanding Store Development Programmes in Post-Property-Crisis UK Food Retailing', *Environment and Planning,* A, 30, 15–35.

Wruck, K.H. (1990) 'Financial Distress, Reorganization and Organizational Efficiency', *Journal of Financial Economics,* 27, 420–44.

Wruck, K.H. (1992) 'Leveraged Buyouts and Restructuring: The Case of Safeway Inc.', *Harvard Business School,* Case 9-192-095.

Yago, G. (1991) 'The Credit Crunch: A Regulatory Squeeze on Growth Capital', *Journal of Applied Corporate Finance,* 4, 96–100.

Yassukovich, S. (1981) 'Could the Euromarkets Leave London?', *Euromoney,* October, 253.

Zazzaro, A. (1997) 'Regional Banking Systems, Credit Allocation and Regional Economic Development', *Economie Appliquée,* 1, 51–74.

Zeckhauser, R., Patel, J. and Hendricks, D. (1991) 'Non-rational Actors and Financial Market Behaviour', *Theory and Decision,* 31, 257–67.

Zelizer, V.A.R. (1994) *The Social Meaning of Money.* New York: Basic Books.

Zimmerman, G.C. (1989) 'The Growing Presence of Japanese Banks in California', *Federal Reserve Bank of San Francisco Economic Review,* 3, 3–17.

LIST OF CONTRIBUTORS

Professor Pietro Alessandrini Faculty of Economics, Universita Degli Studi di Ancona, Italy

Dr Leslie Budd Department of Economics, London Guildhall University, UK

Professor Gordon Clark School of Geography, University of Oxford, UK

Professor Sheila C. Dow Department of Economics, University of Stirling, UK

Professor Richard T. Harrison Department of Management, University of Ulster, UK

Dr Alan C. Hudson Department of Geography, University of Cambridge, UK

Dr Roger Lee Department of Geography, Queen Mary and Westfield College, University of London, UK

Dr Ron Martin Department of Geography, University of Cambridge, UK

Professor Colin M. Mason Department of Geography, University of Southampton, UK

Dr Jane Pollard Department of Geography, University of Birmingham, UK

Dr David J. Porteous Development Banker, South Africa

Professor Barney Warf Department of Geography, Florida State University, United States

Professor Neil Wrigley Department of Geography, University of Southampton, UK

Dr Alberto Zazzaro Faculty of Economics, Universita Degli Studi di Urbino Urbino, Italy

INDEX

Note: Page references in *italics* refer to Figures; those in **bold** refer to Tables

1878 Act (US) 53
1909 Act (US) 53

3i plc 176

Abbey National 49, 66, 67, 281

ABN-Amro 135
active integration 79
AFL-CIO 259
Airline Deregulation Act (1978) (USA) 234
Alliance and Leicester Bank 281
American Airlines 233
American Research and Development 162
American Stores 190, 196, 197, 202
    debt and debt-to-capitalisation *197*
American Telephone & Telegraph (AT&T) 59, 234
Amersham International 268
arbitrage economy 195
Arthur model 108–11
Associated British Ports 268
Association of Private Client Investment Managers and Stockbrokers (APCIMS) 279
Automated Pit Trading (APT) system 128, 130
Automated Teller Machine (ATM) investment 61, 67

Bahamas 9, 111, 139, 142, 144, 148
bank deposits relative to wealth (BD/W) 34
Bank for International Settlements 229
bank multiplier 37
Bank of America 54, 60, 61, 63, 67
Bank of America National Trust and Savings Association 54
Bank of America of California 54
Bank of America/Security Pacific merger 63, 68
Bank of California 52, 53, 55
Bank of England 41, 278
Bank of International Settlements 130, 136
Bank of Italy 53, 54, 81
Bank of Nova Scotia 97

banking
    development 7–8
    place in financial systems and regional development 32–6
    regionalisation 4
    spatial distribution of financial flows 43–8, **44**
    spatial evolution of financial system 40–3
    stages of development (Chick's theory) 32, 36–40, **37**
Barclays Bank 7, 20, 49, 54
Baring, Thomas 125
Barings Bank 16, 122, 123, 124, 136
Basle Banking Accord (1987) 9, 229
BCCI 16
Bielson, A.C. 197
Big Bang (1986) 126, 145
Birmingham 2000 113
Botswana 112
Bretton Woods system 9, 115, 116, 144
    demise of 12, 115, 118, 140, 141, 142, 228
British Aerospace 268
British Airports Authority 268, 272
British Airways 268
British Association Ports 272
British Gas 268, 270, 272, 274, 276
British Petroleum 268
British Steel 268
British Telecommunications 235
    privatisation 268, 270, 272, 24, 279
British Virgin Islands 111
Britoil 268
Building Societies Act (1986) 66
Bundesbank 231
bunds 129
business angels 158, 159, 181
    see also venture capital
Buttrey Food & Drug 197

Cable and Wireless 268
California
    bank employment by occupational group **64**
    changing demand for labour in banking 63–6

California (cont.)
　competitive environment (1980s)　58–60
　globalisation of retailing innovations　68–9
　largest banks　55, **56**
　median hourly income　**65**
　national regulatory context　50–1
　regional regulatory context　52–7
　reorganisation of production in banking　61–3
　spatial organisation of retail banking in　50–7
California Banking Act　53
Canary Wharf　119, 135
capital adequacy ratios　47
Caribbean Basin Initiative　233
Caribbean Data Services (CDS)　233
CATS　230
Cayman Islands　9, 111, 139, 142, 144, 148, 233
centralisation　15
　of transaction handling and credit analysis　68
Chase National Bank　53
Chicago Board of Trade (CBOT)　130, 133
Chicago Mercantile Exchange　129, 231
Chick's theory of banking development　32, 36–40, **37**
Citibank　9
Citicorp　9, 10, 19
Citigroup　19
Clayton Act　192
coin relative to wealth (C/W)　34
commodity money　212
communications, role of　100
　see also information technology; technological innovation
Comox Valley　209
Companies Act (1865) (UK)　263
competitive deregulation　13
Computer Assisted Order Routing and Execution System (Tokyo)　230
corporate control failure　185
credible threat　193
credit, allocation of　45
credit analysis　61
credit card capitalism　208
credit-scoring systems　61
CS First Boston　231
cumulative causation model of uneven regional economics　3
currency boards　9

death of distance　15
debt capital　187, *197*

debt/equity ratio　187
debt-to-capitalisation ratios　192, 196, *197*
decentralising forces　103
Depository Institutions Deregulation and Monetary Control Act (1980) (USA)　234
deregulation　12–13, 234–5
Deutsche Termin Börse (DTB)　128, 129
Developing World debt crisis　16
dispersion forces　103
Drexel Burnham Lambert　194

EBITDA　196
economic and monetary union (EMU)　213, 214
economic geography of money
　as new subdiscipline　3–6
economically targeted investments (ETIs)　257
electricity companies, privatisation, UK　273, 274–6
embedded liberalism　115, 116
　limits of　117–19
employment
　in finance　96–7
　IT and　20, **20**
　labour market externalities　102
　by occupational group, California　**64**
end of geography argument　14–15, 104, 121, 131, 139
endogenous reserves　38
entrepreneurship, financing　157–82
equity guarantee schemes　179
Euro　130, 213
Euro dollar market　9, 125–6, 143, 147–8
Euro market　229
EUROEX　128
European Central Bank　213
European Monetary System (EMS)　213
European Security Dealers Automated Quotation (EASDAQ)　132
European Union (EU)　41–2, 126, 227, 228, 231
　Investment Services Directive　132
　Seed Capital Fund Scheme　180–1
　single currency　9
European Venture Capital Association　168
Eurotop 100　132
Eurotop 300　132–3

Farmers and Merchants Bank　52
FDIC insurance　60, 63
Federal Bank Holding Company Act (1956)
　Douglas Amendment　7, 8
Federal Express　100, 233

Federal Reserve Act (US) 208
Federal Reserve Bank 234
Federal Reserve Board 54, 147
Federal Trade Commission (FTC) 192
Fifth Kondratieff 229
financial centres, development of 95–114
  agglomeration in a particular locality 105–8
    information hinterland 105–6
    inter-regional attachments 106–8
  definition 96–101
    bank head-office locations 97
    cheque clearings 97–8, **98**
    employment in finance 96–7
    foreign bank presence 99
    stock exchange volumes 98–9, **99**
  forces behind 101–5
  modelling path dependence 108–11
    application to the city-pairs 109–11
financial distress 187, 189, *190*
Financial Institutions Reform, Recovery and Enforcement Act (1989) 194
financial intermediation 33
financial interrelations ratio (FIR) 34, *35*
financial leverage 187
Finanzplatzdeutschland 126
First Direct 67
First Interstate Bank 54, 60, 61
Ford, Henry 162
foreign banks **19**
foreign exchange turnover *127*
FOX 134
Friedman, Milton 235
FT-SE 100 254

Garn St-Germain Act (1982) **55**
General Agreement of Tariffs and Trade (GATT) 231
  Uruguay Round 229
geographical consolidation 20
geographies of financial structure and financial exclusion 205
geography of avoidance 197
geography of divestiture 185, 196, 205
geography of divestment 197
geo-political economies 141–2
geopolitics of money 205
German reunification 215
Giannini, A.P. 53
Gibraltar 111
Glass-Steagall Act (1933) 7, 50, 143, 229
global neoclassicism 10, 120, 121–2
globalisation 14, 15
GLOBEX 129–30, 230

gold standard, end of 228
Grand Union 194
Great Depression 16, 248
Green Dollar 209
grocery marketing areas 197, *198*, *200*, 201
Guernsey 112

Halifax 49, 281
Hirschman-Herfindahl Index statistics 196
Hong Kong 9
hypermobility of capital 227–39
  localities 236–7
  post-Fordism and 229–33

industrial capitalism, theory of 263
information technology 123
  impact of 230–1
  investment in, in banking **20**
  see also technological innovation
informational spillovers 102
ING 136
initial public offerings (IPOs) 159, 163
Instinet 230
institutional geographies 7
inter-bank lending 38
Inter-City Cost of Living Index 204
Interest Equalisation Tax (IET) 143
intermediate services, demand for 102
international banking facilities (IBF) 139
  establishment of 145–7
  failure 140, 150–3, 153–4
  impact on off-shore financial centres 148–50, 153
international financial markets, growth of 120–5
international financial service centres (IFSCs) 112
International Gold Standard 9
International Monetary Fund (IMF) 12
Internet banking 61
intra-place competition 119
IOUs 37
Isle of Man 112
Italian banking system 71–92
  bank interest rates by region *89*
  banking performance indicators **77**
  banking system/regional development interaction 75–8
  geographical diffusion of financial innovations 88–91
  geography **76**
  indicators of regional banking structure **77**

Italian banking system (*cont.*)
  percentage of households owning financial assets 91
  problems and inter-regional financial integration 78–83
    inter-regional integration 79–83
    pessimism, optimism and possibilism 78–9
  problems of dimension 84–8
    local banks versus national banks 84–5
    model for regional banking system 85–8
  regional characteristics of firms **74**
  regional economic disparities 72–4
  regional economic indicators **73**

Jersey 111
Johnson, President Lyndon 143
junk bonds 194

Keynesianism, collapse of 233–5
Knickerbocker Trust Company 53
Kroger LBO 190, 192, 193, *194*, 196, 197, 201–2
  annual capital expenditure levels *201*

labour market externalities 102
LDC debt crisis 59
lender-of-last resort 38, 47, 48
leveraged buy-outs (LBOs) in US food retailing 185–205
  debt, leverage and retail firm 186–95
    capital structure decision 186–7
    high leverage and the retail firm 187–95
    perils of high leverage 1889, *189*
  leveraged restructuring 190–5
  spatial outcomes of leveraged restructuring 195–204
    effect of capital structure transformation 196–202
      asset sales and market consolidation 196–201
      reductions in capital expenditure 201–2
    interaction between high-leverage firms and decisions of rivals 202–4
      market entry and market expansion by rival incumbent firms 203
      pricing, predation and competition 203–4
liability management 39
liquidity preference, theory of 33–4

Lloyds Bank 7, 54
Lloyds-TSB 49
local cooperatives 220
Local Exchange Trading Systems (LETS) 21, 209, 219–24
  as a vehicle for change 221–3
local money
  local money and economic geographies 217–23
    LETS as a vehicle for change 221–3
    local money and alternative economic geographies 219–21
    local money and multiple scales of reproduction 218–19
    local resistance vs local sustenance 223–4
  monetary networks 216–17
  money and construction of economic geographies 211–16
    money and local/non-local relations 212–15
    money and social reproduction 215–16
  paradox of 209–11
  re-emergence and growth of 208
locational embeddedness 116
locational inertia 227
locational structure of institutions 6
London as international finance centre 125–33, 135, 278
  Big Bang (1986) 126, 145
  foreign exchange turnover 126–7, *127*, **128**
London International financial futures and Options Exchange (LIFFE) 123, 128, 129, 136
  APT system 130
London Stock Exchange 131, 157
London Underground 246
Lucky 190, 197

Maastricht Treaty (1992) 48
Malaysia (Labuan) 112
management buy-ins (MBIs) 160–1, 166, 170–1
management buy-outs (MBOs) 160–1, 166, 170–1
Marché à Terme International de France (MATIF) 128, 129, 130
MATIF/DTB link 132
Mauritius 112
Maxwell, Robert, scandal 254
McFadden Act (1927) 7, 143
Medicaid 235
Medicare 235

mergers 18–19
Metallgesellschaft 122, 124
Midlands Bank 7, 20, 49, 54, 67
monetary networks 209, 216–17
monetary union 45, 213, 214
money of account 212
mortgages 123
Motor Carriers Act (1980) (USA) 234

National City Bank 53
National Health Service 247
National Power-Power-Gen 268
National Westminster Bank 49, 67
nationalisation of state industries 266
New Deal legislation 9
New York Clearing House Association 146
New York Mercantile Exchange 130
New York Stock Exchange deregulation 118, 145
newly industrialised countries (NICs) 227
Nippon Telegraph and Telephone 235
Nolan Act (1983) 55
non-bank financial intermediary deposits (NBD/W) 34
non-financial corporations 100
North American Free Trade Agreement (NAFTA) 227, 228, 231
North American Securities Dealers Automated Quotation (NASDAW) 132, 157, 232
Northern Rock 281
notes relative to wealth (N/W) 34

off shores on-shore
 developing 144–7
 end of 147–8
 money rules or placing trust 153–4
 not end of 148–50
 path-dependence of place 150–3
off-shore financial centres (OFCs) 9, 43, 111–12, 118, 139
 see also under names
oil shock (1973–4) 117, 143
on-shore/off-shore relations 139, 142–4
 see also off shores on-shore
Orange County, California, bankruptcy of 16, 122, 124
Organisation of Petroleum Exporting Countries 143
OTC markets 123, 124, 129

paradox of risk 123
paradox of thrift 123

passive integration 79
path dependence, definition 95
Pathmark 194, 202
pecking order theory of capital structure 187
pension fund capitalism 241–59, 265–6
 implications 257–9
 pension revision 196
 retreat of the state 246–9
 rise of 250–3
 state income and expenditures 243–6, 245
 urban infrastructure and economic development 253–7
permissive regulatory frameworks and structures 103
personal equity plans (PEPs) 281
petty geography 218
politics of productivity 115
portfolio theory 253–4
pre-modern money 212
pre-packaged Chapter 11s 190
principal-agent problem 257
private banking facilities 6, 68
privatisation 235, 247, 261–83
 declining presence of individual shareholder (UK) *272*
 definition 262
 evolution of share-ownership under capitalism 263–7, **264**, *265*
 geographical limits of private share-ownership 270–7
 growth in number of private shareholders, UK *271*
 household income and savings by UK region 274, **275**
 institutional geography of equity market 277–80
 public and institutional allocations of UK privatisation flotations **271**
 regional distribution of share-ownership, UK **273**, **276**
 share-ownership densities in UK **274**
 of UK state pensions 267–70, *269*
Proctor and Gamble 124
proportion-to-probability mapping *109*
ProShare 279

Quebercois nationalism 108

real estate as regulator of territorial embeddedness 133–6
Real Estate Index Market 134
Real Estate Investment Trusts (REITS) 134

Regulation D 143, 145
Regulation Q 51, 143
regulatory spaces 8–9, 130–41
    monetary flows and 139–41
relationship banking 62
remapping of financial landscape 12–21, 17
retail banking 49–70
Reuters 230
Riegle-Neal Act (1994) 8
Rolls Royce 268
Rothschilds 125
Royal Bank of Scotland 66, 67

S&Ls, deregulation of 55, 57
S&P 500 254
Safeway LBO 190, 192, 193, 196, 197
Sainsbury 202
Salomon Brothers 230
Sanwa Bank of California 55
savings associations, failure of (US) 16, 18
Schwab, Charles 60
Scottish banking system 41
Scottish Financial Enterprise 113
Scottish Power-Hydro-Electric 268, 270
SEAQ 230
Securities Acts Amendments (1975) (USA) 234
Securities and Exchange Commission (SEC) 54
securitisation, development of 39, 123
Security Pacific 60, 61
    merger with Bank of America 63, 68
Selling-Area Markets Inc. 197
Sequence VI 131
Share Shop scheme 279
Silicon valley 175
Singapore 9
single currency 9, 130, 213
small business investment companies (SBICs) 162, 179, **180**
Small Business Investment Company Act (1958) 162
social construct, money as 210
social relation, money as 11–12
socio-institutional and cultural factors 103
Soffex (Swiss Options and Financial Futures Exchange) 128, 230
sovereign risk 153
spatial agglomeration 101
spatial fix argument 227–8
Standard Chartered Bank 54

state credit money 212
stateless monies 141
Stock Exchange Automated Quotation international system (SEQA-1) 132
stock market, world, crash (1987) 16, 122, 126, 131
Stop & Shop 190
subsidies 179
Sumitomo Bank of California 55
Supermarkets General LBO 190, 194
Sydney Stock Exchange 98

taxation 234
technological innovation 13–14, 122
technological spillovers 102
telephone banking 68
tele-servicing 61
    see also information technology
territorial embeddedness 116
    London 125–33, 136–7
    real estate as 133–6
Tesco 202
tiger economies 121
Tobins q ratio 122
Toronto Stock Exchange (TSE) 98, 230
Tradepoint 131
Transamerica Corporation 54
Travellers Group 19
Trustee Savings Bank (TSB) 66, 268

UK banking 66–7
    development 7
    see also under names
Union Bank 55, 61
unit of account function 33
unit of exchange 209
unit trusts 281
univeralism of money 207–9
universal banking model 72
US banking
    development 7–8
    failure (1911–96) 50, *51*
    see also under names
US Bankruptcy Code 189
US Census of retail trade
    (1987) 196
    (1977) 196

venture capital
    economic significance 166–8
    genesis and growth 162–6, *163–5*
    implications of uneven geographical distribution of investments 177–8

locational distribution 168–77
  spatial patterns 168–73
  uneven nature of venture capital investments 173–6
  geography of informal market 176–7
nature of investment 159–62
policy responses to regional venture capital gaps 178–81
sources of 158
types of company 158
in the UK 168–71, 173–6, **174**
in the USA 171–7, *172*, **172**, 174

Vietnam War, financing of 117, 142
virtual money 212

welfare reform 242
Wells Fargo 54, 60, 61, 68
Woolwich Bank 49, 281
World Bank 12
world cities 232
World Trade Organisation (WTO) 227, 228, 231

Index compiled by Annette Musker